Islamic Astronomy and Geography

DAVID A. KING

[illegible]

[illegible]

[illegible]

[illegible]

[illegible]

[illegible]

[illegible]

RICHARD LORCH

Arabic Mathematical Sciences

[illegible]

A. I. SABRA

[illegible]

DAVID A. KING

Astronomy in the Service of Islam

ROSHDI RASHED

Optique et mathématiques

Recherches sur l'histoire de la pensée scientifique en Arabe

*Also in the Variorum Collected Studies Series:*

**DAVID A. KING**
Islamic Astronomy and Geography

**DAVID A. KING**
Astrolabes from Medieval Europe

**JULIO SAMSÓ**
Astronomy and Astrology in al-Andalus and the Maghrib

**J.L. MANCHA**
Studies in Medieval Astronomy and Optics

**GAD FREUDENTHAL**
Science in the Medieval Hebrew and Arabic Traditions

**PAUL KUNITZSCH**
Stars and Numbers
Astronomy and Mathematics in the Medieval Arab and Western Worlds

**RAYMOND MERCIER**
Studies on the Transmission of Medieval Mathematical Astronomy

**DONALD R. HILL**
Studies in Medieval Islamic Technology
From Philo to al-Jazari – from Alexandria to Diyar Bakr

**EDWARD S. KENNEDY**
Astronomy and Astrology in the Medieval Islamic World

**RICHARD LORCH**
Arabic Mathematical Sciences
Instruments, Texts and Transmission

**A.I. SABRA**
Optics, Astronomy and Logic
Studies in Arabic Science and Philosophy

**DAVID A. KING**
Astronomy in the Service of Islam

**ROSHDI RASHED**
Optique et mathématiques
Recherches sur l'histoire de la Pensée Scientifique en Arabe

VARIORUM COLLECTED STUDIES SERIES

# Islamic Astronomical Tables

Benno van Dalen

# Islamic Astronomical Tables

## Mathematical Analysis and Historical Investigation

Routledge
Taylor & Francis Group
LONDON AND NEW YORK

First published 2013 by Ashgate Publishing

2 Park Square, Milton Park, Abingdon, Oxfordshire OX14 4RN
711 Third Avenue, New York, NY 10017

*Routledge is an imprint of the Taylor & Francis Group, an informa business*

First issued in paperback 2018



**British Library Cataloguing in Publication Data**
A catalogue record for this book is available from the British Library.

**The Library of Congress has cataloged the printed edition as follows:** 2013916088

ISBN 978-1-4724-2238-5 (hbk)
ISBN 978-1-138-38265-7 (pbk)

VARIORUM COLLECTED STUDIES SERIES CS1040

# CONTENTS

This volume contains xii + 350 pages

## PUBLISHER'S NOTE

*The articles in this volume, as in all others in the Variorum Collected Studies Series, have not been given a new, continuous pagination. In order to avoid confusion, and to facilitate their use where these same studies have been referred to elsewhere, the original pagination has been maintained wherever possible. Articles VI and IX have necessarily been reset with a new pagination, and with the original page numbers given in square brackets within the text.*

*Each article has been given a Roman number in order of appearance, as listed in the Contents. This number is repeated on each page and is quoted in the index entries.*

*Asterisks in the margins are to alert the reader to additional information supplied at the end of the article.*

# PREFACE

As I write these words it is somewhat more than a quarter century ago that I started my researches in the history of Islamic astronomy under the supervision of Jan Hogendijk and Henk Bos at Utrecht University and with the personal inspiration of some of the greats of the field, especially Edward S. Kennedy (1912–2009) and David A. King (1941–). Working in the tradition of Otto Neugebauer (1899–1990), these two scholars had explained the meaning and the method of computation of numerous mathematical tables from what in Arabic and Persian are called *zījes* (singular pronounced as *zeej*), astronomical handbooks with tables and explanations for their use. These works, of which hundreds are known to have been compiled in the Islamic realm, were written in the tradition of Ptolemy's *Almagest* and *Handy Tables* and could be used by any practising astronomer or astrologer to calculate the positions of the planets and other important astronomical and astrological quantities by means of a small number of simple operations on numbers from the tables.

In 1956, Kennedy had published his 'Survey of Islamic Astronomical Tables' (*Transactions of the American Philosophical Society*, New Series 46-2, pp. 123–177; reprinted 1990), in which he listed the 125 *zījes* that were known at the time, summarized the subject matter usually found in these works, listed the mathematical characteristics of 12 of the most important *zījes*, and gave a preliminary overview of the developments in Islamic astronomy in general. During the following quarter century, Kennedy and King found manuscripts of, or references to, more than 100 further Arabic and Persian *zījes* (many of these were briefly described in David A. King, Julio Samsó & Bernard R. Goldstein, 'Astronomical handbooks and tables from the Islamic world (750–1900): an interim report', *Suhayl* 2 (2001), pp. 9–105). They had used mainframe computers to recompute mathematical tables from Islamic sources, and realized that the emergence of the personal computer during the 1980s offered enormous possibilities for further work on the publication and analysis of medieval astronomical tables. It was in this exact direction that I started working on a master's thesis, and later continued with my doctoral dissertation (*Ancient and Mediaeval Astronomical Tables: mathematical structure and parameter values*, Mathematical Institute, Utrecht University, 1993).[1]

[1] A copy of my dissertation is present in many university libraries around the world that have an exchange programme for Ph. D. theses with Utrecht University.

My research in this period was primarily devoted to the development of statistical estimators of astronomical parameter values underlying mathematical tables. Such parameter values, which are not usually indicated in the accompanying text, can be of great help in determining the interdependence of tables, and hence of whole astronomical works.

Articles I–V in this volume illustrate the type of historical conclusions that can be drawn on the basis of parameter values extracted from astronomical tables by means of statistical estimators and numerical tools. Full explanations of the way in which a least squares estimation can be applied to a table for the equation of time and the Least Number of Errors Criterion to mean motion tables are provided in articles IV and V. Article I introduces an intuitive estimator for a single unknown parameter value which finds an estimate by calculating a weighted average of estimates based on single tabular values, giving the largest weight to those tabular values that are most strongly influenced by changes in the parameter value. In two extensive examples the weighted estimator makes it possible to determine which of some very close historically attested values was used for the calculation of the table.

Articles II and IV tackle two hitherto unexplained tables for the equation of time, a complicated function based on four astronomical parameters. Both for the table in Ptolemy's *Handy Tables* (II) and for that in the Latin version of the *zīj* of al-Khwārizmī (IV) a full survey of the available information in historical sources and the results of earlier research are combined with an estimation of the underlying parameter values by means of the method of least squares and a detailed investigation of the methods of computation used. It is shown that al-Khwārizmī's table as found in the Latin translation of al-Majrīṭī's recension for Cordoba (*ca*. 1000) combines rounded Ptolemaic parameters with newly determined Islamic ones stemming from the observations made under the caliph al-Maʾmūn in Baghdad in 829 and 830. The introduction to article IV summarizes all earlier research on al-Khwārizmī's tables and categorizes them according to Indian, Persian or Ptolemaic characteristics.

Article III introduces a statistical estimator that can be used for estimating a 'translation parameter', such as the longitude of a planetary apogee, without knowing the actual tabulated function. With its help a second instance of the solar equation of Yaḥya ibn Abī Manṣūr (d. 830), author of the earliest surviving Islamic *zīj*, could be identified. This solar equation, not found in the two extant manuscripts of Yaḥyā's *zīj* (see below), is based on a Persian method and a value for the maximum solar equation not found in any other sources.

Article V deals with the mean motion tables in the *Muḥammad Shāh Zīj*, written in the 1730s on the order of the maharajah Jai Singh, who is particularly well-known for the huge masonry observatories that he had built in Delhi, Jaipur and three other Indian cities. The article shows how a highly accurate determination of the daily mean motions underlying the tables in the *zīj* makes it possible to determine the exact way in which Jai Singh's astronomers calculated the mean motion tables from those in the *Tabulae astronomicae Ludovici Magni* of De La Hire, which is known to have been available in Jaipur.

The last four articles in this volume are historical investigations of important *zīj*es or groups of *zīj*es.

Article VI (reset from my original contribution to the *Festschrift* for David Pingree that appeared with Brill in 2004) studies the earliest *zīj* written in Muslim India around the year 1250. It turns out to be part of the influential tradition of the *'Alā'ī Zīj* by al-Fahhād al-Shirwānī (*ca.* 1175), which was translated into Byzantine Greek in the early 14th century. A comparison of the planetary tables in a number of works from the ʿAlāʾī tradition allows the determination of the exact sources on which their authors depended.

Article VII provides a full reconstruction and description of the second manuscript of the already mentioned earliest extant Islamic *zīj*, the *Mumtaḥan Zīj* by Yahyā ibn Abī Manṣūr (*ca.* 830). Until its discovery, research on the *Mumtaḥan Zīj* (by Vernet, Kennedy, Viladrich and others) had been entirely based on the Escorial manuscript. The Leipzig manuscript, which I inspected and was able to identify thanks to a hint in old research notes by David King, turned out to contain much of the same materials found in the Escorial manuscript, but also further early texts and tables attributed to al-Maʾmūn, Yahyā, Ḥabash al-Ḥāsib, al-Battānī, and Ibn al-Aʿlam. It therefore contributes significantly to our knowledge of early Islamic astronomy.

Article VIII (co-authored with Fritz S. Pedersen) revisits the monumental publication (1899–1907) by Carlo Alfonso Nallino of the *Ṣābiʾ Zīj* by the famous observer al-Battānī (*ca.* 900). Whereas Nallino's edition and Latin translation of the Arabic text and his commentary (likewise in Latin) remain very valuable, we showed that his transcription of some of the tables from the *Ṣābiʾ Zīj* is unreliable, because he had tabular values with larger errors corrected by the astronomer Giovanni Schiaparelli without noting the original values found in the unique Escorial manuscript. For the determination of interdependences with earlier and later tables it is therefore absolutely necessary to consult this manuscript and that of the Castilian translation of the *zīj* prepared for Alfonso XII.

Article IX, finally, a newly typeset version of an article from Brill's *Encyclopaedia of Islam, New Edition*, presents an overview of methods for calculating and converting calendar dates based on a survey of a representative set of important and mostly unpublished Islamic *zīj*es.

Recently the prospects of the history of Islamic astronomy have not been very promising. The death of the grand old man of the field, Edward S. Kennedy, in May 2009 at the age of 97 may in fact have signified the end of an age of intensive studies of technical aspects of Arabic and Persian astronomical sources (cf. my obituary in *Historia Mathematica* 37 (2010), pp. 159–163). I hope that the publication of this volume of studies will contribute to the continuation of thorough investigations of the sources of Islamic astronomy and that in the coming decades the field will once again receive the attention it deserves.

BENNO VAN DALEN

*Frankfurt am Main*
*April 2013*

# ACKNOWLEDGEMENTS

For permission to reproduce the various articles in this volume I am grateful to: John Wiley & Sons Ltd, Chichester, UK (I and II); Franz Steiner Verlag GmbH, Stuttgart (III); Professor Julio Samsó, editor of the journal *Suhayl*, and the Instituto 'Millás Vallicrosa', Barcelona (IV and VII); Indian National Science Academy, New Delhi (V); Fritz S. Pedersen (co-author) and the Deutsche Akademie der Naturforscher Leopoldina, Halle (VIII). Articles VI and IX were first published by Brill, Leiden, and were newly typeset for this volume. I am also very grateful to John Smedley and Lindsay Farthing for their extremely helpful advice and continuous support in the preparation of this volume.

Every effort has been made to trace all the copyright holders, but if any have been inadvertently overlooked the publishers will be pleased to make the necessary arrangement at the first opportunity.

I

# A Statistical Method for Recovering Unknown Parameters from Medieval Astronomical Tables

*Contents*



## 1. *Introduction*

In 1956 E. S. Kennedy published a survey of 125 medieval Islamic astronomical handbooks (*zīj*es), most of which were not previously known.[1] Since then another 100 works have come to light.[2] Many of the zījes are extant, but only very few have been published or extensively studied.[3] The determination of parameter values used for the computation of tables in zījes is an important topic, since these values may indicate the origin of the tables or may be useful in identifying later related tables. In many cases the parameter values are not explicitly mentioned in the table headings or in the explanatory text and must be recovered from the tabular values.

This paper concerns the problem of recovering a single unknown parameter from an astronomical table. Until now especially E. S. Kennedy and D. A. King have treated this problem. Kennedy has succeeded in recomputing a number of tables from zījes by means of computer programs. He tried several historical values of the unknown parameter to obtain the best fit with the table. Successful examples of this approach can be found in Kennedy's collected works.[4] For instance, Kennedy (together with Salam) showed that the maximum lunar latitude involved in one of the lunar tables in the sole surviving manuscript of the 9th century Mumtaḥan Zīj, which was compiled by a team of astronomers headed by Yaḥyā ibn Abī Manṣūr, is the value attributed to them in later sources.[5] In 1977 Kennedy showed that a solar equation table in the Ashrafī Zīj, attributed to Yaḥyā and computed by means of an approximation method involving the solar declination, is based on the Ptolemaic value of the obliquity of the ecliptic rather than the Indian or any of the Islamic values.[6] Kennedy started a parameter file, which now contains over 1500 entries: values of the obliquity of the ecliptic, mean motion parameters, planetary apogee positions, eccentricities and maximum equations, and other astronomical parameters. Kennedy also published a list of all geographical coordinates occurring in Islamic sources, arranged according to source, locality, latitude and longitude.[7]

King has made use of computers for recomputing a large number of Islamic astronomical tables. In some cases he verified that the parameters mentioned in the headings and in the explanatory text actually underlied the tables, in other cases he determined the underlying parameters by means of trial-and-error. Furthermore he investigated the resulting error patterns. Examples of King's approach can be found in his papers on timekeeping.[8]

J. D. North gave useful methods for approximating unknown parameters in astronomical tables on several occasions. In his book

*Richard of Wallingford* he gives numerous simple rules for determining the eccentricities and epicycle radii underlying planetary equation tables.[9] In his book *Horoscopes and History*, North studies the problem of determining the way of computation of a given horoscope, as well as finding the latitude for which the horoscope is intended. Again he indicates several methods for approximating unknown parameters from one or two values in tables. In addition, he determines a more accurate approximation to an unknown parameter by computing thc arithmetic mean of approximations computed from single tabular values.[10]

J. P. Hogendijk gave a method for determining the underlying parameters of a table for predicting lunar crescent visibility.[11] By assuming a value of the obliquity of the ecliptic, he computes values for the geographical latitude from pairs of tabular values. If the table was computed according to a so-called "solar criterion", the values for the latitude obtained in this way tend to be equal.

After determining an approximation to the unknown parameter, North and Hogendijk search in its neighbourhood for historically plausible values, i.e. round numbers or parameter values which are attested in the sources. They do not study the errors in the approximations systematically.

The methods described above suffice to give a *rough* approximation to the unknown parameter, and in some cases leave little doubt about the historical value underlying the table under concern. In many other cases, however, more sophisticated methods for approximating unknown parameters are necessary to answer questions like:

- Which neighbouring parameter values yield an equally good or even better recomputation than the approximation? In particular, are there parameter values not attested in the sources which yield a better recomputation than attested values close to the approximation?
- Which of two very close attested parameter values more likely underlies the table? Note, for instance, that two right ascension tables with values to minutes based on the attested obliquities 23°32′30″ and 23°33′ (see Section 4) may be very difficult to distinguish.[12] Furthermore a declination table with entries to minutes

would not display the exact obliquity at 90° if it were based on a value to seconds.

In this paper I shall use statistical methods to find accurate approximations to a single unknown parameter in an astronomical table. These approximations will be determined by means of so-called “statistical estimators”. In principle the estimators make use of all tabular values. As a result they will be more accurate than approximations computed from only one or two tabular values and less sensitive to computational and scribal errors. Therefore the estimators will also be able to distinguish between parameter values which lead to nearly identical tables.

The results of the estimators will usually be given together with a so-called “confidence interval”, which has a fixed probability (usually 95%) of containing the unknown parameter. By means of the confidence interval it can be decided which historically plausible value of the parameter underlies the table in question. If no known historical value is contained within the confidence interval, the table may be based on a hitherto unattested value.

Examples of estimators and confidence intervals will be given for the cases of the obliquity of the ecliptic in a right ascension table and of the solar eccentricity in a solar equation table. Single unknown parameters in other types of astronomical tables can be determined by means of estimators analogous to the ones in the examples. In a subsequent paper I will present a method for recovering multiple unknown parameters. In addition, I am preparing a software package that can be used for analysing various important types of tables.

## 2. *Notation and some astronomical formulae*

The following notation will be used:

| | |
|---|---|
| $T$ | table to be investigated |
| $T(x)$ | tabular value for argument $x$ |
| $\theta$ | unknown parameter underlying the table $T$ |
| $f_\theta$ | tabulated function with parameter $\theta$ |
| $f_\theta(x)$ | functional value for argument $x$ of the tabulated function |

| | |
|---|---|
| $e_\theta(x)$ | tabular error for argument $x$: $e_\theta(x) = T(x)-f_\theta(x)$ |
| $\hat{\theta}$ | statistical estimator for $\theta$ |
| $\lambda$ | ecliptic longitude |
| $\varepsilon$ | obliquity of the ecliptic |
| $\alpha(\lambda)$ | right ascension of a point on the ecliptic |
| $\alpha'(\lambda)$ | normed right ascension of a point on the ecliptic |
| $\bar{a}$ | mean solar anomaly |
| $e$ | solar eccentricity |
| $q(\bar{a})$ | solar equation for mean anomaly $\bar{a}$ |
| Sin | sine for radius 60 ($\text{Sin}\, x = 60 \sin x$) |
| $\langle a,b\rangle$ | open interval, not including $a$ and $b$ |
| $[a,b\rangle$ | half-open interval, including $a$ but not $b$. |

Sexagesimal numbers will be denoted in the usual way: sexagesimal digits are separated by commas, and a semicolon is used to separate the integral and fractional parts; e.g. 1,19;25,55 denotes $1 \cdot 60^1 + 19 + 25 \cdot 60^{-1} + 55 \cdot 60^{-2}$. No distinction will be made between arcs and quantities. For computing the accuracy of approximations to unknown parameters, I will usually deal with the number of sexagesimal fractional digits of a tabular value (denoted by $k$), rather than with its total number of sexagesimal digits.

*Right ascension.* The right ascension is given by the modern formula

$$\begin{cases} \alpha(\lambda) = \arctan(\tan\lambda \cdot \cos\varepsilon) & \lambda \in [0,90\rangle \\ \alpha(90) = 90, \end{cases} \tag{1}$$

where $\lambda$ is the ecliptic longitude and $\varepsilon$ the obliquity of the ecliptic. For $\lambda \in \langle 90,360]$ the right ascension follows from $\alpha(\lambda) + \alpha(180-\lambda) = 180$ or $\alpha(\lambda + 180) - \alpha(\lambda) = 180$. These symmetries are apparent in virtually all tables for the right ascension. Therefore it will usually be sufficient to study the tabular values for $\lambda \in [0,90]$.

In his Almagest Abū'l-Wafā' describes a method for computing the right ascension which is "mathematically equivalent" to (1).[13] For $\cos\varepsilon$ he gives the value 54;59,59,5. This value also occurs in the Mumtaḥan Zīj in a table for multiples of $\cos\varepsilon$, which is attributed to Abū'l-Wafā' and which facilitates the computation of the right ascension according to formula (1).[14] However, Ptolemy and most Islamic astronomers computed the right ascension in a way which is mathematically equivalent to the modern

$$\alpha(\lambda) = \arcsin\left(\frac{\tan\delta(\lambda)}{\tan\varepsilon}\right) \quad \lambda\in[0,90], \qquad (2)$$

where $\delta(\lambda) = \arcsin(\sin\lambda \cdot \sin\varepsilon)$ is the declination of a point on the ecliptic.[15] Tables for $\tan\delta(\lambda)$, which facilitate the computation according to formula (2), are found in the zījes of Ibn Yunus,[16] al-Baghdādī[17] and others.

In several cases the right ascension was tabulated starting from 0° Capricorn. Neugebauer calls this the "normed right ascension" and denotes it by $\alpha'(\lambda)$. We have

$$\alpha'(\lambda) = \alpha(\lambda) + 90, \qquad (3)$$

where $\alpha(\lambda)$ is found from formula (2). The normed right ascension was useful for determining the oblique ascension of the ascendant and hence the longitude of the ascendant itself, which was of prime importance in astrology.[18]

In Islamic astronomical tables the obliquity $\varepsilon$ ranges from the common Indian value 24;0[19] to al-Kāshī's 23;30.[20] In principle it is possible that, when using formula (2), the obliquity underlying the declination table that was used was different from the obliquity assumed in the value for $\tan\varepsilon$. However, even two values of $\varepsilon$ with a very small difference (such as the Ptolemaic values 23;51 and 23;51,20) would lead to a value for $\alpha(90)$ differing from 90 by more than a degree, namely $\alpha(90) = 88;41$. Such an error would be unacceptable to all medieval astronomers. Thus I conclude that only one value of the obliquity is involved in any right ascension table.

*Solar equation.* The solar equation is given by the modern formula

$$q(\bar{a}) = \arctan\left(\frac{e\sin\bar{a}}{60 + e\cos\bar{a}}\right) \quad \bar{a}\in[0,360], \qquad (4)$$

where $\bar{a}$ is the mean solar anomaly and $e$ the solar eccentricity. Ptolemy and the Islamic astronomers used slightly different ways of computation, which were also based on geometrical proofs.[21] Therefore their algorithms involved the quantity $e$ rather than $\theta \stackrel{\text{def}}{=} 60/e$. The latter would simplify formula (4) and will be used for approximating the solar eccentricity in a solar equation table. In medieval tables we find eccentricity values ranging from the Ptolemaic value 2;30[22] to approximately 1;51.[23]

Because of the symmetry $q(\bar{a}) = -q(360-\bar{a})$, the solar equation is usually tabulated only for arguments 0 to 180. The solar equation is also involved in tables which give the true solar longitude as a function of the mean solar longitude or as a function of the days of the year.[24] I will study such tables in a later paper.

For a more extensive description of the right ascension and the solar equation, and of the Ptolemaic solar model in general, the reader is referred to Pedersen's *Survey of the Almagest.*[25]

*Errors.* In the sequel a clear distinction will be made between tabular errors and errors with respect to a recomputation. The *tabular error* $e_\theta(x)$ is defined by $e_\theta(x) = T(x)-f_\theta(x)$, where $T(x)$ is a tabular value of $T$, the table to be investigated, and $f_\theta(x)$ is a functional value of the tabulated function. In general, a tabular error will have an infinite number of sexagesimal fractional digits. An extensive description of tabular errors is given in Appendix A2.

A *recomputation* of $T$ for a specific parameter value $\eta$ is given by values $f_\eta(x)$ rounded to the number of sexagesimals of $T$. The differences between table and recomputation (table minus recomputation) will be called the *errors with respect to the recomputation.* Note that these errors have the same number of sexagesimal fractional digits as the tabular values.

*Example.* Assume that $T$ is a table for the right ascension $\alpha_{23;35}$ based on obliquity $\varepsilon = 23;35$. Furthermore assume that the tabular value for longitude 45° is given by $T(45) = 42;31$. Since $\alpha_{23;35}(45) = 42;30,16,35,23,\ldots$, $T(45)$ contains a tabular error $e_{23;45}(45) = T(45) - \alpha_{23;35}(45) = 0;0,43,24,36,\ldots$ The error in $T(45)$ with respect to the recomputation for $\varepsilon = 23;35$ is $+0;1$, since $\alpha_{23;35}(45)$ is rounded to

42;30. On the other hand, $T(45)$ is not in error with respect to the recomputation for $\varepsilon = 23;33$, because $\alpha_{23;33}(45) \approx 42;30,43$.

## 3. *Preliminaries*

In this section the statistical estimators that can be used for approximating the obliquity of the ecliptic in a right ascension table and the solar eccentricity in a solar equation table will be explained in an informal way. All statistical details have been relegated to the appendices. Readers with a background in statistics may skip the present section and read appendices A3 and A4 instead. Extensive examples of the use of the estimators are presented in Sections 4 and 5.

*Right ascension.* Assume that we have a table $T$ for the right ascension with tabular values $T(\lambda)$, $\lambda = 1,2,3,\ldots,90$.[26] Furthermore, assume that the obliquity of the ecliptic underlying this table is unknown. The tabular values are supposed to be close to the exact right ascension, thus

$$T(\lambda) \approx \arctan(\cos\varepsilon \cdot \tan\lambda) \tag{5}$$

for every $\lambda$. This implies that, for every $\lambda$, $\varepsilon$ can be approximated by means of

$$\varepsilon \approx \arccos\left(\frac{\tan T(\lambda)}{\tan\lambda}\right). \tag{6}$$

However, not every $\lambda$ yields an equally accurate approximation to $\varepsilon$. In particular, the approximations are inaccurate close to arguments 0 and 90, at which the right ascension assumes a value independent of $\varepsilon$. As an example, note that every value of the obliquity in the range 23;30 to 23;51,20 leads to a tabular value $T(4) = 3;40$ (the right ascension is given to minutes in most cases). Thus from $T(4)$ hardly any conclusion concerning the obliquity can be drawn. It will be shown in Appendix A3 that the most accurate approximation to $\varepsilon$

computed from a single tabular value is the one for $\lambda = 47$. However, if the tabular values are given to minutes, this approximation is not sufficient for determining the obliquity with certainty. For instance, the attested values 23;30, 23;30,17, 23;32,29, 23;32,30, 23;33 and 23;33,30[27] all lead to $T(47) = 44;31$.

To obtain a more accurate approximation to the obliquity, an average of the separate approximations can be computed. This average will be a weighted average such that accurate separate approximations give a larger contribution to the average than less accurate ones. It can be shown that the best result is obtained if the weights are taken inversely proportionate to the expected errors in the separate approximations. This turns out to imply that the weights are dependent on the obliquity underlying the table, which brings us in an infinite loop. However, the result of the weighted average is not significantly worse if the obliquity value used for the computation of the weights is only a rough approximation, e.g. $\varepsilon = \arccos(\tan T(47)/\tan 47)$, indicated above as the best approximation from a single tabular value.

The computation of both the weighted average and its accuracy can be simplified drastically by averaging approximations to $\cos\varepsilon$ instead of approximations to $\varepsilon$. Thus I define $\theta$ by $\theta = \cos\varepsilon$ and the rough approximation $\theta_0$ to $\theta$ by $\theta_0 = \tan T(47)/\tan 47$. Now $\theta$ can be well approximated by the weighted average $\hat{\theta}$ given by

$$\hat{\theta} = \frac{1}{W}\sum_{\lambda=1}^{89} \frac{\tan\lambda \cdot \tan T(\lambda)}{(1 + \theta_0^2 \tan^2\lambda)^2}, \tag{7}$$

where

$$W = \sum_{\lambda=1}^{89} \left(\frac{\tan\lambda}{1 + \theta_0^2 \tan^2\lambda}\right)^2$$

is the sum of the weights (the derivation of these formulas can be found in Appendix A3). In the sequel, $\hat{\theta}$ will be called the ***weighted estimator*** for the obliquity of the ecliptic in a right ascension table. An approximation $\hat{\varepsilon}$ to the obliquity itself is obtained by putting $\hat{\varepsilon} = \arccos\hat{\theta}$.

The weighted estimator alone is not sufficient to determine the obliquity used for the computation of the table. In addition, we need to know how much the actual obliquity can be expected to differ from the result of the weighted estimator. This information will be provided by a so-called "95% confidence interval", which has a 95% probability of containing the obliquity. In the present case an approximate 95% confidence interval for $\varepsilon$ is given by

$$\langle \arccos\hat{\theta} + 0.15\sigma^2 - 1.35\sigma, \arccos\hat{\theta} + 0.15\sigma^2 + 1.35\sigma\rangle, \qquad (8)$$

where $\sigma^2$ is the variance of the tabular errors (the concept of variance is explained in Appendix A1). $\sigma^2$ can be approximated by

$$\sigma^2 = \frac{1}{88}\sum_{\lambda=1}^{89}(T(\lambda) - \alpha_{\hat{\varepsilon}}(\lambda))^2, \qquad (9)$$

where $\alpha_{\hat{\varepsilon}}(\lambda)$ are right ascension values computed for an obliquity equal to the approximation $\hat{\varepsilon}$.

Once the confidence interval has been determined, a historical conclusion concerning the unknown parameter can be drawn. Thus we search within the interval for a round number or a value attested in the sources. The obliquity value used for the computation of the table $T$ most likely lies in the middle part of the confidence interval, but may incidentally lie close to one of the bounds.

It will be shown in Appendix A3 that the results obtained from the weighted estimator and the confidence interval (8) are nearly seven times as accurate as those obtained by using (6) for $\lambda = 47$. In the examples in Section 4, this will appear to be crucial for the determination of the obliquity.

For right ascension tables that were accurately computed and correctly rounded, another useful statistical estimator for the obliquity of the ecliptic exists. If the tabular values $T(\lambda)$ are given to minutes, we have for every argument $\lambda$

$$\alpha(\lambda) \in [T(\lambda) - 0;0,30,\ T(\lambda) + 0;0,30\rangle, \qquad (10)$$

since this interval contains all numbers that would be rounded to

$T(\lambda)$.[28] Since $\alpha(\lambda) = \arctan(\tan\lambda \cdot \cos\varepsilon)$, we obtain from (10)

$$\cos\varepsilon \in \left[\frac{\tan(T(\lambda) - 0;0,30)}{\tan\lambda}, \frac{\tan(T(\lambda) + 0;0,30)}{\tan\lambda}\right\rangle \tag{11}$$

for every $\lambda \neq 0,90$. Thus we have 89 intervals containing $\cos\varepsilon$. This implies that $\cos\varepsilon$ is also contained in the intersection $I$ given by

$$I = \bigcap_{\lambda=1}^{89} \left[\frac{\tan(T(\lambda) - 0;0,30)}{\tan\lambda}, \frac{\tan(T(\lambda) + 0;0,30)}{\tan\lambda}\right\rangle. \tag{12}$$

Note that if one or more of the tabular values is in error, $I$ will probably equal the empty set. If $I$ is not empty, all numbers in $I$ are equally valid approximations to $\cos\varepsilon$, since they all lead to a recomputed table identical with $T$.

Usually the width of $I$ is much smaller than the width of the 95% confidence interval given above. However, since for only very few tables all tabular values were accurately computed and correctly rounded, it is hardly ever possible to use the interval $I$ in its plain form. In Appendix A4 I will give the statistical basis for an estimator, called *maximum likelihood estimator*, which extends the use of the interval $I$ to tables that were not accurately computed and correctly rounded. It will be shown that under certain conditions the obliquity of the ecliptic can be approximated by those values of $\varepsilon$ for which the number of errors in the table with respect to recomputation is smallest. Thus we see that Kennedy's approach to the problem of unknown parameters as described in the introduction is in fact based on the maximum likelihood criterion. Although the maximum likelihood estimator yields good results in the examples in this paper, further investigation is necessary to establish its reliability. Especially for tables with a large number of errors it is advisable to use the weighted estimator and not to rely too heavily on the maximum likelihood estimator.

*Solar equation.* The unknown solar eccentricity $e$ underlying a solar equation table $T$ can be approximated in a way completely analogous to the above. Thus let $T(\bar{a})$, $\bar{a} = 1,2,3,\ldots,179$, be the tabular values.

For all $\bar{a}$ we have

$$e \approx \frac{60}{\dfrac{\sin \bar{a}}{\tan T(\bar{a})} - \cos \bar{a}}. \tag{13}$$

Dependent on the value of $e$, $\bar{a} = 94$ or $\bar{a} = 95$ yields the best approximation to the eccentricity based on a single tabular value. However, this approximation may be insufficient: if the tabular values are given to minutes, the attested values 2;4,45[29] and 2;4,35,30[30] yield equal tabular values for $\bar{a} = 87$ to $\bar{a} = 97$.

The computation of the accuracy of the weighted estimator for the eccentricity can be simplified by averaging approximations to $\theta = 60/e$ instead of to $e$. A rough approximation to $\theta$ is given by $\theta_0 = 1/\tan T(90)$. The weighted estimator $\hat{\theta}$ of $\theta$ is given by

$$\hat{\theta} = \frac{1}{W} \sum_{\bar{a}=1}^{179} w_{\bar{a}} \left( \frac{\sin \bar{a}}{\tan T(\bar{a})} - \cos \bar{a} \right), \tag{14}$$

where $w_{\bar{a}}$, $\bar{a} = 1,2,3,\ldots,179$, are the weights of the separate approximations and

$$W = \sum_{\bar{a}=1}^{179} w_{\bar{a}}.$$

The best result is obtained if the weights are chosen according to

$$w_{\bar{a}} = \left( \frac{\sin \bar{a}}{\theta_0^2 + 2\theta_0 \cos \bar{a} + 1} \right)^2.$$

An approximation $\hat{e}$ to the eccentricity itself is found from $\hat{e} = 60/\hat{\theta}$. The variance $\sigma^2$ of the tabular errors can be approximated by

$$\sigma^2 = \frac{1}{178} \sum_{\bar{a}=1}^{179} (T(\bar{a}) - q_{\hat{e}}(\bar{a}))^2, \tag{15}$$

where $q_{\hat{e}}(\bar{a})$ are solar equation values computed for an eccentricity equal to the approximation $\hat{e}$. An approximate 95 % confidence interval for $e$ is given by

$$\langle 60/\hat{\theta} + \sigma^2 - 0.22\sigma, 60/\hat{\theta} + \sigma^2 + 0.22\sigma \rangle. \qquad (16)$$

In the case of the solar eccentricity in a solar equation table, the approximation by means of the weighted estimator is more than nine times as accurate as the result obtained from (13) for the optimal $\bar{a} = 94$ or $\bar{a} = 95$. This implies that the weighted estimator is able to distinguish the attested eccentricities 2;4,45 and 2;4,35,30 in a table with values given to minutes. In the example in Section 5 the weighted estimator will enable the determination of the accuracy used by the author of the table in his calculations.

The solar eccentricity in a solar equation table can also be determined by means of the maximum likelihood estimator. If the table was correctly computed and has values to minutes, then the interval $I$ containing $60/e$ is given by

$$I = \bigcap_{\bar{a}=1}^{179} \left\langle \frac{\sin \bar{a}}{\tan(T(\bar{a}) + 0;0,30)} - \cos \bar{a}, \frac{\sin \bar{a}}{\tan(T(\bar{a}) - 0;0,30)} - \cos \bar{a} \right]. \qquad (17)$$

In practice it will be necessary to check the conditions of the statistical estimators before any conclusions from the results are drawn. For the computation of the confidence intervals (8) and (16) I assumed that the tabular errors are independent, i.e. that there is no obvious relation between any of the tabular errors. The maximum likelihood estimator is based on the assumptions that few tabular values are in error, and that the tabular errors corresponding to correctly computed tabular values are uniformly distributed, i.e. that all values of these tabular errors occur with equal frequency. I will illustrate these conditions by two made-up examples:

The tabular errors −0;24, −0;12, 0;0, +0;12, +0;24 are uniformly distributed but highly dependent, because their differences are constant.

The tabular errors −0;22, −0;2, −0;17, −0;11, −0;5 are independent but not uniformly distributed, since no positive errors occur.

In Appendix A2 tests are indicated which can be used to verify the conditions. In Sections 4 and 5 it is shown how the recovery of an unknown parameter proceeds if the conditions of the statistical estimators are tested.

## 4. *The obliquity of the ecliptic in the Sanjufīnī Zīj*

In the Bibliothèque Nationale in Paris there is a unique copy of a zīj completed in 1366 by Abū Muḥammad ʻAṭā ibn Aḥmad al-Sanjufīnī.[31] The author originated from the region of Samarqand in Central Asia, but his work is dedicated to the viceroy of Tibet, a direct descendant of Genghis Khan.[32] In brief, the zīj is a very interesting mixture of Islamic, Mongolian and Tibetan influences.

At least three different values of the obliquity of the ecliptic are incorporated in the Sanjufīnī Zīj. Firstly, Kennedy and Hogendijk showed that the tables for parallax on folio 42$^v$ and for determining the visibility of the lunar crescent on folio 38$^r$ were computed using the Ptolemaic value 23;51.[33] Secondly, the common Islamic value 23;35 underlies the table of the solar meridian altitude on folio 40$^r$.[34] Thirdly, the declination table on folio 37$^v$ is based on $\varepsilon$ = 23;32,30, a value which has so far been found only in the Zīj of Ibn Isḥāq al-Tūnisī,[35] and which is very close to the 23;32,29 associated with the Alphonsine tables.[36] Since there is as yet no historical reason to assume a relationship between Ibn Isḥāq and the Alphonsine tables on the one hand and al-Sanjufīnī on the other, the value 23;32,30 may in the present context be considered as typical to al-Sanjufīnī. The same holds for the geographical latitude 38°10′, which is explicitly mentioned in the headings of five tables in the Sanjufīnī Zīj. This value underlies the table of the solar meridian altitude[37] and was shown to be involved in the table for determining the visibility of the lunar crescent.[38] Furthermore it corresponds to a locality in Tibet which is mentioned in the manuscript.[39]

Below I will investigate the right ascension table and one of the two oblique ascension tables in the Sanjufīnī Zīj. I will use the estimators presented in Section 3 to determine the values of the obliquity underlying those tables. On the basis of the results, the origin of the tables will be discussed.

*Right ascension table.* The Sanjufīnī Zīj has a table for the normed right ascension (see Section 2) on folio 39$^r$. For the analysis of this table I will use only the first 90 tabular entries, which will be denoted by $T(\lambda')$, $\lambda' = 1,2,3,\ldots,90$.[40] These values contain a single scribal error, noted in the apparatus. A transcription of the corrected table can be found in Table 1. $T(43)$ yields the most accurate approximation to the obliquity computed from a single tabular value. It is correctly recomputed for all $\varepsilon \in \langle 23;33,34,\ 23;38,8\rangle$. Within this interval, $\varepsilon = 23;35$ is the only frequently used attested value. Nevertheless, because of the possibility of an error in $T(43)$, it seems advisable to compute a more accurate approximation.

| | | | | | | | |
|---|---|---|---|---|---|---|---|
| 1 | 1; 5 | 24 | 25;55 | 47 | 49;29 | 70 | 71;33 |
| 2 | 2;11 | 25 | 26;58 | 48 | 50;28 | 71 | 72;29 |
| 3 | 3;16 | 26 | 28; 1 | 49 | 51;27 | 72 | 73;25 |
| 4 | 4;22 | 27 | 29; 4 | 50 | 52;26 | 73 | 74;21 |
| 5 | 5;27 | 28 | 30; 7 | 51 | 53;25 | 74 | 75;17 |
| 6 | 6;32 | 29 | 31;10 | 52 | 54;24 | 75 | 76;12 |
| 7 | 7;38 | 30 | 32;13 | 53 | 55;22 | 76 | 77; 8 |
| 8 | 8;43 | 31 | 33;15 | 54 | 56;20 | 77 | 78; 3 |
| 9 | 9;48 | 32 | 34;17 | 55 | 57;18 | 78 | 78;59 |
| 10 | 10;53 | 33 | 35;19 | 56 | 58;16 | 79 | 79;54 |
| 11 | 11;58 | 34 | 36;21 | 57 | 59;14 | 80 | 80;49 |
| 12 | 13; 3 | 35 | 37;23 | 58 | 60;12 | 81 | 81;44 |
| 13 | 14; 8 | 36 | 38;24 | 59 | 61;10 | 82 | 82;40 |
| 14 | 15;13 | 37 | 39;26 | 60 | 62; 7 | 83 | 83;35 |
| 15 | 16;18 | 38 | 40;27 | 61 | 63; 4 | 84 | 84;30 |
| 16 | 17;23 | 39 | 41;28 | 62 | 64; 1 | 85 | 85;25 |
| 17 | 18;28 | 40 | 42;29 | 63 | 64;58 | 86 | 86;20 |
| 18 | 19;31 | 41 | 43;29 | 64 | 65;55 | 87 | 87;15 |
| 19 | 20;35 | 42 | 44;30 | 65 | 66;52 | 88 | 88;10 |
| 20 | 21;39 | 43 | 45;30 | 66 | 67;48 | 89 | 89; 5 |
| 21 | 22;43 | 44 | 46;30 | 67 | 68;45 | 90 | 90; 0 |
| 22 | 23;47 | 45 | 47;30 | 68 | 69;41 | | |
| 23 | 24;51 | 46 | 48;30 | 69 | 70;37 | | |

Table 1. Transcription of the right ascension table on folio 39$^r$.

The weighted estimator as given in formula (7) in Section 3 yields $\hat{\varepsilon} = 23;34,41$. By means of formula (8) we find that an approximate 95% confidence interval for $\varepsilon$ is given by $\langle 23;34,21,\ 23;35,14 \rangle$, provided that the condition of independence of the tabular errors is satisfied. This appears not to be the case, since the errors with respect to recomputations for obliquity values within the confidence interval occur in small groups in which they have the same sign. The statistical test indicated in Appendix A2 in fact shows that the errors can not be considered to be independent.

To determine a valid 95% confidence interval, I will apply the weighted estimator to a set of values of which the tabular errors do satisfy the condition of independence. In the case of a table computed by means of interpolation, a natural choice for such a set would be the independently-computed tabular values.[41] In the present table, however, no traces of interpolation can be found: there is no set of equidistant tabular values giving significantly smaller tabular errors than the intermediate values, and the differences of the tabular values do not display any regular pattern.

Another possible means of achieving independence of the tabular errors is to omit coherent parts of the table in which the errors are highly dependent, e.g. parts where the tabulated function is almost linear or where many errors with the same sign occur. An example of such a situation is found in the solar equation table in the Shāmil Zīj (see Section 5). In the present case, however, the above-mentioned small groups of errors with respect to recomputation are distributed all over the table, and leaving out a coherent part of the table would not remove the dependence of the tabular errors.

A third possibility is to compute the weighted estimator for subtables $T(\lambda_0 + k \cdot \Delta\lambda)$, $k = 0,1,2,\ldots$, where $\Delta\lambda$ is fixed and $\lambda_0 \in \{1,2,\ldots,\Delta\lambda\}$.[42] Since a number of entries of the original table is left out, the tabular errors in such subtables are expected to show a less obvious mutual relationship, i.e. to be less dependent. In fact, it appears that for the right ascension table in the Sanjufīnī Zīj the tabular errors for both even ($\Delta\lambda = 2$, $\lambda_0 = 2$) and odd ($\Delta\lambda = 2$, $\lambda_0 = 1$) arguments pass the test for independence. For even arguments we have $\hat{\varepsilon} = 23;34,50$ and a 95% confidence interval is given by $\langle 23;34,12,\ 23;35,27 \rangle$, for odd arguments we have $\hat{\varepsilon} = 23;34,45$ and a 95% confidence interval is given by $\langle 23;34,7,\ 23;35,23 \rangle$. As no other frequently used attested value of the obliquity is contained in either confidence interval, I conclude that for this table $\varepsilon = 23;35$.

The same result is found by means of the maximum likelihood estimator. The minimum possible number of errors with respect to recomputation is eight and is reached for $\varepsilon \in \langle 23;34,39,24, 23;34,55,51\rangle$. For $\varepsilon \in \langle 23;34,38,25, 23;34,39,24\rangle$ and $\varepsilon \in \langle 23;34,55,51, 23;35,2,55\rangle$ the number of errors is nine. The conditions for the application of the maximum likelihood estimator are satisfied: the table contains few errors with respect to the recomputation for $\varepsilon = 23;35$, and the tabular errors smaller than 0;0,30 pass the test for uniformity mentioned in Appendix A2.

The nine errors with respect to the recomputation for $\varepsilon = 23;35$ are given in the apparatus. Eight are between 30 and 40 seconds in absolute value and could easily have resulted from small inaccuracies in the intermediate steps of the computation. In this way, only the error for $\lambda' = 17$ remains unexplained. Because of the observed dependence of the tabular errors, I also tried several linear interpolation schemes. It turned out that none of them gave a better fit than exact computation.

*Oblique ascension table.* The oblique ascension $\varrho$ is given by $\varrho(\lambda) = \alpha(\lambda) - \Delta(\lambda)$, where $\Delta(\lambda) = \arcsin(\tan\delta(\lambda) \cdot \tan\varphi)$ is the ascensional difference, dependent both on the obliquity $\varepsilon$ (through the declination $\delta(\lambda) = \arcsin(\sin\lambda \cdot \sin\varepsilon)$) and on the geographical latitude $\varphi$.[43] By means of the relation $\varrho(\lambda) - \varrho(180-\lambda) = 2 \cdot \alpha(\lambda) - 180$, it is possible to recover the right ascension used for the computation of a given oblique ascension table. In the same manner the ascensional difference can be recovered using $\varrho(\lambda) + \varrho(180-\lambda) = 180 - 2\Delta(\lambda)$.

The Sanjufīnī Zīj has two oblique ascension tables, of which I shall study the one on folio 38$^{v}$ (see Plate 1). From its heading it appears that this table was computed for $\varphi = 38;10$. A transcription of the first half of the table is given in table 2, the reconstructed right ascension values appear in table 3. I will denote the tabular values of the oblique ascension table by $T_{\varrho}(\lambda)$, $\lambda = 1,2,3,\ldots,360$, and the tabular values of the reconstructed right ascension by $T_{\alpha}(\lambda)$, $\lambda = 1,2,3,\ldots,90$.

It can be noted that the values of both the right ascension and the ascensional difference used for the computation of the present oblique ascension table were given to seconds (or even more accurately). For

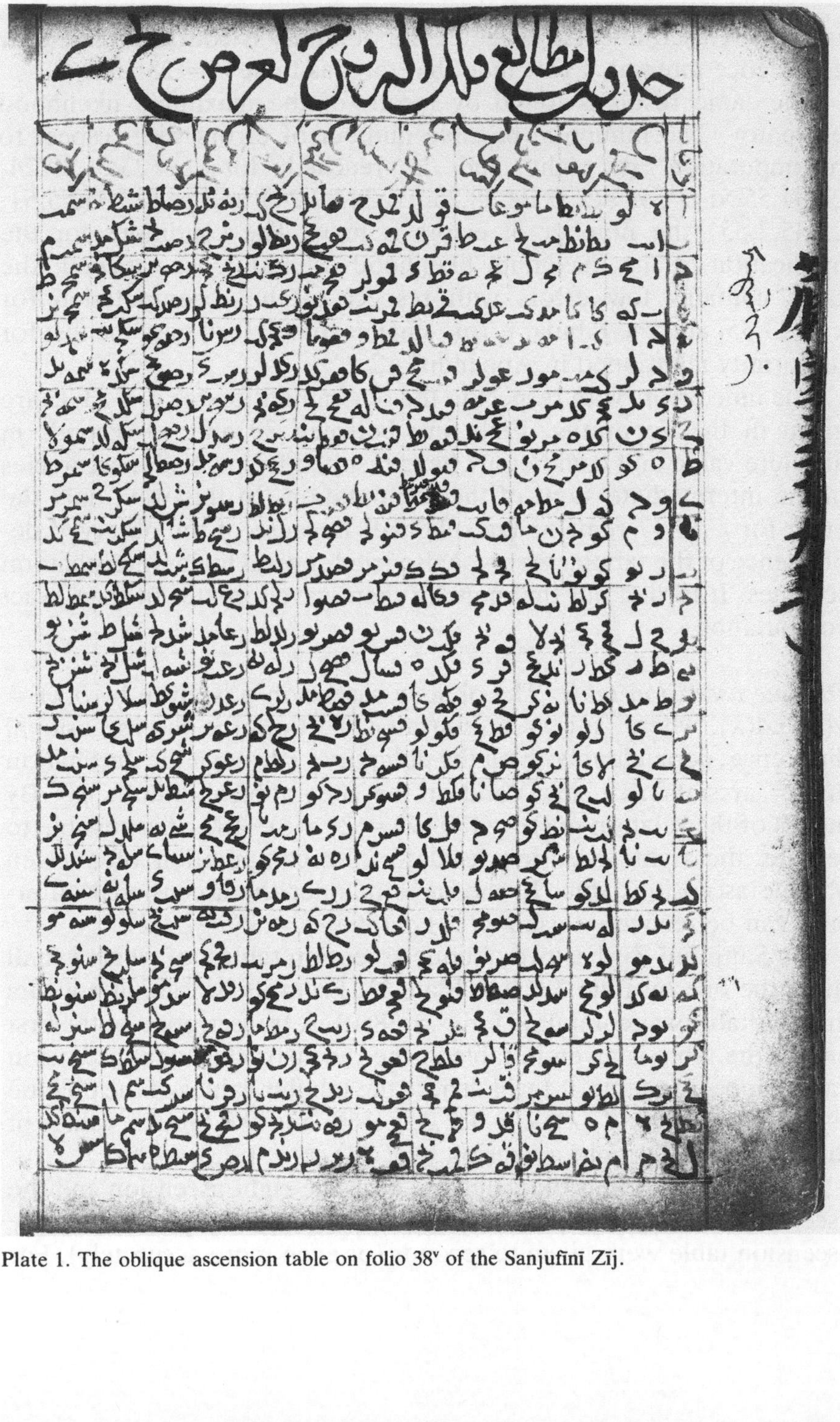

Plate 1. The oblique ascension table on folio 38ᵛ of the Sanjufīnī Zīj.

| | | | | | | | |
|---|---|---|---|---|---|---|---|
| 1 | 0;36 | 46 | 29;51 | 91 | 71; 2 | 136 | 125;21 |
| 2 | 1;12 | 47 | 30;36 | 92 | 72; 9 | 137 | 126;36 |
| 3 | 1;48 | 48 | 31;21 | 93 | 73;15 | 138 | 127;51 |
| 4 | 2;25 | 49 | 32; 8 | 94 | 74;22 | 139 | 129; 6 |
| 5 | 3; 1 | 50 | 32;52 | 95 | 75;29 | 140 | 130;21 |
| 6 | 3;37 | 51 | 33;39 | 96 | 76;37 | 141 | 131;36 |
| 7 | 4;14 | 52 | 34;26 | 97 | 77;45 | 142 | 132;52 |
| 8 | 4;50 | 53 | 35;12 | 98 | 78;54 | 143 | 134; 7 |
| 9 | 5;26 | 54 | 36; 0 | 99 | 80; 3 | 144 | 135;23 |
| 10 | 6; 3 | 55 | 36;48 | 100 | 81;12 | 145 | 136;38 |
| 11 | 6;40 | 56 | 37;37 | 101 | 82;22 | 146 | 137;53 |
| 12 | 7;16 | 57 | 38;27 | 102 | 83;33 | 147 | 139; 8 |
| 13 | 7;53 | 58 | 39;16 | 103 | 84;43 | 148 | 140;23 |
| 14 | 8;30 | 59 | 40; 5 | 104 | 85;53 | 149 | 141;38 |
| 15 | 9; 7 | 60 | 40;55 | 105 | 87; 4 | 150 | 142;53 |
| 16 | 9;44 | 61 | 41;46 | 106 | 88;16 | 151 | 144; 8 |
| 17 | 10;21 | 62 | 42;38 | 107 | 89;28 | 152 | 145;22 |
| 18 | 10;58 | 63 | 43;30 | 108 | 90;40 | 153 | 146;37 |
| 19 | 11;36 | 64 | 44;22 | 109 | 91;51 | 154 | 147;52 |
| 20 | 12;13 | 65 | 45;14 | 110 | 93; 3 | 155 | 149; 6 |
| 21 | 12;51 | 66 | 46; 7 | 111 | 94;16 | 156 | 150;21 |
| 22 | 13;29 | 67 | 47; 2 | 112 | 95;30 | 157 | 151;35 |
| 23 | 14; 7 | 68 | 47;56 | 113 | 96;43 | 158 | 152;50 |
| 24 | 14;46 | 69 | 48;50 | 114 | 97;56 | 159 | 154; 5 |
| 25 | 15;24 | 70 | 49;45 | 115 | 99; 9 | 160 | 155;19 |
| 26 | 16; 3 | 71 | 50;41 | 116 | 100;23 | 161 | 156;33 |
| 27 | 16;41 | 72 | 51;38 | 117 | 101;37 | 162 | 157;47 |
| 28 | 17;20 | 73 | 52;35 | 118 | 102;52 | 163 | 159; 2 |
| 29 | 18; 0 | 74 | 53;31 | 119 | 104; 6 | 164 | 160;16 |
| 30 | 18;40 | 75 | 54;28 | 120 | 105;20 | 165 | 161;30 |
| 31 | 19;19 | 76 | 55;27 | 121 | 106;34 | 166 | 162;44 |
| 32 | 19;59 | 77 | 56;26 | 122 | 107;50 | 167 | 163;59 |
| 33 | 20;40 | 78 | 57;26 | 123 | 109; 4 | 168 | 165;13 |
| 34 | 21;21 | 79 | 58;26 | 124 | 110;19 | 169 | 166;27 |
| 35 | 22; 1 | 80 | 59;26 | 125 | 111;34 | 170 | 167;40 |
| 36 | 22;42 | 81 | 60;27 | 126 | 112;48 | 171 | 168;54 |
| 37 | 23;24 | 82 | 61;28 | 127 | 114; 3 | 172 | 170; 8 |
| 38 | 24; 5 | 83 | 62;30 | 128 | 115;19 | 173 | 171;22 |
| 39 | 24;47 | 84 | 63;32 | 129 | 116;34 | 174 | 172;36 |
| 40 | 25;30 | 85 | 64;34 | 130 | 117;49 | 175 | 173;49 |
| 41 | 26;13 | 86 | 65;38 | 131 | 119; 4 | 176 | 175; 4 |
| 42 | 26;56 | 87 | 66;43 | 132 | 120;20 | 177 | 176;18 |
| 43 | 27;39 | 88 | 67;47 | 133 | 121;35 | 178 | 177;32 |
| 44 | 28;23 | 89 | 68;51 | 134 | 122;50 | 179 | 178;46 |
| 45 | 29; 7 | 90 | 69;56 | 135 | 124; 5 | 180 | 180; 0 |

Table 2. Transcription of the oblique ascension table on folio 38v.

| | | | | | | | |
|---|---|---|---|---|---|---|---|
| 1 | 0;55, 0 | 24 | 22;12,30 | 47 | 44;30,30 | 70 | 68;21, 0 |
| 2 | 1;50, 0 | 25 | 23; 9, 0 | 48 | 45;30,30 | 71 | 69;25, 0 |
| 3 | 2;45, 0 | 26 | 24; 5,30 | 49 | 46;32, 0 | 72 | 70;29, 0 |
| 4 | 3;40,30 | 27 | 25; 2, 0 | 50 | 47;31,30 | 73 | 71;33;30 |
| 5 | 4;36, 0 | 28 | 25;59, 0 | 51 | 48;32,30 | 74 | 72;37,30 |
| 6 | 5;30,30 | 29 | 26;56, 0 | 52 | 49;33,30 | 75 | 73;42, 0 |
| 7 | 6;26, 0 | 30 | 27;53,30 | 53 | 50;34,30 | 76 | 74;47, 0 |
| 8 | 7;21, 0 | 31 | 28;50,30 | 54 | 51;36, 0 | 77 | 75;51,30 |
| 9 | 8;16, 0 | 32 | 29;48, 0 | 55 | 52;37, 0 | 78 | 76;56,30 |
| 10 | 9;11,30 | 33 | 30;46, 0 | 56 | 53;39, 0 | 79 | 78; 2, 0 |
| 11 | 10; 6,30 | 34 | 31;44, 0 | 57 | 54;41,30 | 80 | 79; 7, 0 |
| 12 | 11; 1,30 | 35 | 32;41,30 | 58 | 55;43, 0 | 81 | 80;12, 0 |
| 13 | 11;57, 0 | 36 | 33;39,30 | 59 | 56;45,30 | 82 | 81;17, 0 |
| 14 | 12;53; 0 | 37 | 34;38,30 | 60 | 57;47,30 | 83 | 82;22,30 |
| 15 | 13;48,30 | 38 | 35;36,30 | 61 | 58;50, 0 | 84 | 83;27,30 |
| 16 | 14;44, 0 | 39 | 36;35,30 | 62 | 59;53, 0 | 85 | 84;32,30 |
| 17 | 15;39,30 | 40 | 37;34,30 | 63 | 60;56,30 | 86 | 85;38, 0 |
| 18 | 16;35,30 | 41 | 38;33,30 | 64 | 61;59,30 | 87 | 86;44, 0 |
| 19 | 17;31,30 | 42 | 39;32,30 | 65 | 63; 2,30 | 88 | 87;49, 0 |
| 20 | 18;27, 0 | 43 | 40;31,30 | 66 | 64; 5,30 | 89 | 88;54,30 |
| 21 | 19;23, 0 | 44 | 41;31, 0 | 67 | 65; 9,30 | 90 | 90; 0, 0 |
| 22 | 20;19,30 | 45 | 42;31, 0 | 68 | 66;13, 0 | | |
| 23 | 21;16, 0 | 46 | 43;30,30 | 69 | 67;17, 0 | | |

Table 3. Extracted right ascension table.

if at least one of the two were given to minutes, $T_\varrho(\lambda) + T_\varrho(180-\lambda)$ would have an even number of minutes for all $\lambda$, provided that the standard rounding procedure were used.[44] For the present table, however, $T_\varrho(\lambda) + T_\varrho(180-\lambda)$ is odd for about half of the values of $\lambda$.

If for a particular $\lambda$ neither $T_\varrho(\lambda)$ nor $T_\varrho(180-\lambda)$ is in error, the reconstructed value $T_\alpha(\lambda)$ differs at most 0;0,30 from the exact $\alpha(\lambda)$. If the tabular errors of $T_\varrho(\lambda)$ have a uniform distribution, the tabular errors of $T_\alpha(\lambda)$ will have a triangular distribution.[45] It is not sufficient to approximate the obliquity underlying the right ascension from the single value that yields the most accurate approximation, namely $T_\alpha(47)$: all values of $\varepsilon$ in the interval $\langle 23;31,16, 23;35,51 \rangle$ yield a right ascension value for $\lambda = 47$ which differs from $T_\alpha(47)$ by at most 0;0,30. In particular, the two values 23;32,30 and 23;35, which are attested in the Sanjufīnī Zīj itself, are both contained in this interval.

The weighted estimator yields $\hat{\varepsilon} = 23;32,48$, and a 95% confidence interval is given by $\langle 23;32,23,\ 23;33,13\rangle$, provided that the conditions for its validity are satisfied. However, as in the preceding case, the tabular errors turn out to be insufficiently independent. Therefore I computed the weighted estimator for subtables $T(\lambda_0+k\cdot\Delta k)$, $k = 0,1,2,\ldots$ of the original table. Now for $\Delta\lambda = 3$ the tabular errors of the subtables do pass the test for independence. We have:

| | | | |
|---|---|---|---|
| $\lambda_0 = 0$: | $\hat{\varepsilon} = 23;32,49$ | 95% conf. interval: | $\langle 23;32,10,\ 23;33,27\rangle$ |
| $\lambda_0 = 1$: | $\hat{\varepsilon} = 23;32,26$ | | $\langle 23;31,39,\ 23;33,13\rangle$ |
| $\lambda_0 = 2$: | $\hat{\varepsilon} = 23;33,9$ | | $\langle 23;32,26,\ 23;33,52\rangle$ |

Three different attested values of the obliquity are continued in all three confidence intervals: 23;32,29 (Alphonsine tables), 23;32,30 (Ibn Isḥāq, al-Sanjufīnī) and 23;33 (Mumtaḥan Zīj).[46] Since of these values only 23;32,30 is found elsewhere in the Sanjufīnī Zīj, I conclude that this is the value underlying the extracted right ascension table.

This conclusion is confirmed by the result of the maximum likelihood estimator: the minimum possible number of differences larger than 0;0,30 with a recomputation is five, obtained for $\varepsilon \in \langle 23;32,26,44,\ 23;32,35,14\rangle$. However, if the tabular errors of the reconstructed right ascension in fact have a triangular distribution, then the conditions for the application of the maximum likelihood estimator as given in the previous section are not satisfied, and we must be careful not to overestimate the significance of the result.

The recomputation for $\varepsilon = 23;32,30$ yields five differences larger than 0;0,30 with the extracted table. These differences are given in the apparatus. It is a remarkable fact that none of the differences occurs for an argument which is a multiple of 3. I have not been able to explain this. For instance, a recomputation with linearly interpolated values for every non-multiple of 3 gives 10 differences larger than 0;0,30, 5 more than exact computation.

*CONCLUSIONS. The right ascension table on folio 39ʳ of the Sanjufīnī Zīj was computed using* $\varepsilon = 23;35$. *The right ascension used for the computation of the oblique ascension table on folio 38ᵛ was computed using* $\varepsilon = 23;32,30$.

Preliminary investigations of other tables in the Sanjufīnī Zīj gave the following results:

- The ascensional difference that can be extracted from the oblique ascension table on folio 38$^v$ (see above) was computed using $\varphi$ = 38;10 (in agreement with the heading) and $\varepsilon$ = 23;35. Thus this oblique ascension table involves *two different* values of the obliquity of the ecliptic.
- The second oblique ascension table, which is found on folio 28$^v$, is based on $\varepsilon$ = 23;32,30 (consistently) and $\varphi$ = 32;0.

Since the two oblique ascension tables in the Sanjufīnī Zīj are based on the typical value 23;32,30, it can be concluded that both are either by al-Sanjufīnī himself, or were taken by him from the same source from which he took his declination table. It seems plausible that al-Sanjufīnī had access to a right ascension table based on $\varepsilon$ = 23;32,30. It is unclear why he did not include this table in his zīj.

The ascensional difference used for the computation of the oblique ascension table on folio 38$^v$ may be based on a table for tan$\delta$ from a different source, since it involves a different obliquity. The latitude 38°10′ on which it is based occurs four other times in the zīj. As was indicated before, this latitude corresponds to a locality in Tibet mentioned in the manuscript.[47] Why al-Sanjufīnī also included two tables for latitude 32° is unclear.[48]

## 5. *The solar eccentricity in the Shāmil Zīj*

The Shāmil Zīj is an anonymous Arabic zīj extant in at least eight copies.[49] In spite of the fact that it was obviously frequently used, it has not been investigated systematically. Like the contemporary Baghdādī,[50] Ashrafī[51] and Muṣṭalaḥ[52] Zījes, it was not an original work, but was based largely upon earlier sources now partially or completely lost. I inspected the tables in four of the Shāmil manuscripts indicated in footnote 49 and also those in the Dublin manuscript of the zīj of Athīr al-Dīn al-Abharī (fl. Mardin, ca. 1240).[53] It turned out that the Shāmil Zīj and the zīj of al-Abharī are essentially the same. As Kennedy pointed out, both zījes involve the mean motion parameters found by Abū'l-Wafā' al-Būzajānī (see below).[54] The epoch of the mean motion tables in the Shāmil Zīj is the year 600 of the Yazdigird era (1231 A.D.). From this it may be concluded that the zīj was compiled shortly after this date, which would fit very well with the dates of al-Abharī's scholarly activity. The mean motion tables indicate that they were computed for longitude 84°, and the values of the geographical latitude occurring in the latitude-dependent tables are 37°, 38° and 39°. The conclusion that the Shāmil Zīj was used in north-western Persia is confirmed by the fact that the localities in the geographical table are concentrated in this region.[55] The Chester Beatty manuscript of al-Abharī's zīj has an oblique ascension table stated to be specifically for Mardin (folio 16ᵛ). Its title mentions a latitude of 37°25′ and a longitude of 75°0′.[56] However, the tables for the mean planetary positions are the same as those in the Shāmil Zīj, and their headings likewise mention the longitude 84°.[57] Therefore it seems plausible that al-Abharī simply copied from the Shāmil Zīj.

The above-mentioned Abū'l-Wafā' lived in Baghdad in the 10th century and compiled an astronomical handbook called al-Majistī (Almagest). A substantial part of this work is extant in a Paris manuscript.[58] This unique copy contains important trigonometric material, * which was analysed by Carra de Vaux,[59] but it does not contain any tables. A zīj called Wāḍiḥ Zīj is also attributed to Abū'l-Wafā', but is not extant.[60] The relation between the tables in the Wāḍiḥ Zīj and those belonging to al-Majistī is unknown. More information about Abū'l-Wafā's tables must be obtained from related zījes such as the Shāmil.

An extensive analysis of the manuscripts of the Shāmil Zīj is beyond the scope of this paper. I will restrict myself to applying the estimators introduced in Section 3 to a single table. It will be shown how, by means of these estimators, both the underlying parameter value and the accuracy used in the computation can be determined. Once the table has been "classified" in this way, we will be able to comment on its authorship.

Plate 2 shows one page of the solar equation table found on folios 24ᵛ–27ʳ of Paris Bibliothèque Nationale Ms. Arabe 2528. The table displays the solar equation to seconds for every 6 minutes of the mean solar anomaly. However, especially in the neighbourhood of the maximum (which occurs at 92°) it is clear that the tabular values in between

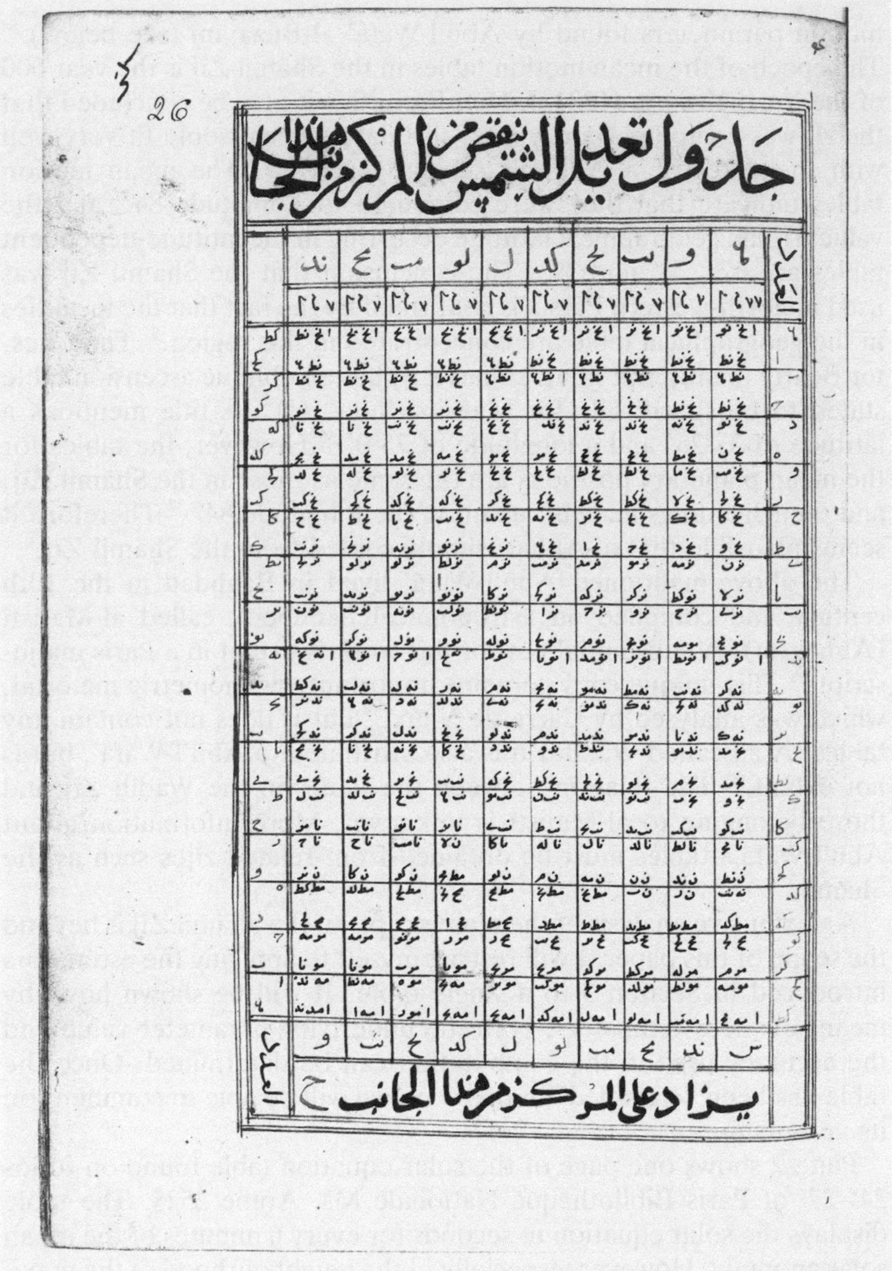

Plate 2. Part of the solar equation table in the Shāmil Zīj (Paris Bibliothèque Nationale Ms. Arabe 2528, folio 26[r]).

| | | | | | | | |
|---|---|---|---|---|---|---|---|
| 1 | 0; 2, 0 | 46 | 1;23,34 | 91 | 1;58,59 | 136 | 1;24,44 |
| 2 | 0; 4, 0 | 47 | 1;24,59 | 92 | 1;59, 0 | 137 | 1;23,13 |
| 3 | 0; 6, 1 | 48 | 1;26,24 | 93 | 1;58,59 | 138 | 1;21,42 |
| 4 | 0; 8, 1 | 49 | 1;27,48 | 94 | 1;58,56 | 139 | 1;20, 8 |
| 5 | 0;10, 1 | 50 | 1;29, 8 | 95 | 1;58,50 | 140 | 1;18,33 |
| 6 | 0;12, 1 | 51 | 1;30,28 | 96 | 1;58,42 | 141 | 1;16,56 |
| 7 | 0;14, 1 | 52 | 1;31,47 | 97 | 1;58,33 | 142 | 1;15,17 |
| 8 | 0;16, 0 | 53 | 1;33, 3 | 98 | 1;58,21 | 143 | 1;13,38 |
| 9 | 0;17,59 | 54 | 1;34,19 | 99 | 1;58, 6 | 144 | 1;11,56 |
| 10 | 0;19,58 | 55 | 1;35,32 | 100 | 1;57,50 | 145 | 1;10,13 |
| 11 | 0;21,57 | 56 | 1;36,45 | 101 | 1;57,31 | 146 | 1; 8,29 |
| 12 | 0;23,55 | 57 | 1;37,55 | 102 | 1;57,10 | 147 | 1; 6,44 |
| 13 | 0;25,53 | 58 | 1;39, 3 | 103 | 1;56,48 | 148 | 1; 4,56 |
| 14 | 0;27,50 | 59 | 1;40,10 | 104 | 1;56,22 | 149 | 1; 3, 9 |
| 15 | 0;29,47 | 60 | 1;41,15 | 105 | 1;55,55 | 150 | 1; 1,19 |
| 16 | 0;31,44 | 61 | 1;42,18 | 106 | 1;55,26 | 151 | 0;59,29 |
| 17 | 0;33,40 | 62 | 1;43,20 | 107 | 1;54,54 | 152 | 0;57,37 |
| 18 | 0;35,36 | 63 | 1;44,20 | 108 | 1;54,20 | 153 | 0;55,44 |
| 19 | 0;37,30 | 64 | 1;45,18 | 109 | 1;53,44 | 154 | 0;53,50 |
| 20 | 0;39,25 | 65 | 1;46,14 | 110 | 1;53, 6 | 155 | 0;51,54 |
| 21 | 0;41,18 | 66 | 1;47, 9 | 111 | 1;52,26 | 156 | 0;49,58 |
| 22 | 0;43,11 | 67 | 1;48, 1 | 112 | 1;51,43 | 157 | 0;48, 1 |
| 23 | 0;45, 3 | 68 | 1;48,52 | 113 | 1;50,59 | 158 | 0;46, 3 |
| 24 | 0;46,54 | 69 | 1;49,41 | 114 | 1;50,12 | 159 | 0;44, 3 |
| 25 | 0;48,45 | 70 | 1;50,27 | 115 | 1;49,24 | 160 | 0;42, 3 |
| 26 | 0;50,35 | 71 | 1;51,12 | 116 | 1;48,33 | 161 | 0;40, 3 |
| 27 | 0;52,24 | 72 | 1;51,56 | 117 | 1;47,40 | 162 | 0;38, 1 |
| 28 | 0;54,12 | 73 | 1;52,36 | 118 | 1;46,45 | 163 | 0;35,58 |
| 29 | 0;55,59 | 74 | 1;53,15 | 119 | 1;45,48 | 164 | 0;33,55 |
| 30 | 0;57,45 | 75 | 1;53,52 | 120 | 1;44,49 | 165 | 0;31,51 |
| 31 | 0;59,30 | 76 | 1;54,27 | 121 | 1;43,48 | 166 | 0;29,47 |
| 32 | 1; 1,15 | 77 | 1;54,59 | 122 | 1;42,45 | 167 | 0;27,42 |
| 33 | 1; 2,58 | 78 | 1;55,30 | 123 | 1;41,40 | 168 | 0;25,36 |
| 34 | 1; 4,40 | 79 | 1;55,59 | 124 | 1;40,33 | 169 | 0;23,30 |
| 35 | 1; 6,21 | 80 | 1;56,25 | 125 | 1;39,24 | 170 | 0;21,23 |
| 36 | 1; 8, 1 | 81 | 1;56,50 | 126 | 1;38,14 | 171 | 0;19,16 |
| 37 | 1; 9,40 | 82 | 1;57,12 | 127 | 1;37, 1 | 172 | 0;17, 9 |
| 38 | 1;11,18 | 83 | 1;57,33 | 128 | 1;35,46 | 173 | 0;15, 1 |
| 39 | 1;12,54 | 84 | 1;57,51 | 129 | 1;34,30 | 174 | 0;12,53 |
| 40 | 1;14,29 | 85 | 1;58, 7 | 130 | 1;33,12 | 175 | 0;10,44 |
| 41 | 1;16, 3 | 86 | 1;58,21 | 131 | 1;31,51 | 176 | 0; 8,36 |
| 42 | 1;17,36 | 87 | 1;58,33 | 132 | 1;30,30 | 177 | 0; 6,27 |
| 43 | 1;19, 8 | 88 | 1;58,43 | 133 | 1;29, 6 | 178 | 0; 4,18 |
| 44 | 1;20,38 | 89 | 1;58,50 | 134 | 1;27,40 | 179 | 0; 2, 9 |
| 45 | 1;22, 6 | 90 | 1;58,56 | 135 | 1;26,13 | 180 | 0; 0, 0 |

Table 4. Transcription of the solar equation table in the Shāmil Zīj.

I



integral numbers of the argument were computed by means of linear interpolation. Note that the interpolated values were rounded according to the standard medieval procedure; thus sexagesimal digits 6, 12, 18 and 24 were rounded down, whereas digits 30, 36, 42, 48 and 54 were rounded up. Because of the regularity of the interpolation pattern it is easy to correct the scribal errors in the tabular values for integral arguments. The errors are given in the apparatus, the corrected values in Table 4.

I will use the tabular entries for integral values of the solar anomaly to find an accurate approximation to the solar eccentricity that was used for the computation of the table. At first sight it is clear that the table was intended for the attested maximum solar equation of 1°59′,[61] which occurs in the table at a mean solar anomaly of 92°. However, all values of the eccentricity between 2;4,34,59 and 2;4,36,1 lead to $T(92) = 1;59,0$. Therefore it is not possible to conclude from the single tabular value $T(92)$ whether 2;4,35 or 2;4,36 or some value in between was used.[62]

If all tabular values for integral arguments are used for the computation of the weighted estimator, we find $\hat{e} = 2;4,35,29$ according to formula (14) in Section 3. Provided that the condition of independence of the tabular errors is satisfied, an approximate 95% confidence interval for the solar eccentricity is given by formula (16) as $\langle 2;4,35,25,\ 2;4,35,34 \rangle$. However, it turns out that for all eccentricity values in this interval a cluster of negative errors with respect to the recomputation occurs for $\bar{a} = 8$ to $\bar{a} = 15$. Apparently, the tabular errors in this region do not satisfy the condition of independence. This is confirmed by the result of the statistical test indicated in Appendix A2.

To obtain a valid 95% confidence interval, I disregard the tabular values for $\bar{a} = 1$ to $\bar{a} = 15$ and compute the weighted estimator $\hat{e}$ according to

$$\hat{e} = \frac{1}{W} \sum_{\bar{a}=16}^{179} w_{\bar{a}} \left( \frac{\sin \bar{a}}{\tan T(\bar{a})} - \cos \bar{a} \right), \tag{18}$$

where again

$$w_{\bar{a}} = \left( \frac{\sin \bar{a}}{\theta_0^2 + 2\theta_0 \cos \bar{a} + 1} \right)^2$$

and

$$W = \sum_{\bar{a}=16}^{179} w_{\bar{a}}.$$

(18) yields $\hat{e} = 2;4,35,30,12$ and the corresponding approximate 95% confidence interval for the eccentricity is given by $\langle 2;4,35,26, 2;4,35,35 \rangle$. Since the tabular errors for $\bar{a} = 16,17,18,\ldots,179$ pass the test for independence, we have now found a valid 95% confidence interval.

The maximum likelihood estimator indicated in Section 3 can also be applied to the solar equation table in the Shāmil Zīj. The minimum possible number of errors with respect to recomputation is 16. Between $e = 2;4,35,29,29$ and $e = 2;4,35,32,56$ recomputation yields 16 and 17 errors alternately. The conditions for the application of the maximum likelihood estimator are satisfied, since the tabular errors which do not exceed 0;0,0,30 in absolute value pass the tests for uniformity mentioned in Appendix A2.

Ptolemy computed his value 2;29,30 of the solar eccentricity $e$ from observations of solstices and equinoxes. Subsequently he computed the maximum solar equation $q_{max}$ in a way equivalent to $q_{max} = \text{arc Sin}\, e$.[63] Al-Battānī used the same procedure and found $e = 2;4,45$.[64] The maximum value in his solar equation table is 1;59,10,[65] but in the explanatory text it is stated that "the maximum is 1;59 approximately".[66] Apparently this rounded value became the basis of later solar equation tables. To compute these tables it was necessary to calculate $e$ using $e = \text{Sin } 1;59,0,\ldots$ Since Sin 1;59 = 2;4,35,29,51,..., it is very well possible that the author of the solar equation table in the Shāmil Zīj used the round value 2;4,35,30. This value makes the computations relatively easy compared to other values in the 95% confidence interval. Furthermore, it seems likely that the author of the table did not consider it necessary to compute the eccentricity with a still higher accuracy, or that he was not able to obtain a value more accurate than 2;4,35,30.

The solar equation table in the Shāmil Zīj shows 17 errors compared to the recomputation for $e$ = 2;4,35,30. Seven of these errors are found for $\bar{a} = 1$ to $\bar{a} = 17$. In this region all tabular errors are negative, whence it looks like the author simply truncated the values of a more accurate solar equation table. The other errors with respect to the recomputation are −0;0,1 (for $\lambda$ = 61,137,148 and 159) or +0;0,1 (for $\lambda$ = 43,49,56,72,76 and 103). In four cases the error could have been caused by small computational errors, in two cases by a scribal mistake (confusion of 6 and 7). As in the case of the ascension tables in the Sanjufīnī Zīj (Section 4), I have not further investigated the way of computation of the table. Researches in this direction are presently being conducted by Glen Van Brummelen of Simon Fraser University (Burnaby, B.C., Canada).[67]

*CONCLUSION. The solar equation table in the Shāmil Zīj was very probably computed using e* = 2;4,35,30. *In particular, the author of the table did not round the eccentricity to* 2;4,35 *or* 2;4,36.

The desired accuracy of Sin 1°59′ could be obtained from any correct Sine table with three sexagesimals after the semicolon, even if linear interpolation on Sin 1° and Sin 2° had to be used. Furthermore, a correct Sine table to seconds with arguments 0;2, 0;4, 0;6, …, 1;58, 2;0, … would also lead to Sin 1°59′ = 2;4,35,30. This is a mere coincidence, since the only two possibilities for the number of thirds in the result are 0 and 30. Now the Sine table in the Paris Ms. 2528 of the Shāmil Zīj (folios 10$^v$–14$^r$) gives Sin $x$ to seconds for arguments 0;2, 0;4, 0;6, … However, this table would not give the value 2;4,35,30 for the eccentricity, since the value for Sin 1°58′ contains an error of −2 seconds which is not the result of a scribal error.[68] In addition, the possibility should not be ruled out that a special computation was conducted to determine the eccentricity with an accuracy which could not be obtained from any available sine table.

The solar equation table on folios 45$^r$–47$^v$ of the Baghdādī Zīj[69] is actually attributed to Abū'l-Wafā'. Apart from a small number of scribal errors, the table is identical with the solar equation table in the Berlin recension of the zīj of Ḥabash.[70] The table has values given to sexagesimal thirds, and a 95% confidence interval for the eccentricity underlying the table is ⟨2;4,35,29,50, 2;4,35,30,2⟩. Therefore the possibility that also this table was computed using 2;4,35,30 can not be excluded. Nevertheless, comparison clearly shows that the solar equation table in the Shāmil Zīj is not a rounded version of the table in the Baghdādī Zīj or otherwise related to it, and therefore it is not yet possible to attribute the Shāmil table to Abū'l-Wafā'.

## 6. *Conclusions*

The statistical estimators described in this paper offer a higher accuracy than methods that have been used till now for approximating unknown parameters. Therefore they may allow the determination of the parameter in cases where other methods fail, or they may yield additional information about the accuracy with which the tabular values were computed. Also from tables with many errors useful conclusions may be drawn.

In principle, the results of the estimators are independent from a priori knowledge concerning the parameter. In this sense, the estimators constitute an *objective* criterion for determining which values of the parameter could have been used for the computation of a particular table.

## 7. *Acknowledgement*

This paper was finished during a stay of 10 months in Frankfurt am Main, which was made possible by a fellowship of the Netherlands Organization for Scientific Research (NWO). I would like to thank Professors D. A. King (Frankfurt), H. J. M. Bos and R. D. Gill (Utrecht) and Dr. J. P. Hogendijk (Utrecht) for their very useful comments on preliminary versions. Furthermore I would like to thank Prof. E. S. Kennedy (Princeton) for providing me with photocopies of tables from the Sanjufīnī Zīj. Plates 1 and 2 were reproduced with kind permission of the Bibliothèque Nationale in Paris.

*Appendices*

## *A1. Some elementary statistical concepts*

A random variable $X$ is one which assumes random values in the set of real numbers $\mathbb{R}$. For every subset $D$ of $\mathbb{R}$ the number $Pr(X \in D)$ denotes the probability that $X$ assumes a value in $D$. We have $0 \leq Pr(X \in D) \leq 1$ for every $D \subseteq \mathbb{R}$. The probability distribution function of $X$ is the function $F_X$ defined by $F_X(x) = Pr(X \leq x)$ for all $x \in \mathbb{R}$.

We distinguish discrete and continuous random variables:

- For a discrete random variable there exists a set $C$ having a finite number of points in every bounded interval such that $Pr(X \in C) = 1$. The probability density function of a discrete variable $X$ is the function $f_X$ such that $f_X(x) = Pr(X=x)$ for all $x \in C$. We have

$$\sum_{x \in C} f_X(x) = 1.$$

- For a continuous random variable the probability distribution function $F_X$ is differentiable on $\mathbb{R}$ minus a set $E$ having a finite number of points in every bounded interval. The probability density function $f_X$ is determined by

$$f_X(x) = \frac{d}{dx} F_X(x)$$

for all $x \in \mathbb{R} - E$ and

$$\int_{\mathbb{R}} f_X(x) dx = 1.$$

Two common examples of continuous distributions are the uniform and the normal distribution. The uniform distribution has a probability density function

$$f_X(x) = \begin{cases} \dfrac{1}{b-a} & a \leq x \leq b \\ 0 & \text{elsewhere,} \end{cases} \tag{19}$$

where $a < b$. Thus it assigns to every subinterval $I$ of the interval $[a,b]$ a probability proportionate to the length of $I$.

The normal distribution has a probability density function

$$f_X(x) = \frac{1}{\sqrt{2\pi}} e^{-\frac{1}{2}x^2}, \quad -\infty < x < +\infty. \tag{20}$$

It can often be used to describe the errors in observations and it occurs naturally as the limiting distribution of other distributions.

The mean or expected value $\mu$ of a random variable $X$ is given by

$$\mu = \sum_{x \in \mathbb{R}} x \cdot f_X(x) \text{ (discrete case) or } \int_{\mathbb{R}} x \cdot f_X(x)dx \text{ (continuous case).}$$

The variance $\sigma^2$ of $X$ is given by

$$\sigma^2 = \sum_{x \in \mathbb{R}} (x-\mu)^2 \cdot f_X(x) \text{ or } \sigma^2 = \int_{\mathbb{R}} (x-\mu)^2 \cdot f_X(x)dx.$$

The standard deviation $\sigma$ of $X$ is the square root of the variance. $\sigma$ measures the average distance of $X$ from its mean.

*Example.* If $X$ is the number of points cast with a fair die, $X$ is a discrete variable with $f_X$ given by $f_X(x) = \frac{1}{6}$ for $x \in \{1,2,3,4,5,6\}$, $f_X(x) = 0$ elsewhere. $X$ has mean $3\frac{1}{2}$, variance $\frac{35}{12}$ and standard deviation $\frac{1}{2}\sqrt{\frac{35}{3}}$. If $X_n$ is the average number of points obtained from $n$ independent casts with a fair die, it can be shown that the distribution of $\sqrt{n}(X_n - 3\frac{1}{2})/\frac{1}{2}\sqrt{\frac{35}{3}}$ tends to the normal distribution as $n$ tends to infinity.[71]

An estimator is a random variable used to approximate a certain unknown quantity. Important properties of an estimator are the bias, i.e. the difference between the expected value of the estimator and the quantity to be approximated, and the variance. Bias and variance may depend on the unknown quantity, but generally approximations to them can be computed. In some cases it may be necessary to carry out a so-called Monte Carlo analysis to obtain these approximations: the estimator is computed a large number of times from computer-generated data, and bias and variance are determined from the errors in the results. In general, an estimator will yield a good approximation to the unknown quantity if bias and variance are both small.

A confidence interval for the unknown quantity is an interval which has a fixed probability (e.g. 95%) of containing the quantity. Such an interval can be computed from the result of the estimator if bias and variance are approximately known.

More details about the above-mentioned statistical concepts can be found in the textbooks by Bickel and Doksum and by Breiman.[72]

### *A2 Tabular errors*

Let $T$ be a table for the function $f_\theta(x)$ with tabular values $T(x)$ and unknown parameter $\theta$. The tabular values are subject to tabular errors $e_\theta(x) = T(x) - f_\theta(x)$. These errors can be divided into three categories: scribal errors, computational errors, and rounding errors, which I will discuss below.

*Scribal errors* result from careless copying of the tabular values, e.g. in Arabic نا = 51 is sometimes carelessly written as ىا, which is also used for 11. Like computational mistakes (e.g. 117 + 248 = 355), scribal errors usually are large and unpredictable and do not have a nice probability distribution. Statistical methods are available which deal with these so-called outliers. However, in the case of medieval manuscripts it is often possible to correct the scribal errors before starting the statistical analysis. Methods for the correction of scribal errors are:

- if linear interpolation was used for the computation of the table: investigation of irregularities in the interpolation pattern;

- investigation of the differences between the tabular values and values recomputed for a first approximation to the unknown parameter. Scribal errors correspond to large irregularities in the error pattern;
- investigation of the fourth differences of the tabular values. If one of the tabular values contains a significant scribal error $e(x)$, then in the fourth differences an error pattern $+e(x), -4e(x), +6e(x), -4e(x), +e(x)$ will be superimposed on the fourth differences of the correctly computed table. Since the fourth differences of correct tabulations of most astronomical functions are very close to zero, the error pattern can be clearly recognized and the tabular error identified.
  *Example.* Let $T$ be a table with values to seconds for the solar equation with eccentricity $e = 2;4,35,30$. If $T$ is correctly computed, the fourth differences of the tabular values $T(61)$ to $T(71)$ are $+1,+1,-3,+4,-3,0$ and $+3$ seconds respectively. However, for the solar equation in the Shāmil Zīj, which is analysed in Section 5, the fourth differences of the tabular values $T(61)$ to $T(71)$ are $0,+12,-47,+70,-47,+11$ and $+3$ seconds. Thus a scribal error of approximately $+12$ seconds is found in $T(66)$. Inspection of the Arabic numeral notation now leads to the exact determination of the error, namely confusion of 9 (ط) and 20 (ك).[73]

*Computational errors* result from intermediate inaccuracies in the computation such as rounding off and interpolation, e.g. if $\cos 23;35 \approx 0;54,59,19$ and $\tan 41 \approx 0;52,9,26$ are rounded to seconds, the resulting right ascension value to minutes is $\arctan(0;54,59 \cdot 0;52,9) = 38;32$ instead of the correct $38;33$. It is difficult to state anything general about the probability distribution of computational errors. If the number of intermediate inaccuracies becomes large, the distribution of the error in the final result tends to the normal distribution.

*Rounding errors* are the errors that result from rounding the exact functional values to the number of sexagesimal places of the table, e.g. $\alpha(45) = 42;30,16,35,23,\ldots$ is rounded to $42;30$ and therefore yields a rounding error of $-0;0,16,35,23,\ldots$ Most medieval astronomers used the same rounding procedure as we use today, i.e. they rounded the digits 29 and lower downwards, 30 and higher upwards. Occasionally evidence of truncation or unsystematic rounding is found.

I will make it plausible that in many tables the rounding error in an arbitrary tabular value has approximately a uniform distribution. To obtain this result I will use an unpublished theorem by Kemperman.[74]

Let $X$ be a random variable with distribution function $F$. Let $F_\sigma$ denote the distribution function of $\sigma X$ (mod 1). Define

$$D(\sigma) = \sup_{0 \leq x, y \leq 1} |F_\sigma(y) - F_\sigma(x) - (y-x)|$$

to measure the distance between $F_\sigma$ and the uniform distribution. See the textbook by Feller for the definition and some elementary properties of the characteristic function.[75]

*THEOREM* (Kemperman, 1975).

$$\lim_{\sigma \to \infty} D(\sigma) = 0$$

*if and only if the characteristic function of F tends to zero at infinity.*

As a corollary of this theorem, the distribution of rounding errors for randomly-chosen arguments approximates the uniform distribution if the number of sexagesimals of the tabular values is relatively large. For in the above let $X$ be the value $f_\theta(Y)$ of the tabulated function at a randomly chosen point $Y$. Then $60^k \cdot X$ (mod 1) can be considered as the rounding error of the tabular value $T(Y)$ if $k$ is the number of sexagesimal fractional digits of the table. Suppose that $X$ has a probability density. Then the characteristic function of $X$ will tend to zero at infinity,[76] and accordingly, by Kemperman's theorem, the distribution of $60^k \cdot X$ (mod 1) tends to the uniform distribution if $k$ approaches infinity.

*Example.* Let $Y$ have the uniform distribution on $\langle 0,90 \rangle$, let the tabulated function be the right ascension $\alpha$ and let $X = \alpha(Y)$. Then we have

$$Pr(X < x) = Pr\left(Y < \arctan\left(\frac{\tan x}{\theta}\right)\right) = \frac{1}{90}\arctan\left(\frac{\tan x}{\theta}\right), \tag{21}$$

and $X$ has a probability density

$$f_X(x) = \frac{1}{90}\,\frac{\theta(1 + \tan^2 x)}{\theta^2 + \tan^2 x} \qquad (0 < x < 90).$$

As a result, the tabular errors of a table for the right ascension can be considered uniformly distributed if the tabular values have a sufficient number of sexagesimals.

The width of the confidence intervals depends largely on the independence of the tabular errors. One may conjecture that for fixed $h$ the rounding errors associated with two functional values $f_\theta(x)$ and $f_\theta(x+h)$ can be considered to be independent if the tabular values have a sufficient number of sexagesimals. I plan to investigate this on a different occasion.

In practice it is necessary to test the tabular errors for uniformity and independence, e.g. by means of the tests given by Knuth.[77] In the examples in Sections 4 and 5 of this paper I used the equidistribution tests (Chi-square and Kolmogorov–Smirnov) to test uniformity,[78] the Serial correlation to test independence.[79] If the tabular errors of a particular table appear to be non-uniform or dependent, the unknown parameter may be estimated from a subtable of which the tabular errors do have the desired properties. For example, in Section 4 the obliquity of the ecliptic was estimated from every second or third value of a right ascension table, and in Section 5 the values for arguments 1 to 15 of a solar equation table were left out in the estimation of the solar eccentricity.

### *A3. The Weighted Estimator for a single unknown parameter*

Let $T$ be a given table for the function $f_\theta(x)$ with tabular values $T(x)$ and a single unknown parameter $\theta$. As before, the tabular errors $e_\theta(x)$

are defined by $e_\theta(x) \overset{\text{def}}{=} T(x) - f_\theta(x)$.[80] Assume that the errors are independent and have a distribution with common mean 0 and variance $\sigma^2$. This implies that all tabular values contain errors of the same order of magnitude.[81] Suppose that we have a function $g\colon \mathbb{R}^2 \to \mathbb{R}$ such that for all $x$ and $\theta$ $g(x, f_\theta(x)) = \theta$.[82] Then every $\hat{\theta}_x \overset{\text{def}}{=} g(x, T(x))$ is an estimator for the unknown parameter. First I will compute the accuracy of these estimators.

By computing the mathematical expectation of the right-hand side of the Taylor series

$$\begin{aligned} \hat{\theta}_x &= g(x, T(x)) \\ &= g(x, f_\theta(x) + e_\theta(x)) \\ &\approx g(x, f_\theta(x)) + e_\theta(x) \cdot \frac{\partial}{\partial y} g(x, f_\theta(x)) \\ &\quad + \tfrac{1}{2} e_\theta^2(x) \cdot \frac{\partial^2}{\partial y^2} g(x, f_\theta(x)) + 0(e_\theta^3(x)), \quad (e_\theta(x) \to 0) \end{aligned} \tag{22}$$

we find that $\hat{\theta}_x$ has approximately an expected value

$$\theta + \tfrac{1}{2}\sigma^2 \cdot \frac{\partial^2}{\partial y^2} g(x, f_\theta(x))$$

and a bias

$$b_{\hat{\theta}_x} \approx \tfrac{1}{2}\sigma^2 \cdot \frac{\partial^2}{\partial y^2} g(x, f_\theta(x)). \tag{23}$$

From (22), (23), and the definition of the variance given in Appendix A1 we find that $\hat{\theta}_x$ has approximately a variance

$$\operatorname{Var} \hat{\theta}_x = \sigma^2 \cdot \left( \frac{\partial}{\partial y} g(x, f_\theta(x)) \right)^2. \tag{24}$$

Here and in the sequel when I use the phrase "the estimator $\hat{\theta}$ has approximately bias $b_{\hat{\theta}}$ and variance Var $\hat{\theta}$" I mean "$\hat{\theta}$ has a probability distribution which can well be approximated by a distribution with bias $b_{\hat{\theta}}$ and variance Var $\hat{\theta}$".

If a particular tabular value contains a scribal error, the results of the estimator computed from it may be highly misleading. But also if no scribal error occurs, the most accurate separate estimator $\hat{\theta}_x$ may be insufficient to determine the unknown parameter. Examples hereof are found in Sections 4 and 5. A more accurate estimator $\hat{\theta}$ is obtained by computing a weighted average of the separate estimators $\hat{\theta}_x$

$$\hat{\theta} = \frac{1}{W} \sum_x w_x \hat{\theta}_x, \tag{25}$$

where $w_x$ denotes the weights and

$$W = \sum_x w_x.$$

The weights will be chosen in such a way that less accurate separate estimators have a smaller influence on the average. It can be shown that in the case of linear parameter estimation the optimal weights of a weighted estimator are inversely proportionate to the variances of the separate estimators. Therefore in the present situation the weights

$$w_x = \frac{1}{\left(\frac{\partial}{\partial y} g(x, f_\theta(x))\right)^2}$$

are approximately optimal.

Consequently $\hat{\theta}$ will approximately have a variance

$$\text{Var}\,\hat{\theta} = \frac{1}{W^2} \sum_x w_x^2 \,\text{Var}\,\hat{\theta}_x \approx \frac{1}{W^2} \sum_x w_x \sigma^2 = \frac{\sigma^2}{W}, \tag{26}$$

where

$$W = \sum_x \frac{1}{\left(\frac{\partial}{\partial y} g(x, f_\theta(x))\right)^2}.$$

The bias $b_{\hat{\theta}}$ is computed according to the formula

$$b_{\hat{\theta}} = \frac{1}{W} \sum_x w_x b_{\hat{\theta}_x}.$$

Throughout this paper $\hat{\theta}$ is called "the weighted estimator for a single unknown parameter".

*The obliquity of the ecliptic in a right ascension table.* The right ascension is given by $\alpha(\lambda) = \arctan(\tan\lambda \cdot \cos\varepsilon)$, where $\varepsilon$ is the obliquity of the ecliptic.[83] Let $T(\lambda)$, $\lambda = 1,2,3,\ldots,89$, be a table for $\alpha(\lambda)$, and assume that the value of $\varepsilon$ used for the computation of this table is unknown. To simplify the calculations, I will compute the weighted estimator for $\theta \stackrel{\text{def}}{=} \cos\varepsilon$ instead of for $\varepsilon$. As is indicated in Section 2, there is no historical reason for doing so, since most Islamic astronomers used formula (2), which involves $\tan\delta$ and $\tan\varepsilon$, rather than formula (1). Assume that the tabular errors $e_\theta(\lambda) \stackrel{\text{def}}{=} T(\lambda) - \alpha(\lambda)$ are independent and have mean 0 and fixed variance $\sigma^2$. Let

$$g(x,y) = \frac{\tan y}{\tan x}.$$

Then we have $g(\lambda, \alpha(\lambda)) = \theta$ for all $\lambda \in \langle 0,90 \rangle$ and $\theta \in \langle 0,1]$. Therefore, for every $\lambda = 1,2,3,\ldots,89$,

$$\hat{\theta}_\lambda \stackrel{\text{def}}{=} g(\lambda, T(\lambda)) = \frac{\tan T(\lambda)}{\tan\lambda} \tag{27}$$

is a reasonable estimator for $\theta$. We have

$$\frac{\partial}{\partial y} g(x,y) = \frac{\pi}{180} \frac{1+\tan^2 y}{\tan x} \text{ and } \frac{\partial^2}{\partial y^2} g(x,y) = \frac{2\pi^2}{180^2} \frac{\tan y(1+\tan^2 y)}{\tan x}. \tag{28}$$

From (23) and (24) we find that $\hat{\theta}_\lambda$ has approximately a probability distribution with bias

$$b_{\hat{\theta}_\lambda} = \tfrac{1}{2}\frac{2\pi^2\sigma^2}{180^2} \frac{\tan \alpha(\lambda)}{\tan \lambda} (1 + \tan^2 \alpha(\lambda)) = \frac{\pi^2\sigma^2}{180^2} \theta(1 + \theta^2 \tan^2 \lambda) \tag{29}$$

and variance

$$\operatorname{Var} \hat{\theta}_\lambda = \left(\frac{\pi}{180} \frac{1 + \tan^2 \alpha(\lambda)}{\tan \lambda}\right)^2 \sigma^2 = \frac{\pi^2\sigma^2}{180^2} \left(\frac{1 + \theta^2 \tan^2 \lambda}{\tan \lambda}\right)^2. \tag{30}$$

The standard deviation of $\hat{\theta}$ is approximately given by

$$\operatorname{Std} \hat{\theta} = \frac{\pi\sigma}{180} \frac{1 + \theta^2 \tan^2 \lambda}{\tan \lambda}. \tag{31}$$

In practice, every right ascension table has at least one sexagesimal fractional digit, which implies that usually $0 < \sigma < 10^{-2}$.[84] Since

$$\frac{b_{\hat{\theta}_\lambda}}{\operatorname{Std} \hat{\theta}_\lambda} = \frac{\pi\sigma\theta}{180} \tan \lambda$$

and $\tan \lambda$ ranges from 0.0175 to 57.29 for $\lambda = 1,2,3,\ldots,89$, the bias of $\hat{\theta}_\lambda$ is negligible compared to its standard deviation. $\operatorname{Std} \hat{\theta}_\lambda$ tends to infinity as $\lambda$ approaches 0 or 90; this corresponds to the fact that the obliquity cannot be estimated from $T(0)$ and $T(90)$, which are both independent of $\varepsilon$. The minimum value

$$\frac{2\pi\sigma\theta}{180}$$

of Std $\hat{\theta}_\lambda$ is assumed for $\lambda = \arctan(1/\theta) \approx 47.5$. However, the use of $T(47)$ or $T(48)$ for estimating $\theta$ would lead to a 95% confidence interval for $\theta$ with a width of at least 3.92 Std $\hat{\theta}_\lambda \approx 6.0 \cdot 10^{-4}$, if $T$ has a single sexagesimal fractional digit.[85] This corresponds to a 95% confidence interval for $\varepsilon$ with a width of at least 0;5, which could easily contain more than one historical value of the obliquity.

To obtain a smaller confidence interval the weighted estimator (25) can be used. Let $\theta_0$ be a reasonable first approximation to $\theta$, e.g.

$$\theta_0 = \frac{\tan T(47)}{\tan 47}.$$

The accuracy of $\theta_0$ has only a slight influence on the accuracy of the weighted estimator. As indicated in the first part of this section, approximately optimal weights are given by

$$w_\lambda = \frac{1}{\operatorname{Var}\hat{\theta}_\lambda} = \left(\frac{\tan\lambda}{1+\theta_0^2\tan^2\lambda}\right)^2.$$

Thus the weighted estimator $\hat{\theta}$ is given by

$$\hat{\theta} = \frac{1}{W}\sum_{\lambda=1}^{89}\frac{\tan\lambda\cdot\tan T(\lambda)}{(1+\theta_0^2\tan^2\lambda)^2}, \tag{32}$$

where

$$W = \sum_{\lambda=1}^{89}\left(\frac{\tan\lambda}{1+\theta_0^2\tan^2\lambda}\right)^2 \approx 13.4.$$

We have

$$b_{\hat{\theta}} = \frac{1}{W}\sum_{\lambda=1}^{89} w_\lambda b_{\hat{\theta}_\lambda} \approx \frac{\pi^2\sigma^2}{W180^2}\,\theta\sum_{\lambda=1}^{89}\frac{\tan^2\lambda}{1+\theta_0^2\tan^2\lambda} \approx 0.0011\sigma^2 \tag{33}$$

and

$$\operatorname{Var}\hat{\theta} = \frac{1}{W^2}\sum_{\lambda=1}^{89} w_\lambda^2 \operatorname{Var}\hat{\theta}_\lambda \approx \frac{\pi^2\sigma^2}{W180^2} \approx 2.3 \cdot 10^{-5}\,\sigma^2. \tag{34}$$

Note that $\operatorname{Std}\hat{\theta} \approx 0.0048\sigma$ and that therefore $b_{\hat{\theta}}$ is negligible compared to $\operatorname{Std}\hat{\theta}$. Furthermore

$$\frac{\min\limits_{\lambda=1,2,3,\dots,89} \operatorname{Std}\hat{\theta}_\lambda}{\operatorname{Std}\hat{\theta}} \approx 6.7,$$

which implies that by means of $\hat{\theta}$ we obtain a 95% confidence interval for the obliquity which is more than 6 times as small as the smallest possible 95% confidence interval obtained from a separate estimator.

An estimator $\hat{\varepsilon}$ for the obliquity itself can be obtained by putting $\hat{\varepsilon} = \arccos\hat{\theta}$. By means of the Taylor series

$$\begin{aligned}\arccos(\hat{\theta} + h) &= \arccos\hat{\theta} - \frac{180}{\pi}\frac{1}{\sqrt{1-\hat{\theta}^2}}h + O(h^2)\\ &= \hat{\varepsilon} - \frac{180h}{\pi\sin\hat{\varepsilon}} + O(h^2) \quad (h \to 0),\end{aligned} \tag{35}$$

it can be shown that $\hat{\varepsilon}$ has approximately a distribution with bias

$$b_{\hat{\varepsilon}} = -\frac{180}{\pi\sin\hat{\varepsilon}}b_{\hat{\varepsilon}} \approx -0.15\sigma^2$$

and standard deviation

$$\operatorname{Std}\hat{\varepsilon} = \frac{180}{\pi\sin\hat{\varepsilon}} \cdot \operatorname{Std}\hat{\theta} \approx 0.68\sigma.$$

To determine the bias and standard deviation of $\hat{\theta}$ and $\hat{\varepsilon}$ for a particular table, it is necessary to approximate the variance $\sigma^2$ of the tabular errors from

$$\sigma^2 \approx \frac{1}{88} \sum_{\lambda=1}^{89} (T(\lambda) - \alpha_{\hat{\varepsilon}}(\lambda))^2, \tag{36}$$

where the $\alpha_{\hat{\varepsilon}}(\lambda)$ are computed values for the right ascension with obliquity equal to the estimate $\hat{\theta}$.

In Appendix A5 it will be shown that $\hat{\theta}$ has a distribution which is approximately normal. Therefore an approximate 95% confidence interval for $\hat{\theta}$ is given by

$$\langle \hat{\theta} - b_{\hat{\theta}} - 1.96 \operatorname{Std} \hat{\theta},\ \hat{\theta} - b_{\hat{\theta}} + 1.96 \operatorname{Std} \hat{\theta} \rangle \tag{37}$$

(cf. footnote 85). An approximate 95% confidence interval for $\varepsilon$ is obtained by taking the arccosine of the bounds of the confidence interval for $\theta$.

A Monte Carlo analysis was carried out to test the validity of (37). Confidence intervals were computed from 200 computer-generated right ascension tables of different types. These types involved tables to minutes and tables to seconds and with additional uniformly distributed errors of up to 5 minutes or seconds. For all tables random values of the obliquity within historically plausible limits were used. The results of the analysis confirmed that (37) constitutes an approximate 95% confidence interval for $\cos \varepsilon$.

*The solar eccentricity in a solar equation table.* The solar equation as a function of the mean anomaly is given by

$$q(\bar{a}) = \arctan \left( \frac{e \sin \bar{a}}{60 + e \cos \bar{a}} \right),$$

where $e$ is the solar eccentricity.[86] Let $T(\bar{a})$, $\bar{a} = 1,2,3,\ldots,179$, be a table for $q(\bar{a})$ and assume that the value of $e$ used for the computation of the table is unknown. Assume that the tabular errors $e_\theta(\bar{a}) \stackrel{\text{def}}{=} T(\bar{a}) - q(\bar{a})$ are independent and have mean 0 and variance $\sigma^2$. As in case of the right ascension, the calculations can be simplified by computing the weighted estimator for

$$\theta \stackrel{\text{def}}{=} \frac{60}{e}$$

instead of for $e$. Again there is no historical reason for doing so, since the medieval astronomers didn't use $\theta$ in their computations (see Section 2). Let

$$g(x,y) = \frac{\sin x}{\tan y} - \cos x.$$

Then we have $g(x,q(x)) = \theta$ for all $x \in \langle 0,180\rangle$ and $\theta > 1$. Now

$$\frac{\partial}{\partial y} g(x,y) = -\frac{\pi}{180}\frac{1 + \tan^2 y}{\tan^2 y}\sin x \tag{38}$$

and

$$\frac{\partial^2}{\partial y^2} g(x,y) = \frac{2\pi^2}{180^2}\frac{1 + \tan^2 y}{\tan^3 y}\sin x. \tag{39}$$

Therefore the separate estimator $\hat{\theta}_{\bar{a}} \stackrel{\text{def}}{=} g(\bar{a}, T(\bar{a}))$ has approximately a distribution with bias

$$b_{\hat{\theta}_{\bar{a}}} = \tfrac{1}{2}\,\frac{2\pi^2\sigma^2}{180^2}\frac{1+\tan^2 q(\bar{a})}{\tan^3 q(\bar{a})}\sin\bar{a} = \frac{\pi^2\sigma^2}{180^2}\frac{(\theta^2 + 2\theta\cos\bar{a} + 1)(\theta + \cos\bar{a})}{\sin^2\bar{a}} \tag{40}$$

and variance

$$\operatorname{Var}\hat{\theta}_{\bar{a}} = \left(-\frac{\pi}{180}\frac{1+\tan^2 q(\bar{a})}{\tan^2 q(\bar{a})}\sin\bar{a}\right)^2\sigma^2 = \frac{\pi^2\sigma^2}{180^2}\frac{(\theta^2 + 2\theta\cos\bar{a} + 1)^2}{\sin^2\bar{a}}. \tag{41}$$

Since

$$\frac{b_{\hat{\theta}_{\bar{a}}}}{\operatorname{Std}\hat{\theta}_{\bar{a}}} = \frac{\pi\sigma}{180}\,\frac{\theta + \cos\bar{a}}{\sin\bar{a}}$$

the standard deviation again plays a more important role than the bias. The minimum standard deviation occurs for

$$\bar{a} = 90 + \arcsin\left(\frac{2\theta}{1+\theta^2}\right),$$

i.e. close to the maximum of the solar equation at

$$\bar{a} = 90 + \arcsin\left(\frac{e}{60}\right),$$

and amounts

$$\frac{\pi\sigma}{180}\,(\theta^2 - 1).$$

Let $\theta_0$ be a reasonable first approximation to $\theta$, e.g.

$$\theta_0 = \frac{1}{\tan T(90)}\,.$$

Since the range of historical values of the eccentricity is rather large, it is advisable not to choose an arbitrary historical value for $\theta_0$. For the weighted estimator

$$\hat{\theta} = \frac{1}{W}\sum_{\bar{a}=1}^{179} w_{\bar{a}}\hat{\theta}_{\bar{a}}$$

we take

$$w_{\bar{a}} = \frac{\sin^2 \bar{a}}{(\theta_0^2 + 2\theta_0 \cos \bar{a} + 1)^2}.$$

We find that $\hat{\theta}$ has approximately a distribution with bias

$$b_{\hat{\theta}} = \frac{1}{W} \sum_{\bar{a}=1}^{179} w_{\bar{a}} b_{\hat{\theta}_{\bar{a}}} \approx \frac{1}{W} \frac{\pi^2 \sigma^2}{180^2} \sum_{\bar{a}=1}^{179} \frac{\theta + \cos \bar{a}}{\theta_0^2 + 2\theta_0 \cos \bar{a} + 1} \tag{42}$$

and variance

$$\operatorname{Var} \hat{\theta} = \frac{1}{W^2} \sum_{\bar{a}=1}^{179} w_{\bar{a}}^2 \operatorname{Var} \hat{\theta}_{\bar{a}} \approx \frac{\pi^2 \sigma^2}{W 180^2}, \tag{43}$$

where

$$W = \sum_{\bar{a}=1}^{179} w_{\bar{a}}.$$

$\sigma^2$ can be estimated by

$$\sigma^2 \approx \frac{1}{178} \sum_{\bar{a}=1}^{179} (T(\bar{a}) - q_{\hat{e}}(\bar{a}))^2, \tag{44}$$

where the $q_{\hat{e}}(\bar{a})$ are computed values for the solar equation with eccentricity equal to $60/\hat{\theta}$.

Since

$$\frac{b_{\hat{\theta}}}{\operatorname{Std} \hat{\theta}} \approx 9\sigma,$$

I



$b_{\hat{\theta}}$ is small compared to Std $\hat{\theta}$. Furthermore

$$\frac{\min_{\bar{a}=1,2,3,\ldots,179} \text{Std}\,\hat{\theta}_{\bar{a}}}{\text{Std}\,\hat{\theta}} \approx 9.5,$$

which means that $\hat{\theta}$ yields a 95% confidence interval more than 9 times as small as the smallest possible 95% confidence interval obtained from a separate estimator.[87]

An estimator $\hat{e}$ for the eccentricity is obtained by putting

$$\hat{e} = \frac{60}{\hat{\theta}}\,.$$

Its approximate bias and standard deviation follow from the Taylor expansion

$$\frac{60}{\hat{\theta} + h} = \hat{e} - \frac{60}{\hat{\theta}^2} h + 0(h^2) \quad (h \to 0). \tag{45}$$

A Monte Carlo analysis as described in the case of the right ascension confirmed that $\langle \hat{\theta} - b_{\hat{\theta}} - 1.96\ \text{Std}\,\hat{\theta},\ \theta b_{\hat{\theta}} + 1.96\ \text{Std}\,\hat{\theta} \rangle$ constitutes a 95% confidence interval for $60/e$.

### *A4 The Maximum Likelihood Estimator for a single unknown parameter*

Let $X_1, X_2, \ldots, X_n$ be independent random variables with probability densities $f_{X_i}(x;\theta)$, dependent on a parameter $\theta$. The likelihood function $L(\theta; x_1, x_2, \ldots, x_n)$ is defined by

$$L(\theta; x_1, x_2, \ldots, x_n) = \prod_{i=1}^{n} f_{X_i}(x_i;\theta) \tag{46}$$

for any set $x_1,x_2,\ldots,x_n$ of realisations of $X_1,X_2,\ldots,X_n$. A maximum likelihood estimator for the parameter $\theta$ is a random variable $u(X_1,X_2,\ldots,X_n)$ such that the likelihood function $\theta \rightarrow L(\theta;x_1,x_2,\ldots,x_n)$ assumes its maximum value for $\hat{\theta} \stackrel{\text{def}}{=} u(x_1,x_2,\ldots,x_n)$ for any set of realisations $x_1,x_2,\ldots,x_n$. This means that a maximum likelihood estimate is a value of the parameter which, of all possible parameter values, makes the actual realisation the most likely.

The maximum likelihood estimator can be applied to the estimation of single unknown parameters from astronomical tables. Let $T$ be a table with correctly rounded tabular values $T(x)$ of a function $f_\theta(x)$ with unknown parameter $\theta$. As is indicated in Appendix A2, it is reasonable to assume that the tabular errors $e_\theta(x)$, defined by $e_\theta(x) = T(x) - f_\theta(x)$, of a correctly rounded table have a uniform distribution on the interval $\langle -\frac{1}{2} \cdot 60^{-k}, +\frac{1}{2} \cdot 60^{-k}]$, where $k$ is the number of sexagesimal fractional digits of the tabular values.[88] Thus the errors have a probability density

$$h(x) = \begin{cases} 60^k & x \in \langle -\frac{1}{2} \cdot 60^{-k}, +\frac{1}{2} \cdot 60^{-k}] \\ 0 & \text{elsewhere.} \end{cases} \tag{47}$$

The maximum likelihood estimates for the unknown parameter are all values $\hat{\theta}$ such that

$$f_{\hat{\theta}}(x) \in [T(x) - \tfrac{1}{2} \cdot 60^{-k}, T(x) + \tfrac{1}{2} \cdot 60^{-k}\rangle \tag{48}$$

for all $x$. This follows from the fact that for these values of $\theta$ $L(\theta;x_1,x_2,\ldots,x_n) = 60^{kn}$, where $n$ is the number of tabular values, while for all other values of $\theta$ $L(\theta;x_1,x_2,\ldots,x_n) = 0$. Note that in this case the maximum likelihood estimate need not be unique.

If the tabular values were correctly rounded but the tabular errors do not have a uniform distribution, the maximum likelihood estimate(s) will again be value(s) $\hat{\theta}$ of the parameter such that for all $x f_{\hat{\theta}}(x) \in [T(x) - \frac{1}{2} \cdot 60^{-k}, T(x) + \frac{1}{2} \cdot 60^{-k}\rangle$. However, in this case not all such values of $\theta$ will be maximum likelihood estimates. It may be difficult to compute the exact maximum likelihood estimate(s).

Most medieval astronomical tables contain at least one incorrect value. For these tables there is no parameter value $\hat{\theta}$ such that

$L(\hat{\theta};x_1,x_2,\ldots,x_n) \neq 0$, i.e. all parameter values are maximum likelihood estimates. I will extend the use of the maximum likelihood estimator by modifying the assumption about the distribution of the tabular errors. Thus I will assume that for certain $a \leq -\frac{1}{2}$ and $b \geq +\frac{1}{2}$ the tabular errors have a probability density

$$h(x) = \begin{cases} \eta \cdot 60^k & x \in \langle -\frac{1}{2} \cdot 60^{-k}, +\frac{1}{2} \cdot 60^{-k}] \\ \zeta \cdot 60^k & x \in [a \cdot 60^{-k}, -\frac{1}{2} \cdot 60^{-k}] \cup \langle +\frac{1}{2} \cdot 60^{-k}, b \cdot 60^{-k}] \\ 0 & \text{elsewhere,} \end{cases} \quad (49)$$

where $\eta + (b-a-1)\zeta = 1$, $\eta < 1$ is close to 1 and $0 < \zeta \ll \eta$. Thus most of the tabular values are expected to be correctly computed. Every tabular error within the interval $\langle -\frac{1}{2} \cdot 60^{-k}, +\frac{1}{2} \cdot 60^{-k}]$ contributes a factor $\eta \cdot 60^k$ to $L(\theta;x_1,x_2,\ldots,x_n)$, every other tabular error a (smaller) factor $\zeta \cdot 60^k$ or zero. Therefore the maximum likelihood estimates are those values $\hat{\theta}$ of the parameter which maximize the number of tabular errors within the interval $\langle -\frac{1}{2} \cdot 60^{-k}, +\frac{1}{2} \cdot 60^{-k}]$, i.e. which minimize the number of differences between the table to be investigated and the recomputation.

*Right ascension.* If all tabular values of a right ascension table $T$ with values to sexagesimal $k^{\text{th}}$s are rounded correctly, the maximum likelihood estimates $\hat{\varepsilon}$ for the obliquity of the ecliptic are the numbers in the interval

$$\bigcap_{\lambda=1,2,3,\ldots,89} \left\langle \arccos\left(\frac{\tan(T(\lambda) + \frac{1}{2} \cdot 60^{-k})}{\tan \lambda}\right), \right.$$
$$\left. \arccos\left(\frac{\tan(T(\lambda) - \frac{1}{2} \cdot 60^{-k})}{\tan \lambda}\right)\right]. \quad (50)$$

I carried out a Monte Carlo analysis on 200 computer-generated right ascension tables for arbitrary values of the obliquity to investigate the width of this interval. For right ascension tables with values to minutes the average width was 0;0,9, for tables with values to seconds, 0;0,0,10.

## *A5 The central limit theorem*

In 1900 Liapounov proved the following theorem:[89]

*CENTRAL LIMIT THEOREM. Let $X_1, X_2, X_3, \ldots$ be independent random variables with means $\mu_i$ and standard deviations $\sigma_i$.* Assume that $\varrho_i^3 \stackrel{\text{def}}{=} E(|X_i - \mu_i|^3)$ *is finite for every* $i = 1,2,3,\ldots$ *and that*

$$\lim_{n\to\infty} \frac{\sqrt[3]{\sum_{i=1}^{n} \varrho_i^3}}{\sqrt{\sum_{i=1}^{n} \sigma_i^2}} = 0. \qquad (51)$$

*Then the sum*

$$S_n = \sum_{i=1}^{n} X_i$$

*is asymptotically normal for $n \to \infty$ with mean*

$$\sum_{i=1}^{n} \mu_i$$

*and variance*

$$\sum_{i=1}^{n} \sigma_i^2.$$

The condition implies that no particular random variable $X_i$ has a disproportionate influence on the sum $S_n$.

For the estimation of the obliquity of the ecliptic from a right ascension table we used the sum $\hat{\theta}$ of the random variables $X_\lambda$, $\lambda = 1,2,3,\ldots,89$, given by

$$X_\lambda = \frac{w_\lambda}{W}\,\hat{\theta}_\lambda.$$ [90]

Here

$$w_\lambda = \frac{\tan^2\lambda}{(1 + \theta_0^{\,2}\tan^2\lambda)^2},$$

where $\theta_0$ is a rough approximation to $\theta$, the parameter to be estimated, and

$$W = \sum_{\lambda=1}^{89} w_\lambda.$$

I will proof that $\hat{\theta}$ has approximately a normal distribution.

For every $\lambda \in [1,89]$ let $T(\lambda)$ be the value of $\alpha(\lambda)$ rounded to $k$ sexagesimal fractional digits. Let $e(\lambda)$ denote the tabular error $T(\lambda)-\alpha(\lambda)$ and $\sigma^2$ its variance. I will assume that the $e(\lambda)$ are independent and have a uniform distribution. From formulas (22), (23) and (28) in Appendix A3 it follows that for every $\lambda$

$$\begin{aligned} X_\lambda - \mu_\lambda &\overset{\text{def}}{=} X_\lambda - E(X_\lambda) \\ &\approx \frac{w_\lambda}{W}\left(\frac{\partial}{\partial y}\,g(\lambda,\alpha(\lambda))\cdot e(\lambda) + \tfrac{1}{2}\frac{\partial^2}{\partial y^2}\,g(\lambda,\alpha(\lambda))\cdot(e(\lambda)^2-\sigma^2)\right) \qquad (52) \\ &= \frac{\pi}{180W}\cdot\frac{w_\lambda(1+\theta^2\tan^2\lambda)}{\tan\lambda}\,(e(\lambda)+c_\lambda(e(\lambda)^2-\sigma^2)) \\ &\approx \frac{\pi}{180W}\cdot\frac{\tan\lambda}{1+\theta^2\tan^2\lambda}\,(e(\lambda)+c_\lambda(e(\lambda)^2-\sigma^2)), \end{aligned}$$

where

$$c_\lambda = \frac{\pi}{180}\,\theta\tan\lambda.$$

According to our assumption, $e(\lambda)$ has a zero probability density outside the interval $[-\sqrt{3}\sigma, +\sqrt{3}\sigma]$. Within this interval the right-hand side of (52) has a zero

$$\frac{\sqrt{1 + 4c_\lambda^2\sigma^2} - 1}{2c_\lambda} \approx c_\lambda\sigma^2,$$

which can be considered equal to zero for practical purposes, since $c_\lambda \leq c_{89} \approx 0.9164$ for every $\lambda$ and $\sigma^2 = \frac{1}{12} \cdot 60^{-2k}$. Thus

$$\begin{aligned} E(|e(\lambda) + c_\lambda(e(\lambda)^2 - \sigma^2)|^3) &\approx -\int_{-\sqrt{3}\sigma}^{0} (\eta + c_\lambda(\eta^2 - \sigma^2))^3 \cdot \frac{1}{2\sqrt{3}\sigma}\, d\eta \\ &\quad + \int_{0}^{+3\sqrt{3}\sigma} (\eta + c_\lambda(\eta^2 - \sigma^2))^3 \cdot \frac{1}{2\sqrt{3}\sigma}\, d\eta \qquad (53) \\ &\approx \tfrac{3}{4}\, 2\sqrt{3}(1 + 2c_\lambda^2\sigma^2)\sigma^3 \approx \tfrac{3}{4}\sqrt{3}\sigma^3. \end{aligned}$$

For all $\lambda \in [1,89]$ we have

$$0.01745 < \frac{\tan\lambda}{1 + \theta^2\tan^2\lambda} \leq \frac{1}{2\theta}.^{91}$$

Thus, if we define $\varrho_\lambda^3 \overset{\text{def}}{=} E(|X_\lambda - \mu_\lambda|^3)$ and take a sum over $n$ values of $\lambda$ in the interval [1,89], we find

$$\begin{aligned} \frac{\sqrt[3]{\sum_\lambda \varrho_\lambda^3}}{\sqrt{\sum_\lambda \operatorname{Var} X_\lambda}} &\approx \frac{\sqrt[3]{\tfrac{3}{4}\sqrt{3}\left(\frac{\pi}{180W}\right)^3 \sigma^3 \sum_\lambda \left(\frac{\tan\lambda}{1 + \theta^2\tan^2\lambda}\right)^3}}{\sqrt{\left(\frac{\pi}{180W}\right)^2 \sigma^2 \sum_\lambda \left(\frac{\tan\lambda}{1 + \theta^2\tan^2\lambda}\right)^2}} \qquad (54) \\ &\leq \sqrt[3]{\tfrac{3}{4}\sqrt{3}}\,\frac{n^{\frac{1}{3}}/2\theta}{n^{\frac{1}{2}} \cdot 0.01745} \approx \frac{34}{\sqrt[6]{n}}. \end{aligned}$$

We conclude that

$$\hat{\theta} = \sum_{\lambda=1}^{89} X_\lambda$$

has approximately a normal distribution if the $e(\lambda)$, $\lambda = 1,2,3,\ldots,89$, are independent and have a uniform distribution. As was indicated in Appendix A2, this will be the case if the number of sexagesimal digits of the tabular values is sufficiently large.

## Critical Apparatus

*The right ascension table on f. 39ʳ of the Sanjufīnī Zīj*

Scribal error (the correction is given in brackets):

$$\lambda' = 66 \quad T(\lambda') = 66;48 \quad (67^\circ)$$

Errors with respect to the recomputation for $\varepsilon = 23;35$ (the recomputed values are given in brackets):

| | | | |
|---|---|---|---|
| $\lambda' = 6$ | $T(\lambda') = 6;32\ (33')$ | $\lambda' = 46$ | $T(\lambda') = 48;30\ (29')$ |
| 16 | 17;23 (22′) | 54 | 56;20 (21′) |
| 17 | 18;28 (27′) | 55 | 57;18 (19′) |
| 20 | 21;39 (40′) | 56 | 58;16 (17′) |
| 21 | 22;43 (44′) | | |

*The oblique ascension table on f. 38ᵛ of the Sanjufīnī Zīj*

The tabular values for arguments 95, 130, and 149 are more or less unreadable. Nevertheless they were reliably restored, since the number of minutes could be identified in each case.

Scribal error:

$$\lambda = 163 \quad T(\lambda) = 119;2 \quad (159^\circ)$$

*Right ascension extracted from the oblique ascension table on f. 38ᵛ of the Sanjufīnī Zīj*

Differences larger than 0;0,30 between the extracted values and recomputed values for $\varepsilon$ = 23;32,30 (the recomputed values to seconds are given in brackets):

| | | | |
|---|---|---|---|
| $\lambda = 5$ | $T_\alpha(\lambda)$ = 4;36, 0 (4;35,9) | $\lambda = 49$ | $T_\alpha(\lambda)$ = 46;32,0 (46;31,22) |
| 7 | 6;26, 0 (6;25,21) | 55 | 52;37,0 (52;37,42) |
| 10 | 9;11,30 (9;10,57) | | |

*Solar equation table in the Shāmil Zīj*

Scribal errors (the corrections in brackets are entirely based on the interpolation pattern, which is very reliable. Note that between $\bar{a}$ = 105 and $\bar{a}$ = 108 several digits were shifted upwards):

| | | | |
|---|---|---|---|
| $\bar{a} = 21$ | $T(\bar{a})$ = 0;41, 0 (18″) | $\bar{a} = 114$ | $T(\bar{a})$ = 1;55,52 (50′12″) |
| 27 | 0;52,29 (24″) | 121 | 1;43,43 (48″) |
| 38 | 1;11,48 (18″) | 128 | 1;25,46 (35′) |
| 47 | 1;24,19 (59″) | 129 | 1;34,36 (30″) |
| 55 | 1;35,37 (32″) | 146 | 1;10,29 (8′) |
| 66 | 1;47,20 (9″) | 149 | 1; 2, 9 (3′) |
| 105 | 1;55,27 (55″) | 155 | 0;51,14 (54″) |
| 106 | 1;55,24 (26″) | 163 | 0;35,18 (58″) |
| 107 | 1;55,20 (54′54″) | 169 | 0;28,30 (23′) |
| 108 | 1;53,44 (54′20″) | 171 | 0;19,56 (16″) |

Errors with respect to the recomputation for $e$ = 2;4,35,30 (the recomputation is given in brackets. The errors for $\bar{a}$ = 148 and $\bar{a}$ = 159 are no scribal errors):

| | | | |
|---|---|---|---|
| $\bar{a} = 2$ | $T(\bar{a})$ = 0; 4, 0 (1″) | $\bar{a} = 56$ | $T(\bar{a})$ = 1;36,45 (44″) |
| 8 | 0;16, 0 (1″) | 61 | 1;42,18 (19″) |
| 9 | 0;17,59 (18′0″) | 72 | 1;51,56 (55″) |
| 10 | 0;19,58 (59″) | 76 | 1;54,27 (26″) |
| 12 | 0;23,55 (56″) | 103 | 1;56,48 (47″) |

| | | | |
|---|---|---|---|
| 14 | 0;27,50 (51″) | 137 | 1;23,13 (14″) |
| 15 | 0;29,47 (48″) | 148 | 1; 4,56 (57″) |
| 43 | 1;19, 8 (7″) | 159 | 0;44, 3 (4″) |
| 49 | 1;27,48 (47″) | | |

## BIBLIOGRAPHY

Bickel, P. J. and Doksum, K. A.
1977: *Mathematical Statistics: Basic Ideas and Selected Topics,* San Francisco.
al-Bīrūnī
*Kitāb al-Qānūn al-Mas'ūdī...,* 3 volumes, Hyderabad 1954–56.
Blochet, E.
1925: *Catalogue des manuscrits arabes des nouvelles acquisitions,* Paris.
Breiman, L.
1973: *Statistics: With a View Towards Applications,* Boston.
Carra de Vaux, B.
1892: "L'Almageste d'Abu'l Wafa al-Buzdjan", *Journal Asiatique* 8–19, pp. 408–471.
Cramèr, H.
1951: *Mathematical Methods of Statistics,* Princeton.
Debarnot, M. Th.
1987: "The Zīj of Ḥabash al-Ḥāsib: A Survey of MS Istanbul Yeni Cami 784/2, From Deferent to Equant", *Annals of the New York Academy of Science* 500, New York, pp. 35–69.
Diaconis, P. and Engel, E.
1986: "Comment to: Good, I. J., "Applications of Poisson's Work"", *Statistical Science* 1, pp. 171–174.
Delambre, J. B.
1819: *Histoire de l'astronomie du moyen âge,* Paris. Reprinted New York 1965.
Feller, W.
1957–66: *An Introduction to Probability Theory and Its Applications,* 2 volumes, New York.
Franke, H.
1988: "Mittelmongolische Glossen in einer arabischen astronomischen Handschrift von 1366", *Oriens* 31, pp. 95–118.
Heiberg, J. L.
1898–1903: *Claudii Ptolemaei: Syntaxis Mathematica,* 2 volumes, Leipzig.
Hogendijk, J. P.
1988: "Three Islamic Lunar Crescent Visibility Tables", *Journal for the History of Astronomy* 19, pp. 29–44.
Irani, R. A. K.
1955: "Arabic Numeral Forms", *Centaurus* 4, pp. 1–12.

Jensen, C.
1971/72: "The Lunar Theories of al-Baghdādī", *Archive for the History of Exact Sciences* 8, pp. 321–328.
Kemperman, J. H. B.
1975: "Sharp Bounds for Discrepancies (mod 1) with Applications to the First Digit Problem", unpublished.
Kennedy, E. S.
1956a: "A Survey of Islamic Astronomical Tables", *Transactions of the American Philosophical Society* N.S. 46/2, pp. 123–177.
1956b: "Parallax Theory in Islamic Astronomy", *Isis* 47, pp. 33–53. Reprinted in Kennedy et al. 1983, pp. 164–184.
1957: "Comets in Islamic Astronomy and Astrology", *Journal of Near Eastern Studies* 16, pp. 44–51. Reprinted in Kennedy et al. 1983, pp. 311–318.
1977: "The solar Equation in the Zīj of Yaḥyā ibn Abī Manṣūr", *PRISMATA: Festschrift für Willy Hartner,* Wiesbaden, pp. 183–186. Reprinted in Kennedy et al. 1983, pp. 136–139.
1985: "Spherical Astronomy in Kāshī's Khāqānī Zīj", *Zeitschrift für Geschichte der arabisch-islamischen Wissenschaften* 2, pp. 1–46.
1987/88: "Eclipse Predictions in Arabic Astronomical Tables Prepared for the Mongol Viceroy of Tibet", *Zeitschrift für Geschichte der arabisch-islamischen Wissenschaften* 4, pp. 60–80.
Kennedy, E. S. and Hogendijk, J. P.
1988: "Two Tables from an Arabic Astronomical Handbook for the Mongol Viceroy of Tibet", *A scientific humanist. Studies in memory of Abraham Sachs,* Philadelphia, pp. 233–242.
Kennedy, E. S. and Kennedy, M. H.
1987: *Geographical Coordinates of Localities from Islamic Sources,* Frankfurt.
Kennedy, E. S. and Muruwwa, A.
1958: "Bīrūnī on the Solar Equation", *Journal of Near Eastern Studies* 17, pp. 112–121. Reprinted in Kennedy et al. 1983, pp. 603–612.
Kennedy, E. S. and Salam, H.
1967: "Solar and Lunar Tables in Early Islamic Astronomy", *Journal of the American Oriental Society* 87, pp. 487–497. Reprinted in Kennedy et al. 1983, pp. 108–113.
Kennedy, E. S. et al.
1983: *Studies in the Islamic Exact Sciences,* Beirut.
King, D. A.
1972: *The Astronomical Works of Ibn Yunus,* unpublished doctoral thesis, Yale University.
1973: "Ibn Yunus' Very Useful Tables for Reckoning Time by the Sun", *Archive for the History of Exact Science* 10, pp. 342–394. Reprinted in King 1986a, no. IX.
1978: "Astronomical Timekeeping in Fourteenth-Century Syria", *Proceedings of the First International Symposium for the History of Arabic Science, Aleppo* 1976, II, pp. 75–84. Reprinted in King 1986a, no. X.
1983: "The Astronomy of the Mamluks", *Isis* 74, pp. 531–555. Reprinted in King 1986a, no. III.
1986a: *Islamic Mathematical Astronomy,* London.
1986b: *A Survey of the Scientific Manuscripts in the Egyptian National Library,* Winona Lake.

Knuth, D. E.
1969: *The Art of Computer Programming, Vol. II, Seminumerical Algorithms,* Reading Massachusetts.
Liapounov, A. M.
1900: "Sur une proposition de la théorie des probabilités", *Bulletin de l'Académie des Sciences de St. Petersbourg* 5, pp. 359–386.
1901: "Nouvelle forme de théorème sur la limite des probabilités", *Mémoires de l'Académie des Sciences de St. Petersbourg* 8, pp. 1–24.
Millás Vallicrosa, J.
1943–50: *Estudios sobre Azarquiel,* Madrid and Granada.
Nallino, C. A.
1899–1907: *Al-Battani sive Albatenii opus astronomicum,* 3 volumes, Milan and Rome. Reprinted Frankfurt 1969.
Neugebauer, O.
1962: "The astronomical tables of al-Khwārizmī", *Det Kongelige Danske Videnskabernes Selskab hist.-fil. Skrifter* 4:2.
1975: *A History of Ancient Mathematical Astronomy,* 3 volumes, Berlin.
North, J. D.
1976: *Richard of Wallingford,* 3 volumes, Oxford.
1986: *Horoscopes and History,* London.
Pedersen, O.
1974: *A Survey of the Almagest,* Odense.
Rico y Sinobas, M.
1864: *Libros del saber de astronomía del rey D. Alfonso X de Castilla,* 5 volumes, Madrid.
Sezgin, F.
1971–: *Geschichte des arabischen Schrifttums,* 9 volumes, Leiden.
de Slane, MacG.
1883–95: *Catalogue des manuscrits arabes,* Paris.
Suter, H.
1892: "Das Mathematiker-Verzeichniss im Fihrist des Ibn Abī Ja'kūb al-Nadīm", *Zeitschrift für Mathematik und Physik* 37 Supplement, pp. 1–87.
1900: "Die Mathematiker und Astronomen der Araber und ihre Werke", *Abhandlungen zur Geschichte der mathematischen Wissenschaften* 10.
1902: "Nachträge und Berichtigungen zu "Die Mathematiker und Astronomen der Araber und ihre Werke"", *Abhandlungen zur Geschichte der mathematischen Wissenschaften* 14, pp. 157–185.
1914: "Die astronomischen Tafeln des Muḥammed ibn Mūsā al-Khwārizmī in der Bearbeitung des Maslama ibn Aḥmed al-Madjrītī und der Lateinischen Übersetzung des Adelard von Bath", *Det Kongelige Danske Videnskabernes Selskab hist.-fil. Skrifter* 3:1.
Toomer, G. J.
1968: "A Survey of the Toledan Tables", *Osiris* 15, pp. 5–174.
1984: *Ptolemy's Almagest,* London.
Van Brummelen, G. R.
*The Numerical Structure of al-Khalīlī's Auxiliary Tables,* to appear.
Yaḥyā ibn Abī Manṣūr
*The Verified Astronomical Tables for the Caliph al-Ma'mūn,* facsimile, Frankfurt 1986.

NOTES

1. Kennedy 1956a.
2. Kennedy gave an update in his lecture "The Zīj Survey: A Progress Report" at the XVIIIth International Congress of History of Science, Hamburg/Munich 1989.
3. Examples of published zījes are Nallino 1899–1907 (al-Battānī), Suter 1914 (Andalusian recension of al-Khwārizmī), Millás Vallicrosa 1943–50 (al-Zarqālī), al-Bīrūnī, and Yaḥyā ibn Abī Manṣūr. Additional extensive studies of zījes are Neugebauer 1962 (al-Khwārizmī), Toomer 1968 (Toledan tables), and Debarnot 1987 (Ḥabash). A large number of short studies on zījes and on tables in general can be found in Kennedy et al. 1983 and King 1986a.
4. Kennedy et al. 1983.
5. Kennedy & Salam 1967, p. 496.
6. Kennedy 1977, p. 184.
7. Kennedy & Kennedy 1987.
8. See King 1973 and King 1978.
9. North 1976, Vol. III, pp. 192–195.
10. See North 1986, pp. 11–16.
11. Hogendijk 1988, p. 31.
12. If such tables were computed correctly, they would differ in only 4 out of 90 entries.
13. Paris Bibliothèque Nationale Ms. Arabe 2494, f. 24$^v$. More information about Abū'l-Wafā' can be found in Section 5. Two formulas or algorithms will be called "mathematically equivalent" if essentially the same intermediate steps occur in their evaluations.
14. Escorial Ms. árabe 927, f. 51$^r$, or Yaḥyā ibn Abī Manṣūr, p. 98.
15. See Heiberg 1898–1903, Vol. I, pp. 82–85 (Greek) or Toomer 1984, pp. 71–73 (English translation); Neugebauer 1962, pp. 46–47; Nallino 1899–1907, Vol. I, pp. 13–14; and King 1972, p. 113. Al-Kāshī used an algorithm equivalent to $\alpha(\lambda) = \arccos(\cos\lambda/\cos\delta(\lambda))$; see Kennedy 1985, pp. 10–11.
16. Leiden Ms. Or. 143, p. 271. King indicates that Ibn Yunus' tables for $\sin\alpha(\lambda)$ and for $\alpha(\lambda)$, which are found on pp. 271 and 268–269 of the Leiden manuscript, were in fact computed from his table for $\tan\delta(\lambda)$; see King 1972, pp. 117–121.
17. Paris Bibliothèque Nationale Ms. Arabe 2486, f. 235$^r$.
18. See Neugebauer 1975, Vol. I, p. 42.
19. Kennedy 1956b, p. 50.
20. Kennedy 1956a, p. 164.
21. See for instance Kennedy & Muruwwa 1958.
22. Heiberg 1898–1903, Vol. I, pp. 234–237 (Greek) or Toomer 1984, p. 155 (English translation).
23. Zīj of Ibn al-Bannā', Escorial Ms. árabe 909, ff. 23$^v$–24$^r$. (The eccentricity underlying this table is the minimum eccentricity of Ibn al-Bannā's solar model.)
24. Examples of such tables are found in Berlin Ms. 5751, pp. 178–179 (zīj of Kushyār) and Millás Vallicrosa 1943–50, pp. 158–165.
25. Pedersen 1974, Chapter 5, pp. 95–101 and 149–151.

26. Usually the right ascension is tabulated for every degree of the ecliptic. However, because of the symmetries indicated in Section 2, the tabular values for $\lambda = 91,92,93,\ldots,360$ do not yield any extra information.
27. See Nallino 1899–1907, Vol. I, p. 160; Delambre 1819, p. 176; and Section 4 of this paper.
28. Similar properties exist for different numbers of sexagesimal fractional digits and different ways of rounding off.
29. Nallino 1899–1907, Vol. I, p. 44.
30. See Section 5 of this paper.
31. Paris Bibliothèque nationale Ms. Arabe 6040, 57 folios, 14th century. The manuscript is an autograph, as is indicated on folio 26$^v$. See Blochet 1925, p. 169. The complete name of the author is given as Abū Muḥammad ʻAṭā ibn Aḥmad ibn Muḥammad Khwāja Ghāzī al-Sanjufīnī al-Samarqandī. The following articles deal with aspects of the Sanjufīnī Zīj: Kennedy 1987/88, Kennedy & Hogendijk 1988 and Franke 1988.
32. See Kennedy & Hogendijk 1988, p. 233.
33. Kennedy & Hogendijk 1988, pp. 237 and 241.
34. This follows from the tabular values 28;15,0 for 0° Capricorn and 51;50,0 for 0° Aries.
35. Hyderabad Andra Pradesh State Library Ms. 298, table 54.
36. Rico y Sinobas 1864, Vol. III, p. 296–297.
37. This follows from the tabular value for 0° Aries given in note 34.
38. Kennedy & Hogendijk 1988, p. 241.
39. Kennedy & Hogendijk 1988, p. 234.
40. Note that $\lambda' = \lambda + 90$. Because of the symmetries indicated just below formula (1) in Section 2, the tabular values for arguments 91 to 360 will usually not yield much extra information. I used them, however, to check for scribal errors.
41. For instance, in the case of linear interpolation, neighbouring tabular values tend to have errors which have the same sign and therefore are not independent.
42. Thus, for example, the estimator could be computed from every second or from every third tabular value.
43. For more information about the oblique ascension, see Kennedy 1956a, p. 140, and Pedersen 1974, pp. 99–101 and 110–115.
44. In case of systematic truncation or rounding up, $T_\varrho(\lambda) + T_\varrho(180-\lambda)$ would have an odd number of minutes for all $\lambda$ if either the right ascension or the ascensional difference were given to minutes. If both were given to minutes, $T_\varrho(\lambda) + T_\varrho(180-\lambda)$ would be even for all $\lambda$.
45. This implies that the tabular errors of the extracted right ascension have a smaller standard deviation than those of the oblique ascension. For instance, a uniform distribution on the interval $[-\frac{1}{2},+\frac{1}{2}]$ has standard deviation $\frac{1}{6}\sqrt{3} \approx 0.29$, a triangular distribution on the same interval has standard deviation $\frac{1}{12}\sqrt{6} \approx 0.20$.
46. See notes 35 and 36, and Kennedy 1956a, p. 145.
47. See note 39.

* 48. Apart from the oblique ascension table on folio 28$^v$, also the parallax table on folio 28$^r$, which is explicitly attributed to al-Sanjufīnī, involves this value. The most important localities to which a latitude of 32° is attributed are Jerusalem and Isfahan; see Kennedy & Kennedy 1987, pp. 678–679.

49. Paris Bibliothèque Nationale: Ms. Arabe 2528 (73 folios, 15th century), Ms. Arabe 2529 (76 folios, 16th century), Ms. Arabe 2540 (folios 29$^v$–97$^v$, 15th century); Florence Laurenziana:

Or. 95 [289] (116 folios, 15th century), Or. 106-1 (folios 1ᵛ–71ʳ, early 14th century, Mosul); Cairo Egyptian National Library (Dār al-Kutūb): Ṭal'at mīqāt 138 (115 pages, ca. 1500), Taimūr riyāḍ 296,1 (pages 1–160, 18th century); London British Library II,2 Ms. 395,3 (Add. 7492 Rich.; 67 folios, ca. 1500). A commentary on the Shāmil Zīj is found in Paris Bibliothèque Nationale Ms. Arabe 2530. See de Slane 1883–95, pp. 451–452 and 454; Suter 1902, pp. 166–167; Kennedy 1956a, p. 129, no. 29; Sezgin 1971–, Vol. V, pp. 324–325; and King 1986b, p. 52, no. B100.

50. The Baghdādī Zīj was compiled shortly before the year 1285 by Jamāl al-Din al-Baghdādī. It is extant in the unique manuscript Paris Bibliothèque Nationale Ms. 2486 (225 folios, 1285 A.D.) and contains material of earlier astronomers such as Ḥabash and Abū'l-Wafā'. See de Slane 1883–95, pp. 440–441; Kennedy 1957, p. 48; and Jensen 1971/72.
51. The Ashrafī Zīj was written in Persian by Muḥammad Sanjar al-Kamālī (Shiraz, south-western Persia, ca. 1300). It gives the mean motion parameters and planetery equations from a large number of earlier zījes and is extant in Paris Bibliothèque Nationale Ms. Suppl. Persan 1488 (251 folios, 1303 A.D.). See Kennedy 1956a, p. 124, no. 4; and Kennedy 1977, p. 183.
52. The anonymous Muṣṭalaḥ Zīj was used in Egypt between the 13th and 15th centuries. It is extant in Paris Bibliothèque Nationale Ms. 2513 (94 folios, 13th century) and Ms. 2520 (175 folios, 14th century). See de Slane 1883–95, pp. 446 and 448–449; Kennedy 1956a, pp. 131–132, no. 47; and King 1983, pp. 535–536.
53. Dublin Chester Beatty Ms. 4076 (61 folios, ca. 1400). Information about al-Abharī can be found in Suter 1900, pp. 145–146, no. 364, and in the article by C. Brockelmann in the *Encyclopaedia of Islam,* 2nd edition, 1960–.
54. Kennedy 1956a, pp. 129 and 133. Information about Abū'l-Wafā' can be found in Suter 1900, pp. 71–72, no. 167, and in the article by A. P. Yushkevich in the *Dictionary of Scientific Biography,* New York 1970–80.
55. The geographical table is given in Kennedy & Kennedy 1987, pp. 471–473.
56. These coordinates are not found in the list given in Kennedy & Kennedy 1987, pp. 216–217. However, al-Battānī's values $\varphi = 37°15'$ and $\lambda = 75°0'$ are very close. It can be shown that the oblique ascension table in the zij of al-Abharī was computed using $\varphi = 37°0'$, and not using $\varphi = 37°15'$ or $\varphi = 37°25'$.
57. Note that the difference in longitude cannot be explained from different meridians of reference. Firstly, both the oblique ascension table in the zīj of al-Abharī and the geographical table in the Shāmil Zīj state that the longitudes are measured from the Fortunate Isles. Secondly, the longitude 75° could in principle refer to north-western Persia (when measured from the "western shore of the encompassing sea"), but in al-Abharī's zīj it occurs only in explicit connection with Mardin. See Kennedy & Kennedy 1987, p. xi for a description of the meridians of reference.
58. Paris Bibliothèque Nationale Ms. 2494 (107 folios, 12th century). See de Slane 1883–95, p. 442.
59. Carra de Vaux 1892.
60. See Kennedy 1956a, p. 134, no. 73; and Suter 1892, pp. 39–40. Suter translates "al-zīj al-wāḍiḥ" as "das Buch der genauen (klaren, zweifellosen) Tafeln".
61. See later in this section.
62. As was indicated in Section 2, *e* was the actual quantity used by Islamic astronomers for the

computation of the solar equation. Therefore the value of $e$ underlying the table can be expected to be a round number.

63. Heiberg 1898–1903, Vol. I, pp. 232–240 (Greek) or Toomer 1984, pp. 153–157 (English translation).
64. Nallino 1899–1907, Vol. III, p. 66 (Arabic) or Vol. I, p. 44 (Latin translation).
65. Nallino 1899–1907, Vol. II, p. 81.
66. Nallino 1899–1907, Vol. III, p. 66, lines 20–21 (Arabic) or Vol. I, p. 44, lines 14–15 (Latin translation).
67. See Van Brummelen, to appear.
68. The same holds for the Sine tables in the other three manuscripts which I inspected.
69. See note 50.
70. Berlin Ahlwardt Ms. 5750, f. $30^r$–$31^r$. This table was analyzed in Kennedy & Salam 1967, pp. 493–494.
71. See Feller 1957–66, Vol. I, p. 229.
72. Bickel & Doksum 1977, Breiman 1973.
73. The determination of scribal errors is facilitated by the list of Arabic numeral forms in Irani 1955.
74. See Kemperman 1975. The theorem is mentioned in Diaconis & Engel 1986, p. 172.
75. Feller 1957–66, Vol. II, p. 472ff.
76. See Feller 1957–66, Vol. II, p. 487.
77. Knuth 1969, pp. 54–66.
78. Knuth 1969, pp. 35–48.
79. Knuth 1969, p. 64.
80. An extensive discussion of tabular errors is found in Appendix A2. Note that, according to the definition of $e_\theta(x)$, also the rounding errors of a correctly computed table are tabular errors.
81. In fact, this turns out to be a reasonable assumption for many tables. As is indicated in Appendix A2, parts of a table not satisfying the assumptions about the tabular errors can be disregarded in the estimation of the unknown parameter.
82. Such a function exists for most functions with a single unknown parameter which were used in zījes. In his Survey Kennedy gave a list of functions occurring in zījes (Kennedy 1956a, pp. 139–145). Neugebauer described the functions that occur in the Ptolemaic planetary model (Neugebauer 1975, Vol. I, pp. 21–230).
83. See Section 2 for more information about the right ascension.
84. $\sigma = 10^{-2}$ corresponds roughly to an average absolute error of 0;0,36. A correctly computed right ascension table with values to minutes and uniformly distributed rounding errors has $\sigma \approx 4.8 \cdot 10^{-3}$.
85. Chebyshev's inequality:

$$\Pr(|X - E(X)| \geq k \cdot \operatorname{Std} X) \leq \frac{1}{k^2}$$

for every random variable $X$ and constant $k > 0$, gives a 95% confidence interval if we take

$$k = \sqrt{\frac{1}{0.05}} \approx 4.47.$$

This interval has a width of 8.94 Std $X$. A much smaller 95% confidence interval is obtained if $X$ has a normal distribution. In this case the width is 3.92 Std $X$. If $X$ has a uniform distribution, its value is contained in an interval having width 3.46 Std $X$. It will be shown in this paper that the weighted estimator $\hat{\theta}$ has approximately a normal distribution. However, this is not likely to hold for a separate $\hat{\theta}_\lambda$. In general, the distribution of $\hat{\theta}_\lambda$ will be quite similar to the distribution of $e_\theta(\lambda)$. This implies that $\hat{\theta}_\lambda$ will approximately have a normal distribution if $e_\theta(\lambda)$ has a normal distribution. As is indicated in Appendix A2, this may be the case if a large number of intermediate rounding errors are made during the computation of $T(\lambda)$. The assumption that a certain tabular error is uniformly distributed implies that the concerned tabular value is not in error. Since many medieval tables contain at least a few errors, it is hardly ever possible to use this assumption.

86. See Section 2 for more information about the solar equation.
87. If

$$i \stackrel{\text{def}}{=} \left( \min_x \operatorname{Std} \hat{\theta}_x \right) / \operatorname{Std} \hat{\theta}$$

and $n$ is the number of tabular values from which the weighted estimator is computed, we have

$$\frac{i}{\sqrt{n}} \approx 0.71$$

for the right ascension and

$$\frac{i}{\sqrt{n}} \approx 0.71$$

for the solar equation. Thus the improvement in the standard deviation seems to be proportional to the square root of the number of tabular values used, as would be the case if all separate estimators had equal variance. This allows us to predict the width of confidence intervals. For instance, if the obliquity is to be estimated from the values $T(5), T(10), \ldots, T(85)$ of a right ascension table, the expected width of the confidence interval is

$$0;0,51 \cdot \frac{\sqrt{89}}{\sqrt{17}} \approx 0;1,57,$$

since 0;0,51 is the expected width if 89 tabular values are used.

88. Tests that can be used to verify the assumption of uniform distribution are also indicated in Appendix A2.
89. Liapounov 1900 and 1901. The theorem is given by Cramèr in Cramèr 1951, p. 215ff.
90. See Appendix A3.
91. The minimum value is assumed for $\lambda = 1$, the maximum for $\lambda = \arctan(1/\theta)$.

**Corrections and Additions**

I 107, line 20: ‘trigonometric’ → ‘trigonometric and spherical-astronomical’.

I 142, note 48: My later research showed that the *Sanjufinī Zīj* is strongly related to an originally Persian *zīj* that was very probably compiled around 1275 at the Islamic Astronomical Bureau of the Mongolian Yuan dynasty in China. This work is extant in various Chinese versions and in a Persian manuscript at the Oriental Institute in St. Petersburg. The tables for a latitude of 32° in the Chinese versions are intended for Nanjing, the place where the Chinese translation was prepared. Cf. Benno van Dalen, ‘Islamic Astronomical Tables in China. The Sources for the Huihui li’, in *History of Oriental Astronomy. Proceedings of the Joint Discussion-17 at the 23rd General Assembly of the International Astronomical Union, organised by the Commission 41 (History of Astronomy) held in Kyoto, August 25–26, 1997* (S.M. Razaullah Ansari, editor), Dordrecht (Kluwer) 2002, pp. 19–31.

# II

# On Ptolemy's Table for the Equation of Time

*Contents*



The second-century Greek scientist *Claudius Ptolemaeus* (or *Ptolemy*) was one of the most important astronomers of Antiquity. He was the first to construct satisfactory models for the apparent motion of the moon and the five visible planets, and he compiled large sets of mathematical tables with which the positions of the heavenly bodies could be predicted at the cost of only a small number of additions and multiplications. Ptolemy's most important astronomical works, the *Almagest* and the *Handy Tables*, had a large influence up until the end of the Middle Ages.

In recent years a large number of publications have appeared about technical aspects of Ptolemy's *Almagest.* In many of these, the origin of material found in the *Almagest* was investigated using a mathematical or statistical analysis of numerical data. The materials investigated include the star catalogue, the observations reported, the values of the astronomical parameters, and the mathematically-computed tables for trigonometrical, spherical astronomical and planetary functions.[1]

Although Glen Van Brummelen [1993] discovered various interesting details of the methods by which Ptolemy computed the table of chords and the planetary interpolation tables in the *Almagest*, our knowledge of the methods of computation used in Ptolemy's tables and in ancient and mediaeval astronomical tables in general is still very limited. Likewise, little is known about the values of the astronomical parameters underlying such tables. Many different values were in use, but especially in the case of tables having multiple unknown parameters the determination of the underlying values has been problematic.[2] Since the method of computation and the underlying parameter values of ancient and mediaeval astronomical tables are often specific to the astronomer who constructed them, the determination of the mathematical properties of a table can be extremely useful when one is trying to determine its origin.

In this article an extensive mathematical analysis of the table for the equation of time in Ptolemy's second important astronomical work, the *Handy Tables,* will be presented. The equation of time is the correction that must be applied to the time read from a sundial in order to obtain the time indicated by a watch adjusted to local time. It is a complicated function depending on five different astronomical parameters, and can be tabulated for different independent variables. Besides Ptolemy's *Handy Tables*, numerous extant astronomical handbooks from Islamic, Byzantine and Western European origin contain a table for the equation of time. Most of these tables are essentially different, i.e. they are based on different values for the underlying parameters or they were computed according to different algorithms. The use of interpolation and the application of approximation methods cannot be ruled out in advance.

Only occasionally can detailed information about the mathematical properties of tables for the equation of time be found in the primary sources. Usually the information presented is insufficient for determining the precise method according to which a table was computed; in many cases not even all of the underlying parameter values are mentioned. Thus the only way to obtain this information is to perform a mathematical or statistical analysis of the tabular values. Up to the present day practically no such analyses have been published, and only one or two tables for the equation of time have been successfully recomputed.[3] Since the effect of changes to the underlying parameter values on the equation of time is hard to predict, and since the independent variable of a table for the equation of time cannot be ascertained in a straightforward way, the problem of determining the mathematical structure of a table for the equation of time is non-trivial.

In this article I will show how the method of least squares can be used to determine the underlying parameter values of a table for the equation of time or, more generally, of any astronomical table having multiple unknown parameters. Furthermore, I will present a number of special properties from which other information concerning the mathematical structure of a table for the equation of time can be derived; for instance, the accuracy of the underlying auxiliary tables.

In the analysis of Ptolemy's table for the equation of time I will show how an application of the above-mentioned methods combined with historical considerations allow us to determine practically every detail of the method of computation. I will show that Ptolemy used approximation methods at various stages of his computation without letting the errors get out of hand. His table for the equation of time was *not* computed using other tables in the *Almagest* or the *Handy Tables,* contrary to what is suggested by the fourth-century commentator Theon of Alexandria. Although the table for the equation of time in the second-century Greek papyrus P. London 1278 displays fewer and less accurate values than the table in the *Handy Tables*, I will show that the two tables are related.

In Section 1 of this article the equation of time will be explained from a technical point of view. Mathematical properties of the equation of time and methods to analyse tables for this function can be found in Section 2. A complete analysis of Ptolemy's table for the equation of time, preceded by a survey of the relevant information found in the primary and secondary sources, is presented in Section 3, with the table in the papyrus P. London 1278 treated in Section 3.5. The reader is referred to the Appendix for information about the statistical background of my analysis of Ptolemy's table for the equation of time.

## *1 The equation of time: theoretical exposition*

Astronomers from Antiquity and the Middle Ages knew that true solar time (as can be read from a sundial) and astronomical or mean solar time (which was used to determine the positions of the planets from tables) are not generally equal. Although the difference is about half an hour at most, it has to be taken into account when an accurate position of a fast celestial body like the moon is needed, e.g. in the calculation of the time of an eclipse. To convert true solar time to mean solar time or vice versa one has to add or subtract a quantity which varies throughout the year. Many Greek and Byzantine astronomers called this quantity *ἡ παρὰ τὴν ἀνισότητα τῶν νυχθημέρων διαφορά* (the difference due to the inequality of the days and nights),[4] but Ptolemy did not use any particular designation. The Arabic name was *taʿdīl al-ʾayyām bi-layālīhā* (correction of the days and their nights), which was translated into Latin as *equatio dierum cum noctibus suis.* In modern astronomy the correction is called the *equation of time.* Ptolemy's *Handy Tables* and numerous mediaeval Islamic astronomical handbooks contain a table for the equation of time as a function of the solar position.

To understand the concept of the equation of time and the way in which the equation can be computed, we will have to consider both annual and daily effects.[5] The tropical year is defined by Ptolemy as the period of time in which the true sun *S*, which moves on the ecliptic at a variable apparent speed, returns to a particular equinox or solstice [Heiberg 1898–1903, vol. 1, pp. 192–193; Toomer 1984, p. 132]. It can be noted that Ptolemy considers the length of the tropical year (and also the tropical longitude of the solar apogee) to be a constant. The so-called "ecliptical mean sun" $\bar{S}$, which moves on the ecliptic, and the "equatorial mean sun" *M*, which moves on the equator, both have a uniform motion and a period of precisely one tropical year.[6] $\bar{S}$ and *M* will be well defined as soon as we fix their position at a certain point in time with respect to the position of the true sun *S*. Let $\lambda$ denote the ecliptical longitude of the true sun, $\bar{\lambda}$ the ecliptical longitude of the ecliptical mean sun, and $\mu$ the right ascension of the equatorial mean sun.[7] $\lambda$, $\bar{\lambda}$, and $\mu$ are all measured from the vernal point. Note that, since $\bar{S}$ and *M* have a uniform motion and the same period, $\bar{\lambda}$ and $\mu$ increase linearly at the same rate. Therefore there is a constant *c* such that $\bar{\lambda}+c=\mu$ at any moment. $\lambda$ and $\bar{\lambda}$ are related through $\bar{\lambda}=\lambda+q$, where *q* is a variable quantity called the "solar equation". A more extensive discussion of *c* and formulae for *q* will be given below.

Now we will consider the daily effects. Time is usually reckoned from midday, i.e. from the upper culmination of the sun. Thus true solar time is defined as the hour angle $h(S)$ of the true sun, mean solar time is defined as the hour angle $h(M)$ of the equatorial mean sun.[8] The equation of time $E_d$ (where the subscript *d* indicates that the equation is measured in degrees) can now be defined as the difference $h(S)-h(M)$ between true and mean solar time (cf. Figure 1). $E_d$ is not a constant for two reasons. Firstly, the true sun has a variable speed, whereas the equatorial mean sun moves uniformly. Secondly, the true sun moves on the ecliptic, the equatorial mean sun on the equator, and in general equal arcs of the equator and the ecliptic do not pass the meridian in equal time spans.

From $E_d=h(S)-h(M)$ it follows that at any moment $E_d$ equals the difference in the right ascension of the equatorial mean sun and the true sun:

$$E_d = \mu - \alpha(\lambda) = \bar{\lambda} - \alpha(\lambda) + c \quad . \tag{1}$$

Here $\alpha(\lambda)$ denotes the right ascension of the true sun *S*, which

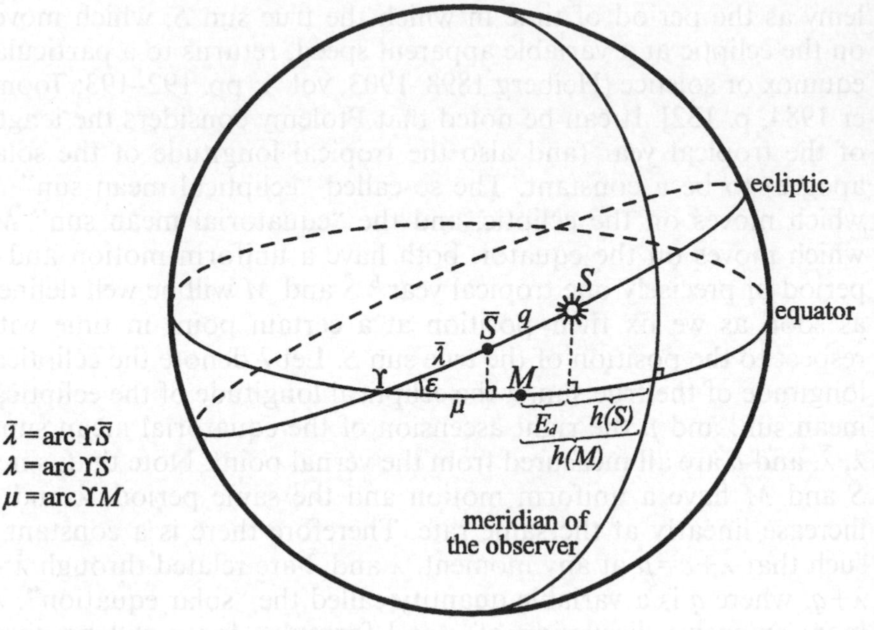


Figure 1. Geometrical explanation of the equation of time.

depends on the obliquity of the ecliptic $\varepsilon$. For $\lambda \in [0,90\rangle$ we have the modern formula

$$a(\lambda) = \arctan(\cos\varepsilon \cdot \tan\lambda)\ . \tag{2}$$

For $\lambda \in [90, 360]$ the right ascension can be determined using the symmetry relations

$$a(180-\lambda) = 180-a(\lambda) \quad \text{and} \quad a(180+\lambda) = 180+a(\lambda)\ . \tag{3}$$

As was mentioned before, we have $\overline{\lambda}=\lambda+q$, where the solar equation $q$ depends on the longitude $\lambda_A$ of the solar apogee and on the solar eccentricity $e$. As a function of $\lambda$, $q$ can be expressed as

$$q(\lambda) = \arcsin\left(\frac{e}{60}\sin(\lambda-\lambda_A)\right) \tag{4}$$

(cf. the explanations of the Ptolemaic solar model mentioned in footnote 5). When the solar equation is expressed as a function of $\bar{\lambda}$, we will use the symbol $\bar{q}$. We have

$$\bar{q}(\bar{\lambda}) = \arctan\left(\frac{e \cdot \sin(\bar{\lambda}-\lambda_A)}{60+e \cdot \cos(\bar{\lambda}-\lambda_A)}\right) . \tag{5}$$

Note that $q=\bar{q}=0$ (and hence $\lambda=\bar{\lambda}$) whenever $\lambda=\lambda_A$ or $\lambda=\lambda_A+180°$. Furthermore, it follows directly from formulae (4) and (5) that the solar equation satisfies the symmetry relations

$$q(\lambda) = q(2\lambda_A+180-\lambda), \quad q(\lambda) = -q(180+\lambda) ,$$
$$\text{and} \quad \bar{q}(\bar{\lambda}) = -\bar{q}(2\lambda_A+360-\bar{\lambda}) . \tag{6}$$

The constant $c$ occurring in formula (1) determines the synchronization of the ecliptical mean sun and the equatorial mean sun according to $\bar{\lambda}+c=\mu$ (see above). In modern astronomy $c$ is usually taken to be zero; this implies that $\bar{S}$ and $M$ pass the vernal equinox simultaneously. Note, however, that at the moment when $\bar{S}$ and $M$ pass the vernal equinox, the true sun has an ecliptical position $\lambda=-\bar{q}(0)$ which is generally not equal to zero.

Ptolemy and mediaeval astronomers used several other methods to fix $c$, some of which involved convenient choices of the so-called "epoch", the starting point of the tables for planetary motion. For example, Ptolemy defined the equation of time to be the difference between the true solar time and the mean solar time that had passed since epoch (see the works mentioned in note 5). Thus he implicitly assumed that mean and true solar time at epoch were equal. At the epoch of his *Handy Tables* the equation of time happened to be close to its yearly minimum. Thus $c$ had to be chosen in such a way that this minimum became almost equal to $0;0,0^h$. As a result, all equation of time values were non-negative.[9] In the sequel $c$ will be called "epoch constant".

In Ptolemy's *Handy Tables* and in most mediaeval astronomical handbooks which contain a table for the equation of time, the tabulated quantity is $E_h$, the equation of time expressed in hours, rather than $E_d$. To convert $E_d$ to $E_h$, $E_d$ was usually divided by 15, in agreement with the identity $24^h = 360°$. However, since the sun's daily motion in the ecliptic amounts to approximately 0;59,8°/day, a more accurate conversion factor is $(360;0+0;59,8)/24 \approx 15;2,28°/$ hour. This number was for instance used by the 15th-century Persian mathematician and astronomer al-Kāshī in his Khāqānī Zīj [Kennedy 1988, p. 5]. The maximum error introduced by taking the factor 15 instead of 15;2,28 is only 5 seconds of time. We will write $E_h = \frac{1}{D} E_d$, where $D$ ordinarily can only take the values 15 and 15;2,28. My impression is that all early Islamic astronomers used the conversion factor 15.

Thus we find the following formula for the equation of time expressed as a function of the true solar longitude $\lambda$:

$$E_h(\lambda) = \frac{1}{D}(\lambda + q(\lambda) - a(\lambda) + c) \ . \tag{7}$$

When the equation of time is expressed as a function of the mean solar longitude $\bar{\lambda}$, we will use the symbol $\bar{E}_h$. We have

$$\bar{E}_h(\bar{\lambda}) = \frac{1}{D}(\bar{\lambda} - a(\bar{\lambda} - \bar{q}(\bar{\lambda})) + c) \ . \tag{8}$$

The general shape of the equation of time both as a function of the true solar longitude and as a function of the mean solar longitude can be seen from Figure 2.

Ptolemy's *Handy Tables* and many mediaeval astronomical handbooks contain a table for the equation of time, usually in the same section as the tables for the mean solar motion and the solar equation. Both the true solar longitude and the mean solar longitude occur as the independent variable of tables for the equation of time [Kennedy 1988]. I will now argue that, as far as the computation of a table for the equation of time or its application is concerned, it makes little difference whether the independent variable is the true or the mean solar longitude.

To calculate the equation of time as a function of the true solar longitude, one need only perform a simple addition and/or subtraction of right ascension and solar equation values. The right

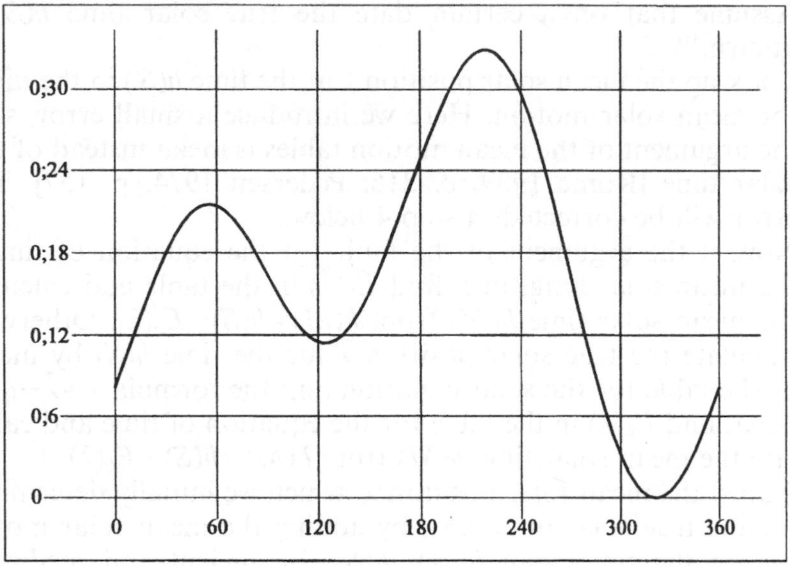


Figure 2. Graphical representation of the equation of time.

ascension values can readily be taken from a sufficiently accurate table for $a(\lambda)$. If a table for $q(\lambda)$ is available, the solar equation values can readily be taken from it except if the longitude of the solar apogee does not occur in the argument column (in that case interpolation is needed). If only a table for $\bar{q}(\bar{\lambda})$ is available, inverse interpolation in this table or a separate calculation of $q(\lambda)$ is necessary. Note that the solar equation tables in most zījes display $\bar{q}(\bar{\lambda})$ rather than $q(\lambda)$.

To calculate the equation of time as a function of the mean solar longitude, one always needs to perform interpolation in a right ascension table. If a table for $\bar{q}(\bar{\lambda})$ is available, the solar equation can readily be taken from it except if the longitude of the solar apogee does not occur in the argument column (in that case interpolation is needed). If only a table for $q(\lambda)$ is available (we have noted above that this is an unlikely situation), inverse interpolation or a separate calculation of $\bar{q}(\bar{\lambda})$ will be necessary.

The equation of time can be applied as follows:

1. Assume that on a certain date the true solar time $h(S)$ is known.[10]
2. Look up the mean solar position $\bar{\lambda}$ at the time $h(S)$ in the tables for mean solar motion. Here we introduce a small error, since the argument of the mean motion tables is mean instead of true solar time [Rome 1939, p. 216; Pedersen 1974, p. 157]. This error will be corrected in step 4 below.
3. Now, if the argument of the table for the equation of time is the mean solar longitude, find $\bar{E}_h(\bar{\lambda})$ in the table and calculate the mean solar time $h(M)$ from $h(M)=h(S)-\bar{E}_h(\bar{\lambda})$. Otherwise, calculate the true solar position $\lambda$ for the time $h(S)$ by means of the table for the solar equation and the formula $\lambda=\bar{\lambda}-\bar{q}(\bar{\lambda})$. Next, find $E_h(\lambda)$ in the table for the equation of time and calculate the mean solar time $h(M)$ from $h(M)=h(S)-E_h(\lambda)$.
4. Adjust the mean solar position $\bar{\lambda}$, which we initially determined for the true solar time $h(S)$, by adding the mean solar motion during the time span $h(M)-h(S)$. In ancient and mediaeval practice it was not necessary to repeat steps 2 and 3 using the corrected mean solar position, since the difference in the equation of time would not exceed a second of time.

## *2 Analysis of tables for the equation of time*

The investigation of the mathematical structure of a table for the equation of time can increase our knowledge of the methods by which ancient and mediaeval astronomers computed their tables and of the many different values for the astronomical parameters that they employed. Since the mathematical structure of an astronomical table is often specific to the astronomer who calculated the table, its investigation may also enable us to establish the author of tables of which the origin is unknown.

In order to ascertain the complete mathematical structure of a given table for the equation of time, we need to know at least the following:

- the independent variable (mean or true solar longitude);
- the values of the underlying parameters: the obliquity of the

ecliptic $\varepsilon$, the longitude $\lambda_A$ of the solar apogee, the solar eccentricity $e$, the epoch constant $c$, and the conversion factor $D$ (15 or 15;2,28);
- the accuracy of the underlying tables for the solar equation and the right ascension;
- whether interpolation was used for the computation of part of the tabular values.

If the desired information is not presented explicitly in the source containing the table for the equation of time or in other primary sources, then we may use the following properties and mathematical methods to extract the information from the tabular values:

1. The use of interpolation for the computation of part of the tabular values can often be recognized from groups of (practically) constant first or second order tabular differences; see Section 3.3.1 for an example. Tabular values calculated independently (i.e. without the use of interpolation) will be called *"nodes"*, other values, *"internodal values"*. In the analysis of a table the internodal values can usually be disregarded. However, sometimes they can be used to correct scribal errors in the nodes.

   For every type of linear interpolation the tabular differences between consecutive nodal values differ at most by a single unit. The type in which the internodal values are determined by calculating values on straight lines between the nodal values, will be referred to as "exact linear interpolation". In the case of exact linear interpolation, the tabular differences between every two nodes are *evenly* distributed. In Section 3.1.1 we will see a type of linear interpolation for which the tabular differences between every two nodes are *unevenly* distributed.
2. In an accurate trigonometric table all possible final digits occur approximately equally often (cf. the Appendix). This implies that if the final sexagesimal digit of (nearly) all values or all nodes in a table for the equation of time is a multiple of 4, we can safely conclude that for the table concerned $D$=15. In that case the values for $E_d$ were calculated to one sexagesimal place less than those for $E_h$, e.g. $E_h$=0;18,20 derives from $E_d$=4°35′.

Note that if only the final sexagesimal digit of the *nodes* is a multiple of 4, the interpolation must have been performed after the division by 15.

3. Using the symmetry relations (3) for the right ascension and (6) for the solar equation we can derive the following properties of the equation of time $E_h$ and $\bar{E}_h$:

$$E_h(\lambda_A) = E_h(\lambda_A+180) \quad \text{and} \quad \bar{E}_h(\lambda_A) = \bar{E}_h(\lambda_A+180) \tag{9}$$

$$\sum_{i=1}^{2n} E_h\left(\frac{180i}{n}\right) = \frac{2nc}{D} \quad \text{for every } n, \quad \text{and} \quad \int_0^{2\pi} \bar{E}_h(\bar{\lambda})\, d\bar{\lambda} = \frac{2\pi c}{D} \tag{10}$$

$$E_h(\lambda)+E_h(180-\lambda)+E_h(\lambda+180)+E_h(360-\lambda) = \frac{4c}{D} \quad \text{for every } \lambda \tag{11}$$

$$a(\lambda) = \lambda+c-\tfrac{1}{2}D\cdot(E_h(\lambda)+E_h(\lambda+180)) \quad \text{for every } \lambda \tag{12}$$

$$q(\lambda) = \tfrac{1}{2}D\cdot(E_h(\lambda)-E_h(\lambda+180)) \quad \text{for every } \lambda \tag{13}$$

(in (10), $n$ denotes half the number of available equation of time values). Using (9) and, if necessary, interpolation between tabular values, a rough estimate of $\lambda_A$ can be obtained [Kennedy 1988, pp. 3–4]. Using (10) and (11) accurate estimates for $c$ can be calculated. Using (12) and (13) the underlying right ascension and solar equation can be extracted from a table for the equation of time as a function of the true solar longitude; for examples see Sections 3.3.4 and 3.3.5.

4. Using a least squares estimation as explained in the Appendix, accurate estimates of $\varepsilon$, $\lambda_A$, $e$, and $c$ can be calculated from the tabular values of a table for the equation of time. With most of the estimations in this article, so-called 95% confidence intervals for the parameter values will be presented, i.e. intervals that are expected to contain the actually used parameter values in 19 out of 20 cases.

Before the parameters can be estimated the formula according to which the table was computed must be known. Since this is not always the case, it may be necessary to perform the estimation for various possible formulae, for instance for all four possible combinations of the independent variable ($\lambda$ or $\bar{\lambda}$) and the conversion factor $D$ (15 or 15;2,28). The plausibility of a certain hypothesis concerning the method of computation can be judged from the minimum obtainable standard deviation of the tabular errors (cf. the Appendix). If this standard deviation is much larger than the one for a correct table, we have probably used an incorrect formula. Thus for tabular values that are given to $k$ sexagesimal fractional digits the minimum obtainable standard deviation of the tabular errors should not be much larger than $17 \cdot 60^{-(k+1)}$.

Using this property a least squares estimation distinguishes very clearly between the true and mean solar longitude as the independent variable. A decision about the conversion factor can be made from the historical plausibility of the estimates of the parameters. Note that a least squares estimation can only be applied if certain conditions concerning the tabular errors (namely that they are independent and have mean zero and identical standard deviations) are satisfied. This will not always be mentioned explicitly.

In Section 3.3 an extensive analysis of the table for the equation of time in Ptolemy's *Handy Tables* will be carried out. Thereby it will be demonstrated how by means of the methods indicated above the underlying parameter values and the precise method of computation can be determined.

## *3 Ptolemy's table for the equation of time*

### *3.1 Historical context*

Ptolemy is the most influential astronomer of antiquity.[11] He lived in the second century A.D. in Alexandria (Egypt), where he made a large number of observations which are recorded in his works.

Ptolemy expanded Hipparchus's solar and lunar models and was the first to develop satisfactory models for the other planets. He compiled sets of tables for easy determination of planetary positions at arbitrary points in time. Ptolemy's most important astronomical works, the more theoretical *Almagest* (originally entitled *Μαθηματικὴ Σύνταξις*, Mathematical Composition) and the practical *Handy Tables* (see below), were in use up to the end of the Middle Ages.

In the second half of the fourth century Theon of Alexandria (father of the female mathematician Hypatia) wrote various commentaries on Ptolemy's work. Theon was involved in higher education, possibly at the Alexandria Museum. His commentaries were written to comply with the needs of his students, some of whom "were not able to follow a multiplication or division of [sexagesimal] numbers" [Tihon 1978, pp. 199 or 301]. Little is known about original scientific achievements of Theon.

In most extant manuscripts containing Ptolemy's *Handy Tables* we also find a copy of Theon's *Small Commentary* on these tables (more information about the primary sources mentioned in this paragraph is presented in Section 3.1.1 below). Furthermore, in his *Great Commentary* Theon describes extensively how the *Handy Tables* were computed and how they were derived from the tables in the *Almagest*. Since the contents of the *Handy Tables* is in good agreement with Ptolemy's own *Introduction,* which has only come down to us separately, we can nevertheless conclude that the *Handy Tables* were written by Ptolemy and not by Theon. Although Theon's descriptions of the *Handy Tables* are in many cases correct, we will see in this article that he did not know the exact way in which the table for the equation of time had been computed.

*3.1.1 Primary sources.* Historical information with regard to the equation of time as used by Ptolemy is found in the following primary sources: the *Almagest,* the *Handy Tables,* Ptolemy's own *Introduction to the Handy Tables,* and Theon of Alexandria's *Commentary on the Almagest, Great Commentary on the Handy Tables,* and *Small Commentary on the Handy Tables.*[12] Much of this information was summarized by A. Rome [1939; see Section 3.1.2 below]. Short descriptions of the sections from the above-mentioned primary sources which contain information about the equation of time will now be presented.

In Section 9 of Book III of the *Almagest* Ptolemy examines the problem of the "inequality of the days" [Heiberg 1898–1903, vol. 1, pp. 258–263; Toomer 1984, pp. 169–172]. He explains the reasons for this inequality and determines the maximum deviation. Furthermore he shows how a given time span in true solar days can be converted to mean solar days. Values for the mean and true solar position at the beginning of the Egyptian year 1 Nabonassar are presented. No mention is made of a table for the equation of time.

Theon of Alexandria's *Commentary on the Almagest* follows the Almagest closely and does not give important new information. An example is presented that utilizes the initial values mentioned in the *Almagest.*[13]

The set of astronomical tables known as the *Handy Tables* (*Πρόχειροι Κανόνες*) is extant in a large number of Greek manuscripts. Three manuscripts of the Bibliothèque Nationale in Paris were used by Halma in his unreliable edition and French translation published in 1822–1825. Stahlman consulted the manuscript Vatican gr. 1291 for his English edition of the tables [1959]. Neugebauer gave a comprehensive technical description of the *Handy Tables* in *A History of Ancient Mathematical Astronomy* [Neugebauer 1975, vol. 2, pp. 969–1028]. Recently Tihon published an extensive description of the oldest manuscripts of the *Handy Tables* and of the tables occurring in these manuscripts [Tihon 1992]. As was indicated above, most of the contents of the *Handy Tables* originates from Ptolemy, although some of the tables are obviously later (Byzantine) additions. In most copies of the *Handy Tables* we find a table displaying in two adjacent columns the normed right ascension and the equation of time. It is this table that we will analyse in the remainder of this article. Since the equation of time depends both on the right ascension and on the solar equation, we will also investigate the table for the solar equation in the *Handy Tables.*

In his own *Introduction to the Handy Tables*[14] Ptolemy first gives a partial table of contents, which includes the table for the equation of time. Later he explains how to convert true solar to mean solar time using the table for the equation of time [Heiberg 1907, pp. 162–163; Halma 1822–1825, vol. 1, p. 5].

In the *Great Commentary on the Handy Tables*[15] Theon of Alexandria discusses the layout, computation and use of the *Handy Tables* extensively. In Chapter 2 of Book 1 he describes the layout of the table for the right ascension and the equation of time and the type of interpolation which was used to derive the right ascension values from those in the *Almagest* [Mogenet & Tihon 1985, pp. 97 or 162]. Both descriptions are in full agreement with what we find in the extant manuscripts of the *Handy Tables.* Next he describes how the equation of time could be computed from the true solar position using the right ascension table and inverse interpolation in the solar equation table, and following Ptolemy's method in the *Almagest* [*idem,* pp. 98–100 and 163–164]. In Chapter 9 of the *Great Commentary* Theon explains the conversion from local true solar time given in seasonal hours to Alexandria mean solar time [*idem,* pp. 117–120 and 177–179; see the commentary on pp. 276 ff. for an explanation of the concepts local time and seasonal hours]. The third and last step of this conversion is the application of the equation of time.

In Chapter 5 of the *Small Commentary on the Handy Tables,* which Theon wrote after the *Great Commentary,* the above-mentioned conversion from true to mean solar time is illustrated with a numerical example [Tihon 1978, pp. 216–218 or 309–310]. For a particular date in the year 360 A.D. Theon arrives at a true solar position of 175° and mentions that the equation of time amounts to approximately 7 minutes (the value occurring in the table for the equation of time in the *Handy Tables* is $6^m55^s$).

In the sources mentioned above we find the following specific information with regard to the independent variable and the underlying parameters of Ptolemy's table for the equation of time:

- The *independent variable* is the true solar position. This can be concluded from the instructions for the application of the table for the equation of time which we find both in Ptolemy's *Introduction to the Handy Tables* and in Theon's *Commentaries.*
- *For the obliquity of the ecliptic* $\varepsilon$ Ptolemy uses the value 23;51,20

throughout the *Almagest*. Since the equation of time depends on the obliquity through the right ascension, the right ascension in its turn through the solar declination,[16] we have to consider the solar declination and right ascension tables in Ptolemy's works as well. In the *Almagest* both the declination table and the right ascension values for ecliptical arcs of 10 degrees can be verified to have been computed using $\varepsilon$=23;51,20. The declination table in the *Handy Tables* displays values to minutes only, which were rounded from the values in the *Almagest* [Stahlman 1959, pp. 105–106].[17]

In his *Great Commentary* Theon indicates correctly that the right ascension table in the *Handy Tables* was computed from the values given in the *Almagest* by means of a particular type of linear interpolation which I have called "distributed linear interpolation" [see also Rome 1939, p. 219; Mogenet & Tihon 1985, pp. 97 or 162]. As usual the tabular values between two consecutive independently calculated values were determined in such a way that the resulting tabular differences differed at most by a single unit in their final sexagesimal digit. Thus between the independently calculated values $\alpha(10)$=10;55 and $\alpha(20)$=21;42 Ptolemy obtained three tabular differences of 1;4, seven of 1;5. But instead of the tabular differences being distributed evenly (as in the case of exact linear interpolation), the larger differences were placed closer to the nearest solstice, the smaller differences, closer to the nearest equinox. Such a distribution is in accordance with the fact that the right ascension increases faster in the neighbourhood of the solstices. In fact it can be checked that for the right ascension this type of linear interpolation is more accurate than the usual type which involves even distribution of the tabular differences [Stahlman 1959, pp. 31–32 and 41–42].

- In the *Almagest* Ptolemy finds the value 2;29$\frac{1}{2}$ for the *solar eccentricity e*, which he rounds to 2;30. He uses the rounded value in his worked examples, but it is doubtful whether it also underlies his table for the solar equation as a function of the mean solar longitude.[18] In the *Great Commentary* Theon states that the solar equation table in the *Handy Tables* [Leiden BPG 78, ff. 94$^{r}$–96$^{v}$; Stahlman 1959, pp. 249–254] was derived from the one in the *Almagest* using the same type of linear interpolation which was used for the computation of the right ascension (see above). In fact, the interpolation pattern concerned can be recognized in nearly every internodal block of the table. However, the nodal values $q(6)$, $q(12)$,...,$q(90)$, $q(93)$,..., $q(177)$ differ from the corresponding values in the *Almagest* in 9 out of 44 cases, and fit the rounded eccentricity value 2;30 better than the *Almagest* values [see also Mogenet & Tihon 1985, pp. 253–264].

No information about the solar eccentricity can be found in the *Small Commentary*.

- In all sources that I consulted the *solar apogee* is stated to be in 5°30′ Gemini, i.e. $\lambda_A$=65;30.
- No explicit information about the *epoch constant* can be found in either of the above-mentioned primary sources. The following, however, can be said about the epochs of the planetary motions in Ptolemy's works:

In the *Almagest* the tables for the planetary motions are based on the Era Nabonassar.[19] Ptolemy gives $\bar{\lambda}_0$=330;45 and $\lambda_0$= 333;8 as epoch values for the mean and true solar longitudes respectively,[20] whence $\bar{\lambda}_0 - a(\lambda_0) \approx$ 330;45−335;8=−4;23°. Since the equation of time at epoch was close to minimum, an epoch constant $c$ equal to 4;23 leads to equation of time values of which the absolute value must (nearly) always be *added* to the mean solar time to obtain true solar time.[21] The table that is found in the second-century Greek papyrus London 1278 has this property [Neugebauer 1958, pp. 97–102 and 109–111; Section 3.5 of this article].

In the *Handy Tables* the tables for the planetary motions are based on the Era Philip.[22] Thus we obtain as epoch values $\bar{\lambda}_0$=227;40 (from the solar mean motion table), $\lambda_0$= $\bar{\lambda}_0 - \bar{q}(\bar{\lambda}_0 - 65;30) \approx$ 226;54 (using the solar equation table), and $\bar{\lambda}_0 - a(\lambda_0) \approx$ 227;40−224;22=3;18 (using the right ascension table).[23] Since in this case the equation of time at epoch was close to maximum, an epoch constant $c$ equal to −3;18 leads to equation of time values of which the absolute value must (nearly) always be *subtracted* from the mean solar time to obtain true solar time. As we will see, the table for the equation of time in the *Handy Tables* in fact has this property.

- In all used sources a division by 15 is said to have been used to convert equinoctial degrees into hours. In the *Almagest* it is mentioned that the mean solar day equals 360°+0;59° approximately [Heiberg 1898–1903, vol. 1, pp. 258; Toomer 1984, p. 170], but this fact does not seem to have been utilized to carry out a more accurate conversion. Thus we expect $D$=15.

Neither in the *Almagest* and the *Handy Tables* nor in Theon's commentaries on these works can explicit information about Ptolemy's method of *rounding* be found. I investigated some of Ptolemy's tables and found that modern rounding was used in the computation of the right ascension values in the *Almagest* and the declination table in the *Handy Tables*. The recomputation of the solar equation tables in both sources is problematical (see above), but the use of modern rounding can be shown to be more probable than upward rounding or truncation. From recent research by Glen Van Brummelen it can be concluded that the calculation of the table of chords in the *Almagest* and its so-called sixtieths involves modern rounding as well [Van Brummelen 1993, pp. 60–73]. Dr. Alexander Jones kindly drew my attention to the fact that many of the planetary mean motion tables in the *Handy Tables* were calculated from those in the *Almagest* by rounding the tabular values in the modern way to a single sexagesimal fractional digit. I verified that the mean motion tables calculated independently also involve modern rounding.

Since in several of the above-mentioned tables the use of modern rounding could be established and in none of them are obvious traces of upward rounding or truncation present, we will assume that Ptolemy made use of modern rounding for the computation of his table for the equation of time. At various critical stages of the analysis in Section 3.3 I checked the possibility of other types of rounding, but in no case did this lead to better results. Therefore these results will not be mentioned.

3.1.2 *Secondary sources.* Editions and translations of the primary sources containing information about the equation of time as used by Ptolemy were mentioned in the previous section. Only few discussions of the material in these primary sources can be found in the secondary literature.

In his article "Le problème de l'équation du temps chez Ptolémée" Rome describes the usage of the table for the equation of time in the *Handy Tables*, but does not add any information about the mathematical structure of the table to what is found in the primary sources discussed above [Rome 1939, pp. 217–218]. Rome devotes a section to the astronomer Serapion [pp. 223–224] who according to Theon indicated that a correction is necessary in order to compensate for the fact that the minimum equation of time occurs at 0° Scorpio, whereas the solar position at the epoch 1 Philip of the tables for planetary motion in the *Handy Tables* was 17° Scorpio [Mogenet & Tihon 1985, pp. 123 or 182]. Since several sources mention a geographer Serapion who lived before Ptolemy, we have to consider the possibility that Ptolemy adopted the table for the equation of time from the work of an earlier astronomer.[24] An extensive discussion of this matter is beyond the scope of this article, but an interesting remark concerning the error in the equation of time at epoch will be made in Section 3.5.

In his edition of Theon's *Commentary on the Almagest* Rome makes some remarks about the table for the equation of time in the *Handy Tables* in the footnotes of the section on the "inequality of the days" [Rome 1931–1943, vol. 3, pp. 922–924]. He investigates the tabular differences to check Ptolemy's statement that the maximum variation in the value of the equation of time is 31″ per day and finds that only one difference of 32″ occurs.[25] He also mentions that if Ptolemy had computed the table for the equation of time according to the procedure described by Theon in his *Great Commentary* (see Section 3.1.1 above), all tabular values should be multiples of 4″, which is not the case.

In his edition of the *Handy Tables* Stahlman does not discuss the technical details of the table for the equation of time at length. He suggests that the table was computed by means of linear interpolation within intervals of 3 or 6 degrees [Stahlman 1959, pp. 43–44]. In his *A Survey of the Almagest* Pedersen gives a thorough definition of the equation of time based on Ptolemy's material in the *Almagest* [Pedersen 1974, pp. 154–158], but he does not address the table for the equation of time in the *Handy Tables*.

In *A History of Ancient Mathematical Astronomy,* Neugebauer gives an extensive analysis of the equation of time as defined by

Ptolemy and includes various examples of its calculation [Neugebauer 1975, vol. 1, pp. 61–68]. He explains how the different epochs of the tables for planetary motion lead to additive values for the equation of time in the *Almagest* and the papyrus P. London 1278, but to subtractive values in the *Handy Tables* [Neugebauer 1975, vol. 2, pp. 984–986; Neugebauer 1958, pp. 109–111]. Neugebauer does not analyse the table for the equation of time in the *Handy Tables* mathematically.

From this overview of secondary literature it can be concluded that practically nothing is known about the method of computation and the underlying parameter values of Ptolemy's table for the equation of time in the *Handy Tables.* We will therefore start our analysis of the tabular values in Section 3.3 from scratch and will compare our findings with the information from the primary sources listed in Section 3.1.1 above. It will turn out that the actual method of computation of the values for the equation of time does not agree with Theon's description in the *Great Commentary* and that the value used for the longitude of the solar apogee is different from the value given in the primary sources.

### *3.2 Description of the table*

The manuscript *Biblioteca Publica Graeca (BPG) 78* in the University Library of Leiden contains a copy of Ptolemy's *Handy Tables* on folios 50$^v$–155$^r$. This copy is written in uncial Greek, can be dated to the early 9th century and is one of the earliest extant manuscripts of the *Handy Tables.* An extensive description of the manuscript can be found in Tihon 1978, pp. 105–106; a list of the tables that the manuscript contains is presented in Tihon 1992, pp. 58–61.[26]

On folios 75$^r$–76$^v$ of Leiden BPG 78 we find a table entitled "*Ορθῆς σφαίρας σὺν μεσουρανήματι πανταχῆι*" (literally: [Ascensions] of the right sphere with culmination everywhere; see Plate 1). Each page displays seven columns: one for the argument and two columns with tabular values for each of three zodiacal signs, starting with Capricorn on folio 75$^r$. The argument is the solar position, which runs from 1° to 30° with steps of 1°. For each

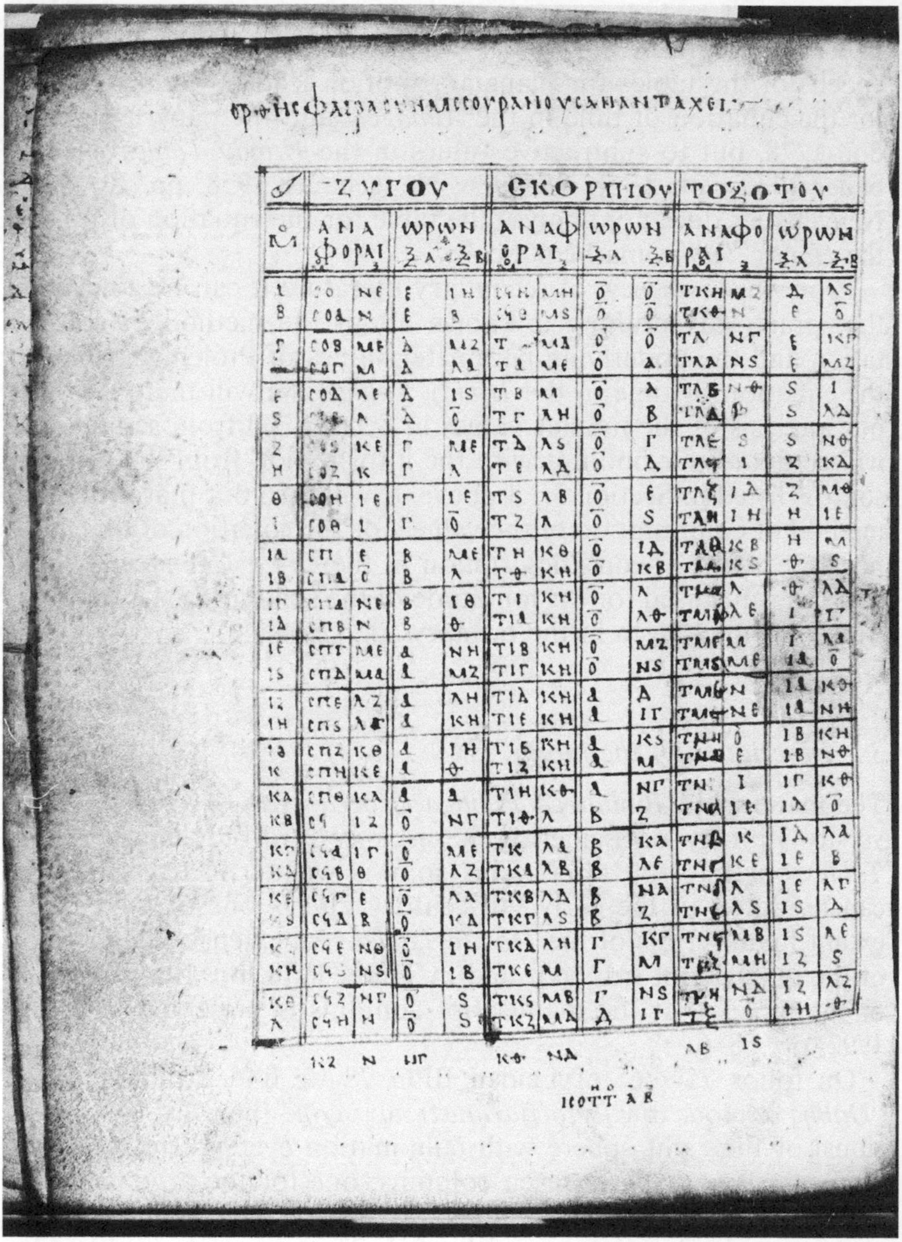


Plate 1. Fragment of the table for the equation of time in Leiden BPG 78 [folio 76ᵛ]⊥.

zodiacal sign the first column (headed "ascensions", ἀναφοράι) displays the "normed right ascension" (i.e. the right ascension with the winter solstice as reference point instead of the vernal equinox), the second column (headed "sixtieths of hours", ὡρῶν ἑξηκοστὰ) the equation of time.[27]

The right ascension values are given to minutes. As was indicated in Section 3.1.1 their mathematical structure was described by Theon of Alexandria in his *Great Commentary on the Handy Tables.* The equation of time is tabulated in minutes and seconds of an hour. It has a maximum $0;33,23^h$ for 18° Aquarius, a secondary minimum $0;6,12^h$ for 0° Gemini, a secondary maximum $0;16,21^h$ for 9° Leo, and a minimum $0;0,0^h$ for 0–3° Scorpio. From the relative positions of the extremes it follows that the tabular entries must be *added* to the true solar time to obtain mean solar time (cf. Figure 2). This implies that they were computed according to a variant of either formula (7) or formula (8), namely

$$E_h^{HT}(\lambda) = \frac{1}{D}(a(\lambda) - \lambda - q(\lambda) + c) \tag{14}$$

or

$$\bar{E}_h^{HT}(\bar{\lambda}) = \frac{1}{D}(a(\bar{\lambda} - \bar{q}(\bar{\lambda})) - \bar{\lambda} + c) \ , \tag{15}$$

depending on the independent variable. Note that in Tables 1, 2 and 6 the argument is the true solar longitude and is thus reckoned from Aries instead of from Capricorn as in the original table.

### 3.3 *Mathematical analysis*

3.3.1 *Interpolation.* First I investigated the tabular differences of Ptolemy's table for the equation of time. If the tabular values are denoted by $T(\lambda)$, $\lambda = 1,2,3,\ldots,360$, the first order tabular differences can be defined as $\Delta^{(1)}(\lambda) \stackrel{\text{def}}{=} T(\lambda+1) - T(\lambda)$ for every $\lambda$. A typical sample of the first order tabular differences of Ptolemy's table for the equation of time is found in Table 1.[28] It can be noted that for most multiples of 6° of the argument there are obvious jumps in the differences.[29] Furthermore in nearly all cases the six differences between tabular values for consecutive multiples of 6° differ by 1 second at most. We conclude that Ptolemy used linear interpolation within intervals of 6°. Stahlman mentions the possibility of interpolation within intervals of 3 degrees [Stahlman 1959, pp. 43–44].[30] However, it can be shown that the tabular values for arguments $(3+6n)°$ are much closer to values computed by means of linear interpolation within intervals of 6 degrees than to values computed by means of the (less exact) method according to which the values for multiples of 6 degrees will be shown to have been determined.

It can be noted that the tabular differences were distributed between the nodal values in an irregular way. In Table 1 only the differences within the intervals 36–42, 54–60 and 72–78 are evenly distributed; the remaining differences seem to be unevenly distributed in an inconsistent way. In any case the interpolation method

| true solar long. $\lambda$ | tabular differ-ences $\Delta^{(1)}$ | true solar long. $\lambda$ | tabular differ-ences $\Delta^{(1)}$ | true solar long. $\lambda$ | tabular differ-ences $\Delta^{(1)}$ | true solar long. $\lambda$ | tabular differ-ences $\Delta^{(1)}$ |
|---|---|---|---|---|---|---|---|
| 36 | −0; 0,14 | 48 | −0; 0, 6 | 60 | 0; 0, 8 | 72 | 0; 0, 9 |
| 37 | −0; 0,15 | 49 | −0; 0, 6 | 61 | −0; 0, 1 | 73 | 0; 0, 9 |
| 38 | −0; 0,14 | 50 | −0; 0, 5 | 62 | 0; 0, 4 | 74 | 0; 0, 9 |
| 39 | −0; 0,15 | 51 | −0; 0, 6 | 63 | 0; 0, 3 | 75 | 0; 0, 9 |
| 40 | −0; 0,14 | 52 | −0; 0, 5 | 64 | 0; 0, 4 | 76 | 0; 0, 9 |
| 41 | −0; 0,13 | 53 | −0; 0, 6 | 65 | 0; 0, 4 | 77 | 0; 0, 9 |
| 42 | −0; 0,12 | 54 | −0; 0, 3 | 66 | 0; 0, 5 | 78 | 0; 0,11 |
| 43 | −0; 0,11 | 55 | −0; 0, 4 | 67 | 0; 0, 5 | 79 | 0; 0,11 |
| 44 | −0; 0,10 | 56 | −0; 0, 3 | 68 | 0; 0, 6 | 80 | 0; 0,11 |
| 45 | −0; 0,10 | 57 | −0; 0, 4 | 69 | 0; 0, 5 | 81 | 0; 0,11 |
| 46 | −0; 0,10 | 58 | −0; 0, 3 | 70 | 0; 0, 6 | 82 | 0; 0,12 |
| 47 | −0; 0,10 | 59 | −0; 0, 4 | 71 | 0; 0, 5 | 83 | 0; 0,11 |

Table 1. First differences of Ptolemy's table for the equation of time

occurring in the tables for the right ascension and the solar equation in the *Handy Tables* (see Section 3.1.1 above) was *not* applied to the table for the equation of time. I have not been able to explain the irregular distribution of the tabular differences fully. Possible explanations are that the values for the equation of time on which Ptolemy based the interpolation were given to more than two sexagesimal fractional digits or that he used a type of linear interpolation different from the ones we have seen so far. Note that a completely irregular pattern (tabular differences +6,+6,+6,−6,−6,−7″) occurs around the secondary maximum at 9° Leo (which is the only extreme value which does not occur at a multiple of 6 degrees). Later we will see that a range of tabular values in the neighbourhood of this maximum are corrupt.

We will assume that the tabular values for multiples of 6 degrees were calculated independently and will refer to these values as the *nodes.* They are displayed in the second and fifth columns of Table 2, where the argument (in the first and fourth columns) is degrees of the ecliptic reckoned from the vernal point.[31] Two obvious scribal errors in the nodes can be corrected: $T(108)=0;14,3$ and $T(210)=0;0,6$. A systematic discussion of scribal errors is presented in Section 3.3.6. In the remainder of the mathematical analysis of Ptolemy's table for the equation of time only the nodal values will be used.

3.3.2 *Independent variable.* For both possibilities of the independent variable I performed a least squares estimation of the parameter values underlying Ptolemy's table for the equation of time (see the Appendix for an explanation of the least squares estimator). If we assume that the independent variable of the table is the mean solar longitude, the minimum possible standard deviation of the residuals is 21″ if all nodes are used, 24″ if only tabular values for multiples of 30° are used. If we assume that the independent variable is the true solar longitude, the minimum possible standard deviations of the residuals are 4″ and 20‴ (thirds) respectively. These results are not affected by the choice of the conversion factor (15 or 15;2,28). In the Appendix it is indicated that the tabular errors of an accurate table for the equation of time with values to seconds have a standard deviation of approximately 17‴. Because out of the four standard deviations found above only the fourth one is reasonably close to this value, we conclude that the independent variable of Ptolemy's table for the equation of time is the true solar longitude. This is in agreement with the information in the historical sources discussed in Section 3.1.1. Hereafter the tabular values for multiples of 30° will be called *supernodes.* The large difference between the minimum possible standard deviation if all nodes are used and if only supernodes are used will be explained below.

*3.3.3 Preliminary estimates of the parameters.* To get a first impression of the underlying parameter values and the accuracy of Ptolemy's table for the equation of time, we compute confidence intervals for the parameters using a least squares estimation. Since we have already seen that the supernodes are much more accurate than the ordinary nodes, we will base our estimations on the supernodes only.[32] If we take the conversion factor equal to 15 the results are as follows:

| *parameter* | *95% confidence interval* |
|---|---|
| obliquity of the ecliptic $\varepsilon$ | ⟨23;51,50, 23;52,26⟩ |
| solar eccentricity $e$ | ⟨ 2;29,51, 2;30, 0⟩ |
| longitude of the solar apogee $\lambda_A$ | ⟨65;57,25, 66; 0,38⟩ |
| epoch constant $c$ | ⟨ 3;34, 3, 3;34, 9⟩ |

For the historically less plausible conversion factor 15;2,28 we find as confidence intervals ⟨23;53,45, 23;54,20⟩, ⟨2;30,16, 2;30,24⟩, ⟨65;57,28, 66;0,34⟩, and ⟨3;34,39, 3;34,44⟩ respectively. For conversion factor 15 the results correspond better to the historically plausible values $\varepsilon$=23;51,20 (Ptolemy's obliquity of the ecliptic), $e$=2;30 (Ptolemy's solar eccentricity) and $\lambda_A$=66;0 (a round number close to Ptolemy's value 65;30 for the longitude of the solar apogee). Apparently Ptolemy chose the epoch constant $c$ in such a way that the minimum equation of time value became 0;0,0. The fact that the minimum value occurs for $\lambda$=210 (i.e. 0° Scorpio), whereas the solar position at the epoch 1 Philip was 17° Scorpio, explains why the estimated epoch constant is different from the value 3;18 calculated in Section 3.1.1.[33]

I recomputed Ptolemy's table for the equation of time using con-

| true solar long. | Ptolemy's equation of time | error | true solar long. | Ptolemy's equation of time | error |
|---|---|---|---|---|---|
| 6 | 0;20,33 | +3 | 186 | 0; 4, 0 | +2 |
| 12 | 0;18, 4 | +4 | 192 | 0; 2,30 | −3 |
| 18 | 0;15,38 | +4 | 198 | 0; 1,28 | +5 |
| 24 | 0;13,24 | +7 | 204 | 0; 0,37 | +7 |
| 30 | 0;11,13 | −1 | 210 | 0; 0, 0 | |
| 36 | 0; 9,35 | +7 | 216 | 0; 0, 2 | +7 |
| 42 | 0; 8,10 | +8 | 222 | 0; 0,22 | +6 |
| 48 | 0; 7, 7 | +6 | 228 | 0; 1,13 | +6 |
| 54 | 0; 6,33 | +9 | 234 | 0; 2,35 | +9 |
| 60 | 0; 6,12 | −1 | 240 | 0; 4,13 | |
| 66 | 0; 6,34 | +7 | 246 | 0; 6,34 | +7 |
| 72 | 0; 7, 6 | +4 | 252 | 0; 9, 6 | +4 |
| 78 | 0; 8, 0 | +3 | 258 | 0;11,58 | +3 |
| 84 | 0; 9, 7 | +1 | 264 | 0;15, 2 | +2 |
| 90 | 0;10,23 | | 270 | 0;18, 9 | |
| 96 | 0;11,42 | −1 | 276 | 0;21,15 | −1 |
| 102 | 0;12,57 | −3 | 282 | 0;24,11 | −2 |
| 108 | 0;14, 3 | −4 | 288 | 0;26,50 | −4 |
| 114 | 0;14,53 | −7 | 294 | 0;29, 4 | −8 |
| 120 | 0;15,37 | +1 | 300 | 0;31, 4 | +1 |
| 126 | 0;16, 3 | +12 | 306 | 0;32,15 | −9 |
| 132 | 0;16, 2 | +16 | 312 | 0;33, 6 | −7 |
| 138 | 0;15,18 | | 318 | 0;33,23 | −5 |
| 144 | 0;14,27 | −4 | 324 | 0;33, 5 | −7 |
| 150 | 0;13,27 | +1 | 330 | 0;32,26 | |
| 156 | 0;11,59 | −7 | 336 | 0;31, 6 | −6 |
| 162 | 0;10,31 | −3 | 342 | 0;29,30 | −4 |
| 168 | 0; 8,52 | −4 | 348 | 0;27,33 | −4 |
| 174 | 0; 7,11 | −3 | 354 | 0;25,21 | −3 |
| 180 | 0; 5,33 | | 360 | 0;23, 0 | |

Table 2. Preliminary recomputation of Ptolemy's table for the equation of time

version factor 15, the above-mentioned historically plausible values for $\varepsilon$, $e$, and $\lambda_A$, and $c$=3;34,6 (least squares estimate of $c$ for conversion factor 15). The errors in the final sexagesimal digit (Ptolemy's table minus recomputation) are shown in the third and sixth columns of Table 2. At first sight the results appear to be disastrous. However, it can be noted that in fact the supernodal values are much more accurate than the ordinary nodes (as could be expected because of the minimum standard deviations found above). Furthermore, the errors in between the supernodes are nearly all positive in the first and third quadrants, but negative in the second and fourth. By comparing with the symmetry relations (3) for the right ascension and (6) for the solar equation we conclude that the errors in the equation of time derive from systematic errors in the underlying right ascension values. If the underlying solar equation values contained similar systematic errors, the errors in the first and second quadrants of the table for the equation of time would all be positive and those in the third and fourth quadrant negative, or the other way around.

Note that obvious outliers (i.e. errors that do not fit into the overall error pattern) are found for arguments $\lambda=126$, 132, 138 and 192. It turns out to be impossible to correct these errors on the basis of the interpolation pattern or by consulting other manuscripts containing Ptolemy's table for the equation of time. The possibility of scribal errors in the Greek transmission will be investigated in Section 3.3.6 after establishing the method according to which the table was computed. The outliers will be disregarded in the estimation of the underlying parameter values.

*3.3.4 The extracted solar equation.* Using formula (13) we can extract the solar equation underlying Ptolemy's table for the equation of time as a function of the true solar longitude. The only condition for the validity of the extraction is that the tabulations of the right ascension and the solar equation used for the computation of the equation of time satisfy the symmetry relations (3) and (6). This can safely be assumed because Ptolemy utilizes the symmetry in the *Almagest*. Since the extracted values depend on the conversion factor $D$, we have to distinguish the cases $D=15$ and $D=15;2,28$.

The result of the extraction when the conversion factor $D$ is taken equal to 15 can be found in the second column of Table 3 (the first column displays the argument, the true solar longitude). Note that the values for $\lambda=12$, 126, 132 and 138 are probably outliers, since they were derived from outlying values of the equa-

| true solar long. | extracted solar equation | recomputed solar equation | error |
|---|---|---|---|
| 6 | −2; 4, 7,30 | −2; 4, 4,33 | −0; 0, 2,57 |
| 12 | −1;56,45, 0 | −1;55,54,18 | −0; 0,50,42 |
| 18 | −1;46,15, 0 | −1;46,27,53 | 0; 0,12,53 |
| 24 | −1;35,52,30 | −1;35,51,30 | −0; 0, 1, 0 |
| 30 | −1;24, 7,30 | −1;24,12. 9 | 0; 0, 4,39 |
| 36 | −1;11,37,30 | −1;11,37,30 | 0; 0, 0, 0 |
| 42 | −0;58,30, 0 | −0;58,15,49 | −0; 0,14,11 |
| 48 | −0;44,15, 0 | −0;44,15,53 | 0; 0, 0,53 |
| 54 | −0;29,45, 0 | −0;29,46,54 | 0; 0, 1,54 |
| 60 | −0;14,52,30 | −0;14,58,22 | 0; 0, 5,52 |
| 66 | 0; 0, 0, 0 | 0; 0, 0, 0 | 0; 0, 0, 0 |
| 72 | 0;15, 0, 0 | 0;14,58,22 | 0; 0, 1,38 |
| 78 | 0;29,45, 0 | 0;29,46,54 | −0; 0, 1,54 |
| 84 | 0;44,22,30 | 0;44,15,53 | 0; 0, 6,37 |
| 90 | 0;58,15, 0 | 0;58,15,49 | −0; 0, 0,49 |
| 96 | 1;11,37,30 | 1;11,37,30 | 0; 0, 0, 0 |
| 102 | 1;24,15, 0 | 1;24,12, 9 | 0; 0, 2,51 |
| 108 | 1;35,52,30 | 1;35,51,30 | 0; 0, 1, 0 |
| 114 | 1;46,22,30 | 1;46,27,53 | −0; 0, 5,23 |
| 120 | 1;55,52,30 | 1;55,54,18 | −0; 0, 1,48 |
| 126 | 2; 1,30, 0 | 2; 4, 4,33 | −0; 2,34,33 |
| 132 | 2; 8, 0, 0 | 2;10,53,15 | −0; 2,53,15 |
| 138 | 2;15,37,30 | 2;16,15,52 | −0; 0,38,22 |
| 144 | 2;19,45, 0 | 2;20, 8,53 | −0; 0,23,53 |
| 150 | 2;22,22,30 | 2;22,29,44 | −0; 0, 7,14 |
| 156 | 2;23,22,30 | 2;23,16,51 | 0; 0, 5,39 |
| 162 | 2;22,22,30 | 2;22,29,44 | −0; 0, 7,14 |
| 168 | 2;20, 7,30 | 2;20, 8,53 | −0; 0, 1,23 |
| 174 | 2;16,15, 0 | 2;16,15,52 | −0; 0, 0,52 |
| 180 | 2;10,52,30 | 2;10,53,15 | −0; 0, 0,45 |

Table 3. The extracted solar equation

tion of time. Correct values of the equation of time only contain a rounding error of at most 30‴ in absolute value. Consequently solar equation values extracted from correct values of the equation of time contain errors with a maximum absolute value of $\frac{1}{2}$ $D \cdot 30'''$, i.e. approximately $7\frac{1}{2}''$. It can be shown that the standard deviation of these errors is approximately 3″.[34] Hereafter, when I speak about the errors in an extracted table, I will mean the differences between the extracted table and a particular recomputation which are larger than $7\frac{1}{2}''$ in absolute value.

First note that the fact that the extracted solar equation equals 0;0,0 for true solar longitude 66° (and to a lesser extent also the fact that the maximum 2;23,22,30 occurs at 156° and that the table is almost symmetric around 66° and 156°) confirms our preliminary estimate 66°0′ for the solar apogee. If $\lambda_A$ were equal to 65;30, the correct solar equation value for argument 66 would be 0;1,15, which is significantly different from 0;0,0 since the expected maximum error in the extracted values is $7\frac{1}{2}''$. A least squares estimation using all extracted values minus the four above-mentioned outliers yields a minimum possible standard deviation of 7″ and 95% confidence intervals ⟨2;29,52, 2;30,0⟩ for the eccentricity $e$, ⟨65;58,30, 66;1,19⟩ for the longitude $\lambda_A$ of the apogee. If we extract the solar equation using $D$=15;2,28 the minimum possible standard deviation and the 95% confidence interval for the longitude of the apogee are the same; the confidence interval for the eccentricity is ⟨2;30,17, 2;30,25⟩.

Independent of the value used for the conversion factor we can conclude that Ptolemy calculated the equation of time using the value 66°0′ for the longitude of the solar apogee. In fact, in both cases this value lies in the middle of the confidence interval, whereas the value mentioned in all historical sources (65°30′) is far removed from the interval. Below I will argue that the value 66°0′ is historically plausible, although not attested. Since for $D$=15;2,28 the confidence interval for the solar eccentricity does not contain any historically plausible (i.e. either attested or round) value, we conclude that the conversion factor used is 15, the solar eccentricity 2;30. In fact, for this choice of $D$ and $e$ the extracted solar equation values contain only 3 errors, whereas for all other historically plausible choices (involving $e$=2;29,30 and $D$=15;2,28) we find at least 18 errors in 26 extracted values.

In the *Almagest* and in the *Handy Tables* Ptolemy gives tables for the solar equation as a function of the *mean* solar anomaly $\bar{a}(\bar{a}=\bar{\lambda}-\lambda_A)$.[35] In both cases the tabular values are given to minutes. In order to compute the table for the equation of time Ptolemy needed the solar equation as a function of the *true* solar anomaly $a$ ($a=\lambda-\lambda_A$). In his *Great Commentary* Theon suggests that Ptolemy used inverse linear interpolation in the solar equation table in the *Almagest* in order to obtain the desired values [Mogenet & Tihon 1985, pp. 99–100 or 163–164]. However, it can be shown that inverse linear interpolation in any solar equation table with values to minutes is not sufficiently accurate to produce the extracted solar equation table.[36] Thus we conclude that Ptolemy had in fact independently computed values for the solar equation as a function of the true solar longitude.

Note that such solar equation values are not very difficult to obtain. For each argument Ptolemy had to look up a value in the table of chords, divide that value by 24 (multiply by 0;2,30), and perform an inverse linear interpolation in the table of chords. The algorithm (equivalent to our formula 4) is indicated in the *Almagest* section on the solar model and a worked example is given in the calculation of the mean solar position at an observed autumnal equinox. It can be checked that the table of chords in the *Almagest* is sufficiently accurate to produce the solar equation values which we extracted from the table for the equation of time. In fact, using Ptolemy's table of chords the solar equation can be computed with a maximum absolute error of 30‴.[37]

The computation of the solar equation values is equally easy for $\lambda_A=65°30'$ and $\lambda_A=66°0'$; one only has to start with a different value from the table of chords. The degree of difficulty of the inverse interpolation in the final step of the computation is not affected by the choice of $\lambda_A$. Ptolemy's use of $\lambda_A=66°0'$ for the equation of time may be explained from the existence of a tabulation for the solar equation as a function of the true solar anomaly. If such a tabulation gave the solar equation only for every integer value of the anomaly (or for multiples of 3° or 6°), the use of $\lambda_A=66°0'$ instead of $\lambda_A=65°30'$ would avoid the need for interpolation. We can assume that Ptolemy was aware that the use of $\lambda_A=66°0'$ instead of 65°30′ only introduces negligible errors in the equation of time values (the largest of which amounts to 5 seconds of time).

Recomputed solar equation values for $\lambda_A=66;0$ and $e=2;30$ are displayed in the third column of Table 3, differences between the extracted solar equation and the recomputation in the fourth column. Apart from the four outliers that we expected, there seem to be outliers for arguments 18, 42 and 144 as well. All other extracted values are correct, i.e. they differ from the corresponding recomputed values by no more than $7\frac{1}{2}''$. Since the outliers do not satisfy the conditions for the least squares estimation (cf. the Appendix), we must repeat the estimation without the newly found outliers to obtain valid confidence intervals. If we use conversion factor 15 to extract the solar equation, the minimum possible standard deviation is $4''$ (slightly more than the standard deviation of the expected triangular distribution of the errors in the extracted values) and 95% confidence intervals are ⟨2;29,56, 2;30,1⟩ for the eccentricity and ⟨65;58,46, 66;0,19⟩ for the longitude of the apogee. For conversion factor 15;2,28 the minimum standard deviation and the confidence interval for the longitude of the apogee are the same; a 95% confidence interval for the eccentricity is given by ⟨2;30,21, 2;30,26⟩. This confirms our conclusions.

Since the solar equation is symmetric around the apogee and perigee, we can compare the extracted solar equation values for arguments $\lambda_A+6k$ and $\lambda_A-6k$ ($k=1,2,3,\ldots,14$) to obtain more insight into the origin of the errors these values contain. It turns out that for all seven outliers the symmetrically corresponding value is correct. This implies that the errors were not made in the computation of the solar equation, but in a later stage of the determination of the equation of time values. The errors may be partly due to scribal mistakes.

*3.3.5 The extracted right ascension.* From the error pattern shown by our preliminary recomputation in Section 3.3.3 we concluded that systematic errors are present in the right ascension values that were used for the computation of Ptolemy's table for the equation of time. We will now try to discover the origin of these errors by extracting the right ascension using formula (12). Here a problem is that we need a value for the epoch constant $c$. As we have seen, no value of $c$ can be found in the available historical sources. From the table for the equation of time itself it can be concluded that $c$ was chosen in such a way that the minimum equation equals zero. Since the minimum is assumed for $\lambda=210$ we expect

$$c \approx 210-a(210)+q(210-66) \approx 210-207;50,7+1;24,12 = 3;34,5$$

(cf. formula (14) in Section 3.2). As we have seen in Section 3.3.3, the least squares estimate of $c$ based on the supernodes only is $\hat{c}=$ 3;34,6. Since it is impossible to decide on an accurate value of $c$ on the basis of the above information, we will examine the extracted right ascension for values of the epoch constant in the range 3;34,0 to 3;34,15; we need not consider values outside this range, since they would lead to an unacceptable number of errors in both the extracted right ascension and the equation of time itself. If not mentioned separately, the results given below hold for all values of $c$ in the range. Note that the extracted right ascension values for $\lambda=12$, 126, 132 and 138 are probably outliers, since they were derived from outlying values of the equation of time.

First it can be noted that for the Ptolemaic obliquity value 23;51,20 and for all values of the epoch constant $c$ in the range 3;34,0 to 3;34,15, the extracted right ascension values show the same error pattern as the equation of time: the supernodes contain only small errors; in between the supernodes the errors are positive in the first quadrant and negative in the second. This situation cannot be remedied by varying the obliquity of the ecliptic: for no combination of values for the obliquity and the epoch constant is the standard deviation of the errors in the extracted right ascension smaller than 42″, whereas the expected standard deviation for an extracted table is only 3″ (Section 3.3.4). We conclude that the right ascension values were *not* all computed according to the precise formula (2) or an equivalent one.

As was noted in Section 3.1.1, in the *Almagest* Ptolemy only gives correct right ascension values to minutes for ecliptic arcs of 10°. In the *Handy Tables* the intermediate values were determined by means of so-called "distributed linear interpolation". Therefore it seems plausible that the right ascension underlying the table for the equation of time also has interpolated values for non-multiples of 10°. Since we assume that the equation of time values for non-multiples of 6° were calculated by means of interpolation, this would imply that only the extracted right ascension values for arguments which are common multiples of 6 and 10 are accurate. As we have seen, this is indeed the case.

The extracted right ascension values show little agreement with the values occurring in the *Handy Tables* in the column adjacent to the equation of time. In fact, the standard deviation of the differences is at least 79″ for any value of *c*. Therefore we will consider seven historically plausible ways of recomputing the right ascension extracted from Ptolemy's table for the equation of time, all based on obliquity 23;51,20. Whenever linear interpolation is involved the values for multiples of 10° are taken as nodes. The seven possibilities are:

A. Exact computation to minutes of every single right ascension value.
B. Exact computation to seconds of every single right ascension value.
C. Exact linear interpolation between correct nodes to seconds, followed by rounding to minutes (see Section 2 for the definition of exact linear interpolation).
D. Exact linear interpolation between correct nodes to seconds, followed by rounding to seconds.
E. Exact linear interpolation between the *Almagest* values without rounding (the resulting values are to seconds).
F. Exact linear interpolation between the *Almagest* values followed by rounding to minutes.
G. Distributed linear interpolation between the *Almagest* values as described by Theon (the resulting values occur in the *Handy Tables* in the column adjacent to the equation of time).

We have already noted that right ascension values computed according to methods A, B and G are in poor agreement with the extracted right ascension values; we will use these methods mainly for comparison. Methods C and D are historically less plausible, since Ptolemy only gives right ascension values to minutes in the *Almagest* and *Handy Tables*.

For each of the possibilities A to G and for four values of the epoch constant in the range that we consider, Table 4 displays the number of errors $n$ (i.e. the number of differences between the extracted values and recomputed values larger than $7\frac{1}{2}''$ in absolute value), the mean difference $\mu$ in seconds, and the standard deviation $\sigma$ of the differences, also in seconds (for the correct method of computation this standard deviation is expected to be not much

| Type | c=3;34,0 | | | c=3;34,5 | | | c=3;34,10 | | | c=3;34,15 | | |
|---|---|---|---|---|---|---|---|---|---|---|---|---|
| | $n$ | $\mu$ | $\sigma$ | $n$ | $\mu$ | $\sigma$ | $n$ | $\mu$ | $\sigma$ | $n$ | $\mu$ | $\sigma$ |
| A | 18 | +18 | 76 | 18 | +13 | 75 | 19 | +8 | 74 | 19 | +3 | 74 |
| B | 22 | +18 | 71 | 23 | +13 | 70 | 24 | +8 | 69 | 23 | +3 | 69 |
| C | 10 | +9 | 17 | 10 | +4 | 15 | 12 | −1 | 15 | 12 | −6 | 16 |
| D | 13 | +7 | 15 | 13 | +2 | 13 | 15 | −3 | 13 | 17 | −8 | 15 |
| E | 4 | +8 | 10 | 3 | +3 | 7 | 6 | −2 | 7 | 11 | −7 | 9 |
| F | 10 | +9 | 17 | 10 | +4 | 15 | 12 | −1 | 15 | 12 | −6 | 16 |
| G | 17 | +16 | 81 | 17 | +11 | 80 | 19 | +6 | 79 | 19 | +1 | 79 |

Table 4. Error statistics for recomputations of the extracted right ascension

larger than 3 seconds; cf. Section 3.3.4). Since in all cases the four outliers were excluded, the total number of extracted right ascension values that were considered is 26.

From Table 4 we conclude without reservation that the right ascension underlying Ptolemy's table for the equation of time was computed according to method E, i.e. exact linear interpolation between the *Almagest* values without rounding. This follows both from the given numbers of errors and from the standard deviations. Even the methods which are similar to method E (in particular methods D and F) yield significantly worse results. Note that our conclusion is independent of the value of the epoch constant *c*, although the differences between the methods are most obvious in the middle of the considered range.

Assuming the use of method E we find that the number of errors in the extracted right ascension is minimized for *c*=3;34,7,30. Since we cannot choose a historically more plausible value (for *c*=3;34 the number of errors in the extracted right ascension is reasonably small, but the table for the equation of time itself has as many as 25 errors in 60 values), we assume for the time being that the equation of time was computed using *c*=3;34,7,30. Disregarding the four outliers, the extracted right ascension contains only two errors for this value of the epoch constant: a difference of $16\frac{1}{2}''$ for argument 42 and a difference of $22\frac{1}{2}''$ for argument 144. Since these differences are significantly larger than all others, and because the extracted solar equation has errors for the same arguments, we conclude that both errors result from outliers in the table for the equation of time. The second column of Table 5 displays the extracted right ascension for epoch constant $c$=3;34,7,30, the third column the right ascension recomputed according to method E and the fourth column the differences between the two. The argument in the first column is the true solar longitude. Note

| true solar long. | extracted right ascension | recomputed right ascension | error |
|---|---|---|---|
| 6 | 5;30, 0, 0 | 5;30, 0 | 0; 0, 0, 0 |
| 12 | 11; 0, 7,30 | 11; 1, 0 | −0; 0,52,30 |
| 18 | 16;34, 7,30 | 16;34, 0 | 0; 0, 7,30 |
| 24 | 22;11, 0, 0 | 22;11, 0 | 0, 0, 0, 0 |
| 30 | 27;50, 0, 0 | 27;50, 0 | 0; 0, 0, 0 |
| 36 | 33;38, 0, 0 | 33;38, 0 | 0; 0, 0, 0 |
| 42 | 39;29,52,30 | 39,29,36 | 0; 0,16,30 |
| 48 | 45;28,22,30 | 45;28,24 | −0; 0, 1,30 |
| 54 | 51;34,22,30 | 51;34,24 | −0; 0, 1,30 |
| 60 | 57;44, 0, 0 | 57;44, 0 | 0; 0, 0, 0 |
| 66 | 64; 4,22,30 | 64; 4,24 | −0; 0, 1,30 |
| 72 | 70;27,22,30 | 70;27,24 | −0; 0, 1,30 |
| 78 | 76;55,37,30 | 76;55,36 | 0; 0, 1,30 |
| 84 | 83;27, 0, 0 | 83;27, 0 | 0; 0, 0, 0 |
| 90 | 89;59,52,30 | 90; 0, 0 | −0; 0, 7,30 |
| 96 | 96;33, 0, 0 | 96;33, 0 | 0; 0, 0, 0 |
| 102 | 103; 4,22,30 | 103; 4,24 | −0; 0, 1,30 |
| 108 | 109;32,30, 0 | 109;32,36 | 0; 0, 6, 0 |
| 114 | 115;55,30, 0 | 115;55,36 | −0; 0, 6, 0 |
| 120 | 122;16, 0, 0 | 122;16, 0 | 0; 0, 0, 0 |
| 126 | 128;28, 7,30 | 128;25,36 | 0; 2,31,30 |
| 132 | 134;34,22,30 | 134;31,36 | 0; 2,46,30 |
| 138 | 140;31, 0, 0 | 140;30,24 | 0; 0,36, 0 |
| 144 | 146;22,22,30 | 146;22, 0 | 0; 0,22,30 |
| 150 | 152;10, 0, 0 | 152;10, 0 | 0; 0, 0, 0 |
| 156 | 157;49, 0, 0 | 157;49, 0 | 0; 0, 0, 0 |
| 162 | 163;26, 0, 0 | 163;26, 0 | 0; 0, 0, 0 |
| 168 | 168;59, 0, 0 | 168;59, 0 | 0; 0, 0, 0 |
| 174 | 174;29,52,30 | 174;30, 0 | −0; 0, 7,30 |
| 180 | 180; 0, 0, 0 | 180; 0, 0 | 0; 0, 0, 0 |

Table 5. Recomputation of the right ascension underlying Ptolemy's equation of time

that all extracted right ascension values necessarily are multiples of $7\frac{1}{2}''$, the recomputed values, multiples of $12''$. Therefore there is only a limited number of possibilities for the final digits of the differences.

We can compute an estimate $\hat{c}$ of the epoch constant which minimizes the sum of the squares of the differences between the extracted right ascension and values computed according to method E. If the six outliers are left out, the result is $\hat{c}$=3;34,6,26 and an approximate 95% confidence interval for $c$ is given by ⟨3;34,5,11, 3;34,7,42⟩. The minimum obtainable standard deviation of the errors in the extracted right ascension values is $3''$, which is equal to the expected standard deviation of a correct extracted table.[38]

*3.3.6 Final recomputation.* In this Section a final recomputation of Ptolemy's table for the equation of time will be given and an attempt to explain some of the outliers will be made. If we assume that Ptolemy had accurate solar equation values (see Section 3.3.4), that he determined the right ascension according to method E in Section 3.3.5, and that he used the same method of rounding that we use today (see my remarks in Section 3.1.1), the choice $c$= 3;34,7,30 for the epoch constant leads to the minimum possible number of errors in the table for the equation of time, namely 10. Of these errors seven may be called true outliers: they will not disappear for any value of $c$ in the range 3;34,0 to 3;34,15. (As was noted before, we need not consider values of $c$ outside this interval, since for such values the total number of errors in the nodes amounts to at least 25.) These true outliers occur for arguments 42, 126, 132, 138, 144, 192 and 198, and caused the errors that we found in the extracted right ascension and solar equation.

The errors in the final recomputation for epoch constant $c$= 3;34,7,30 are displayed in Table 6, where the first and fourth columns contain the true solar longitude, the second and fifth, Ptolemy's equation of time values, and the third and sixth, the errors.

I have checked the possibility that the errors in Ptolemy's table for the equation of time result from scribal errors in the Greek transmission of the table by inspecting photographs of five other manuscripts of the *Handy Tables,* namely Vatican gr. 1291, Laurentianus gr. 28/26 and Laurentianus gr. 28/48 (Florence), Mar-

| true solar long. | Ptolemy's equation of time | error | true solar long. | Ptolemy's equation of time | error |
|---|---|---|---|---|---|
| 6 | 0;20,33 | | 186 | 0; 4, 0 | |
| 12 | 0;18, 4 | | 192 | 0; 2,30 | −7 |
| 18 | 0;15,38 | | 198 | 0; 1,28 | +1 |
| 24 | 0;13,24 | | 204 | 0; 0,37 | |
| 30 | 0;11,13 | | 210 | 0; 0, 0 | |
| 36 | 0; 9,35 | | 216 | 0; 0, 2 | |
| 42 | 0; 8,10 | +2 | 222 | 0; 0,22 | |
| 48 | 0; 7, 7 | | 228 | 0; 1,13 | |
| 54 | 0; 6,33 | | 234 | 0; 2,35 | |
| 60 | 0; 6,12 | | 240 | 0; 4,13 | |
| 66 | 0; 6,34 | | 246 | 0; 6,34 | |
| 72 | 0; 7, 6 | | 252 | 0; 9, 6 | |
| 78 | 0; 8, 0 | | 258 | 0;11,58 | |
| 84 | 0; 9, 7 | | 264 | 0;15, 2 | |
| 90 | 0;10,23 | | 270 | 0;18, 9 | −1 |
| 96 | 0;11,42 | | 276 | 0;21,15 | |
| 102 | 0;12,57 | | 282 | 0;24,11 | |
| 108 | 0;14, 3 | | 288 | 0;26,50 | |
| 114 | 0;14,53 | | 294 | 0;29, 4 | −1 |
| 120 | 0;15,37 | | 300 | 0;31, 4 | |
| 126 | 0;16, 3 | +20 | 306 | 0;32,15 | |
| 132 | 0;16, 2 | +23 | 312 | 0;33, 6 | |
| 138 | 0;15,18 | +5 | 318 | 0;33,23 | |
| 144 | 0;14,27 | +3 | 324 | 0;33, 5 | |
| 150 | 0;13,27 | | 330 | 0;32,26 | |
| 156 | 0;11,59 | | 336 | 0;31, 6 | |
| 162 | 0;10,31 | | 342 | 0;29,30 | |
| 168 | 0; 8,52 | | 348 | 0;27,33 | |
| 174 | 0; 7,11 | | 354 | 0;25,21 | −1 |
| 180 | 0; 5,33 | | 360 | 0;23, 0 | |

Table 6. Final recomputation of Ptolemy's table for the equation of time

cianus gr. 325 (Venice) and Ambrosianus gr. H 57 sup (Milan).[39] It turned out that in four of these five manuscripts the tabular value for true solar longitude 42 is 0;8,8, equal to the recomputed value. The same four manuscripts have $T(138)$=0;15,19 instead of 0;15,18, but here the recomputed value is 0;15,13. Two manuscripts have $T(192)=0;2,31$ instead of 0;2,30, whereas the recomputed value is 0;2,37. Two other manuscripts display $T(198)=0;1,26$ instead of 0;1,28, whereas the recomputed value is 0;1,27.

We conclude that in our final recomputation the error for argument 42 can be explained from an error in the transmission of the table. Furthermore the error for argument 192 could easily result from a scribal mistake (B Λ instead of B ΛZ). I have not been able to find an explanation for the errors for arguments 126 to 144, which seem to be correlated. Note that these errors have been made before the interpolation was performed, since the interpolation pattern is regular.[40]

### *3.4 Conclusions*

The mathematical analysis in the previous section has enabled us to recover practically every detail of the method that Ptolemy used for the computation of his table for the equation of time. From primary sources and from work by Rome and Neugebauer (see Section 3.1.2) we know the following:

- The independent variable of Ptolemy's table for the equation of time is the true solar longitude.
- The epoch constant *c* was chosen in such a way that the minimum equation of time value was equal to zero. Furthermore, the tabular values had to be added to true solar time to obtain mean solar time.

From this information we concluded that Ptolemy must have calculated his table for the equation of time according to the general formula $E_h^{HT}(\lambda) = (a(\lambda)-\lambda-q(\lambda)+c)/D$ (where $a(\lambda)$ is the right ascension of the true sun, $q(\lambda)$ the solar equation as a function of the true solar position and $D$ the conversion factor). Our analysis confirmed that the table has the true solar longitude for the independent variable and furthermore led to the following new results:

- The conversion factor $D$ is equal to 15.
- The underlying solar equation $q(\lambda)$ is based on the value 2;30

for the eccentricity, not on 2;29,30. Instead of the value 65;30 for the longitude of the solar apogee, which is the only value mentioned in the historical sources, Ptolemy used the rounded value 66;0. The underlying solar equation was not determined by means of inverse linear interpolation in a table for the solar equation as a function of the mean solar longitude, as suggested by Theon of Alexandria, but was computed independently and was accurate to seconds (Section 3.3.4).

- The right ascension $\alpha(\lambda)$ underlying Ptolemy's table for the equation of time is based on the values for multiples of 10° given in the *Almagest*. The intermediate values were determined by means of exact linear interpolation without rounding. The underlying value of the obliquity is 23;51,20 (Section 3.3.5).
- Neither numerical nor historical considerations make it possible to decide on a definite value for the epoch constant. The number of errors in both the extracted right ascension table and our final recomputation of the table for the equation of time is minimized for $c$=3;34,7,30. A 95% confidence interval for $c$ based on all tabular values for multiples of 6° excluding the seven outliers that we have found is ⟨3;34,4,45, 3;34,7,4⟩ (Sections 3.3.5 and 3.3.6).
- Linear interpolation was used to determine the tabular values for non-multiples of 6°. The tabular differences seem to be distributed irregularly between the nodal values (Section 3.3.1).

From these results we see that in computing his table for the equation of time Ptolemy was a very practical astronomer. He simplified the calculations in three different ways: by rounding the longitude of the solar apogee in order to avoid interpolation between his solar equation values, by using linear interpolation between right ascension values for every 10°, and by using linear interpolation between equation of time values for multiples of 6°. We can assume that Ptolemy was aware that the resulting errors were small (it can be checked that the maximum difference between Ptolemy's equation of time values and precisely calculated values for Ptolemy's parameter values is at most 15 seconds of time).

From our results it also becomes clear that one must be very careful when comparing Ptolemy's tables with recomputations based on modern formulae. Incorrect conclusions concerning the underlying parameters or the dependence on other tables may easily be drawn if one ignores the possibility that Ptolemy made use of approximation methods that simplified the computations.

We have noted that in his *Great Commentary* Theon gives correct descriptions of the mathematical structure of the solar declination table and the right ascension table in Ptolemy's *Handy Tables.* He also correctly describes the interpolation pattern in the solar equation table, but fails to mention that the nodal values are not all equal to the corresponding values in the *Almagest.*

Our analysis reveals that there are several divergences between Theon's description in the *Great Commentary* [Mogenet & Tihon 1985, pp. 99–100 or 163–164] and the actual computation of Ptolemy's table for the equation of time. Firstly, Theon does not mention the interpolation within intervals of 6 degrees and even gives his numerical example for argument 271, which is not a node. Secondly, he uses the right ascension which occurs in the column adjacent to the equation of time instead of exact interpolation within intervals of 10 degrees. Thirdly, he uses inverse interpolation in the table for the solar equation as a function of the mean anomaly instead of accurate values from a table for the solar equation as a function of the true anomaly. It seems certain that Theon did not know precisely how Ptolemy's table for the equation of time had been computed.

Although we have shown that the table for the equation of time in the *Handy Tables* was computed differently from what we expected on the basis of available historical information and other tables from the *Almagest* and the *Handy Tables,* we can conclude that there is no reason to question Ptolemy's authorship of the table. We have seen that the particular type of linear interpolation which was used in the right ascension and various other tables in the *Handy Tables* was not used in the right ascension underlying the table for the equation of time, or in the table for the equation of time itself. Therefore it seems plausible that Ptolemy already had accurate equation of time values before he compiled the *Handy Tables* and for some reason left these unchanged when he modified many of the other tables that he included in the *Handy Tables.*

### *3.5 The table for the equation of time in the papyrus P. London 1278*

The second-century Greek papyrus P. London 1278, kept in the British Museum, contains six fragments of numerical tables. Neugebauer made a careful analysis of the papyrus and suggested that the tables it contains may be recensions of tables found in Ptolemy's *Handy Tables* [Neugebauer 1958, pp. 109–112]. Neugebauer concluded that the fragments 6$^v$, 3$^r$, 3$^v$, 1$^r$ and 2$^r$ in that order constitute a badly damaged trigonometric table in the Ptolemaic tradition, displaying the sine of the right ascension, the right ascension itself, and the equation of time for every degree of the ecliptic, starting with Aries [*idem,* pp. 97–103]. It turns out that the right ascension values are identical to those in the *Handy Tables* (see Section 3.1.1) apart from a difference of 90° resulting from the fact that the *Handy Tables* tabulate the normed instead of the ordinary right ascension. The values for the sine of the right ascension can also be found in the *Handy Tables* in connection with the determination of the length of daylight [Neugebauer 1958, pp. 102–103; Stahlman 1959, pp. 265–266]. The values for the equation of time, however, are different from those in the *Handy Tables* analysed in Section 3.3 of this article. From the relative positions of the extremes, it can be seen that the equation of time in P. London 1278 was computed for the Era Nabonassar. This was the era utilized in the *Almagest,* whereas the *Handy Tables* make use of the Era Philip [Sections 1 and 3.1.1 of this article; Neugebauer 1958, pp. 109–111]. The minimum equation of time in P. London 1278 occurs in Aquarius and the equation must always be subtracted from true solar time to obtain mean solar time. The equation of time values are given to an accuracy of minutes of an hour, whereas the *Handy Tables* give values to seconds. Because of the eras involved Neugebauer suggested that the table for the equation of time in P. London 1278 is a version intermediate between the *Almagest* and the *Handy Tables*. We will now further investigate this matter by analysing the equation of time values found in the papyrus.

Since the papyrus contains only 72 equation of time values to minutes, occurring in scattered groups, it would normally be difficult to determine the underlying parameter values and mathematical structure. However, by means of a least squares estimation as explained in the Appendix and by comparing the separate tabular values with those in the *Handy Tables*, it will be possible to draw conclusions about both. We will see that there is reasonable evidence to assume that the equation of time found in the papyrus in the British Museum is directly related to the equation of time in the *Handy Tables.*

Table 7 displays the 72 legible equation of time values in P. London 1278. To obtain a first impression of the underlying parameters and the mathematical structure of the table, I applied a least squares estimation using all available values. Assuming that the conversion factor $D$ equals 15 and that the independent variable of the table is the true solar longitude, we find a minimum possible standard deviation of the tabular errors of 20″ and the following approximate 95% confidence intervals for the underlying parameters:

| *parameter* | *95% confidence interval* |
|---|---|
| obliquity of the ecliptic $\varepsilon$ | ⟨23;20,22, 23;38, 2⟩ |
| solar eccentricity $e$ | ⟨ 2;26,34, 2;32,13⟩ |
| longitude of the solar apogee $\lambda_A$ | ⟨65;44,48, 66;53,54⟩ |
| epoch constant $c$ | ⟨ 4;26, 6, 4;28,48⟩ |

For $D=15;2,28$ the results are essentially the same. Since in this case it will turn out to be impossible to distinguish between the two values of the conversion factor, and since there is no reason so far to believe that $D=15;2,28$ was used for the computation of tables for the equation of time by Greek astronomers, we will take $D$ equal to 15 in the remainder of this section. Because the minimum possible standard deviation that we found is hardly larger than the expected standard deviation of the tabular errors of a correct table with values to minutes (namely 17″; cf. the Appendix), we can conclude that the table for the equation of time in P. London 1278 fits into the Ptolemaic tradition and was probably computed according to formula (7). For two reasons I consider it less probable that the table was computed according to formula (8), i.e. that the independent variable of the table is the mean solar longitude. Firstly, even though the minimum obtainable standard deviation of the tabular errors would be even smaller (namely

| λ | Aries | Taurus | Gemini | Cancer | Leo | Virgo |
|---|---|---|---|---|---|---|
| 1 | | | | 0;21 | | |
| 2 | | | | 0;21 | | |
| 3 | | | | 0;21 | | |
| 4 | | | | 0;21 | | |
| 5 | | | | 0;21 | | |
| 6 | | | 0;25 | 0;20 | | |
| 7 | | 0;23 | 0;25 | 0;20 | | |
| 8 | | 0;23 | 0;25 | 0;20 | | |
| 9 | | 0;23 | | 0;20 | | |
| 10 | | 0;23 | | 0;20 | | |
| 11 | | 0;24 | | 0;19 | | |
| 12 | | 0;24 | | 0;19 | | |
| 13 | | 0;24 | | 0;19 | | |
| 14 | | 0;24 | | | | |
| 15 | | 0;24 | | | | |
| 16 | | | | | | |
| λ | Libra | Scorpio | Sagittarius | Capricornus | Aquarius | Pisces |
| 1 | | 0;32 | 0;28 | 0;13 | | |
| 2 | | 0;32 | 0;27 | 0;13 | | |
| 3 | | 0;32 | 0;27 | 0;12 | | |
| 4 | | 0;32 | 0;26 | 0;12 | | |
| 5 | | 0;32 | 0;26 | 0;11 | | |
| 6 | | 0;32 | 0;25 | 0;11 | 0; 0 | |
| 7 | | 0;32 | 0;25 | 0;10 | 0; 0 | |
| 8 | | 0;32 | 0;24 | 0;10 | 0; 0 | |
| 9 | | 0;32 | 0;24 | 0; 9 | 0; 0 | |
| 10 | | 0;32 | 0;24 | 0; 9 | 0; 0 | |
| 11 | | 0;32 | 0;23 | 0; 8 | 0; 0 | |
| 12 | | 0;32 | 0;23 | 0; 8 | 0; 0 | |
| 13 | | 0;32 | 0;22 | | | |
| 14 | | | 0;22 | | | |
| 15 | | | 0;21 | | | |
| 16 | | | | | | |

Table 7. The equation of time values in the papyrus London 1278

18″), the approximate 95% confidence intervals would no longer contain the historically plausible values 2;29,30 or 2;30 for the eccentricity and 65;30 or 66;0 for the longitude of the apogee, or any other plausible values. Secondly, we have seen that the table for the equation of time in the *Handy Tables* had the true solar longitude as its independent variable, and that none of the relevant sources related to Ptolemy suggests the possibility that the mean solar longitude might be the independent variable of a table for the equation of time (Sections 3.1.1 and 3.3.2). Consequently, I will assume in the sequel that the independent variable of the table for the equation of time in P. London 1278 is the true solar longitude.

The approximate 95% confidence intervals suggest that, as in the case of the *Handy Tables,* the underlying value for the eccentricity is $e$=2;30 (or possibly $e$=2;29,30), the longitude of the apogee $\lambda_A = 66°$. The confidence interval for the obliquity contains only values much smaller than the attested value $\varepsilon$=23;51,20. Since the same phenomenon occurs for the equation of time in the *Handy Tables,*[41] it seems possible that the right ascension used for the calculation of the equation of time values in P. London 1278 contains errors similar to those in the right ascension underlying the equation of time in the *Handy Tables* or even that the table for the equation of time in P. London 1278 is directly related to the one in the *Handy Tables.*

In the remainder of this Section we will investigate the possibility that the equation of time values in P. London 1278 were computed by subtracting the values in the *Handy Tables* from a constant and rounding the results to minutes, i.e.

$$T_{PL}(\lambda) = r_1(C - T_{HT}(\lambda)) \qquad (16)$$

for every $\lambda$, where $C$ is a constant, $T_{PL}(\lambda)$ are the values in P. London 1278, $T_{HT}(\lambda)$ those in the *Handy Tables*, and $r_1$ indicates (modern) rounding to minutes. Thus we will consider the distribution of $\Sigma(\lambda) \stackrel{\text{def}}{=} T_{PL}(\lambda) + T_{HT}(\lambda)$ for all $\lambda$ for which P. London 1278 displays values. It turns out that $\Sigma(\lambda)$ lies in the range from 0;31,31 to 0;32,30 for 64 values of $\lambda$, and that furthermore

$\Sigma(100)$ = 0;32,32, $\Sigma(309)$ = 0;32,41,
$\Sigma(241)$ = 0;32,36, $\Sigma(310)$ = 0;32,49,
$\Sigma(248)$ = 0;31,24, $\Sigma(311)$ = 0;32,58,
$\Sigma(308)$ = 0;32,32, and $\Sigma(312)$ = 0;33, 6.

We note that for $C$=0;32 (in which case the sum of the epoch constants of both tables equals 8;0, namely 15 · 0;32), 64 out of the 72 available equation of time values in P. London 1278 were correctly computed according to formula (16) if the rounding was performed in the modern way (the same would hold for $C$=0;32,30 if truncation were used). The values $\Sigma(\lambda)$ outside the range 0;31,31 to 0;32,30 for $\lambda$=100, 241 and 248 could be attributed to small rounding errors. The values $\Sigma(\lambda)$ for $\lambda$=308, 309,...,312 are clearly correlated and could be explained by assuming that the author of the table in P. London 1278 set $T_{PL}(\lambda)$ to 0;0 whenever $r_1(0;32-T_{HT})<0$.

*Conclusions.* The table for the equation of time in the papyrus London 1278 was computed in accordance with the theory developed by Ptolemy in the *Almagest*. It is probable that the table was directly computed from the equation of time in the *Handy Tables* using formula (16), where $C$ was taken equal to 0;32, the rounding was performed in the modern way, and the resulting negative values in the sign Aquarius were set to zero. Thus, like the equation of time in the *Handy Tables*, the table in P. London 1278 is based on parameter values $\varepsilon$=23;51,20, $e$=2;30 and $\lambda_A$= 66;0.

It seems reasonable to expect that the author of P. London 1278 took $C$ equal to the maximum equation of time in the *Handy Tables* in order to obtain values that must always be subtracted from true solar time to obtain mean solar time. A possible explanation why he used the value 0;32 instead of the maximum 0;33,23, is as follows.

In Section 1 it was noted that Ptolemy assumed that mean and true solar time at epoch were equal. As Theon explained in his *Great Commentary on the Handy Tables,* Ptolemy made the minimum equation of time in his table equal to zero in order to obtain an equation that must always be added to the true solar time [Section 3.1.1 of this article; Mogenet & Tihon 1985, pp. 123 or 182]. Thus he neglected the fact that the solar position at the epoch 1 Philip of the *Handy Tables* was 17° Scorpio, whereas the minimum equation of time occurred for 0° Scorpio. Consequently, for every true solar position, the actual equation of time was approximately

1′4″ smaller than the value given in the *Handy Tables* (this difference can be found in the table for the equation of time as the value for argument 17° Scorpio).

For the Era Nabonassar the same problem occurred: the true solar position at epoch was 3° Pisces, whereas the minimum equation of time was assumed for 18° Aquarius. To tabulate the equation of time correctly, the epoch constant must be chosen in such a way that the equation for 3° Pisces became zero. This implies that, when computing a table for the Era Nabonassar from the equation of time in the *Handy Tables* by means of formula (16) above, $C$ should be taken equal to the tabular value for $\lambda$=333, i.e. to 0;31,47, which, when rounded to minutes, becomes $C$=0;32. Instead of accepting the negative values that would thus arise for longitudes 308 to 329, the author of the table for the equation of time in P. London 1278 apparently preferred to make all these values equal to zero.

Thus we see that the tables for the equation of time in the *Handy Tables* and in P. London 1278 present two different solutions to the problem that the actual maximum or minimum equation does not occur precisely at epoch. In the case of the *Handy Tables* a small constant was added to all tabular values in order to make them non-negative; in the case of P. London 1278 all negative values were simply set to zero.

Note that the relationship between the tables for the equation of time in the *Handy Tables* and in P. London 1278 need not necessarily be the one described above. Another possibility is that both were computed from a non-extant table for the equation of time with values to seconds for the Era Nabonassar. The errors that we found in the *Handy Tables* (Section 3.3.6) cannot be used to obtain more detailed information concerning this matter, since either they are too small to show up in the values in P. London 1278 or they occur in regions where the papyrus has no tabular values at all.

*Acknowledgement*

It is a pleasure to thank Prof. Anne Tihon (Louvain-la-Neuve) and Dr. Alexander Jones (Toronto) for their helpful advice with regard to various details of my analysis of Ptolemy's table for the equation of time. The Universiteitsbibliotheek of Leiden kindly gave me permission to include a photograph of folio 76$^v$ of their manuscript BPG 78.

This article is a reworked version of Sections 3.1 and 3.2 of my doctoral thesis (van Dalen 1993). During the work on my thesis I particularly appreciated the support and invaluable suggestions and comments of Professors Henk Bos and Richard Gill and Dr. Jan Hogendijk (Utrecht) and Prof. David King (Frankfurt am Main). My doctoral research was financially supported by the Netherlands Organization for Scientific Research (NWO).

## *Appendix*

In this appendix the statistical background of the analyses of the tables for the equation of time in Ptolemy's *Handy Tables* and in the Greek papyrus P. London 1278 will be briefly described. A more extensive discussion of tabular errors and their probability distribution can be found in van Dalen 1993, Section 1.2. More information about the application of least squares estimators to astronomical tables in ancient and mediaeval sources is given in *idem*, Section 2.4; general information about least squares estimation can for instance be found in Draper & Smith 1981, Chapter 10.

*Tabular errors.* Let $T(x), x \in X$ denote tabular values for a given function $f$. For every argument $x \in X$ I define the *tabular error* $e(x)$ by $e(x)=T(x)-f(x)$. A tabular value $T(x)$ is said to be *correct* if $T(x)=r_k(f(x))$, where $r_k$ denotes a presumed rounding procedure (to $k$ sexagesimal fractional digits). Thus for a correct tabular value the tabular error is identical to the error made in rounding the functional value to the number of sexagesimal digits of the table. This implies that in general tabular errors are non-zero, even if the tabular values concerned are correct.

The following can be conjectured [van Dalen 1993, Section 1.2.4]: *if the number of sexagesimal digits of the tabular values of a correct table is sufficiently large, then the tabular errors (i.e. the rounding errors) can be assumed to be independent and to have a uniform distribution.*

This implies, for instance, that in a correct table for the equation of time with values to seconds of an hour, all possible final digits occur approximately equally often (cf. property 2 in Section 2). Furthermore, the standard deviation of the tabular errors can be approximated by the standard deviation of a uniform distribution on the interval $[-\frac{1}{2}\cdot 60^{-2}, +\frac{1}{2}\cdot 60^{-2}]$ and is thus approximately equal to 0;0,0,17. When performing a least squares estimation of the underlying parameter values (see below), the minimum obtainable standard deviation of the tabular errors is expected to be approximately equal to this value.

*Outliers* are tabular errors that are significantly larger than most other errors in the table concerned. Outliers can for instance result from computational errors or scribal mistakes.

*Least squares estimation.* Let $T(x)$, $x \in X$, denote tabular values for the function $f_\theta$ which depends on the parameter vector $\theta$. Let the objective function $\Phi(\theta)$ be defined as the sum of the squares of the tabular errors: $\Phi(\theta)=\Sigma_{x \in X}(T(x)-f_\theta(x))^2$. A least squares estimate for the parameter vector $\theta$ is a vector $\hat{\theta}$ that minimizes $\Phi(\theta)$. In order to determine a least squares estimate of the underlying parameter values of a table for the equation of time, we need an iterative optimization procedure such as the method of Gauss-Newton (see for instance van Dalen 1993, pp. 55–60).

Once the estimate $\hat{\theta}$ has been obtained, we can approximate the standard deviation $\sigma$ of the tabular errors using $\sigma^2 \approx \Phi(\hat{\theta})/n$. I refer to this approximation as the "minimum obtainable standard deviation" of the tabular errors. If the minimum obtainable standard deviation is much larger than the standard deviation calculated on the basis of the assumption that the tabular errors have a uniform distribution (see above), then it is probable that $f_\theta$ is *not* the tabulated function. If the minimum obtainable standard deviation is small enough and the tabular errors can be assumed to be independent and to have zero means and equal variances, then confidence intervals for all underlying parameters of the table under consideration can be computed from the least squares estimates found.

## BIBLIOGRAPHY

Britton, John P.

1969: "Ptolemy's Determination of the Obliquity of the Ecliptic", *Centaurus* 14, pp. 29–41.

1992: *Models and Precision: The Quality of Ptolemy's Observations and Parameters,* New York (Garland).

Dalen, Benno van

1988: *A Statistical Method for the Analysis of Medieval Astronomical Tables,* University of Utrecht (Netherlands), Mathematical Institute, preprint no. 517.

1989: "A Statistical Method for Recovering Unknown Parameters from Medieval Astronomical Tables", *Centaurus* 32, pp. 85–145.

1993: *Ancient and Mediaeval Astronomical Tables: mathematical structure and parameter values* (doctoral thesis), University of Utrecht (Netherlands).

Draper, Norman R. & Smith, Harry

1981: *Applied Regression Analysis* (2nd ed.), New York (Wiley).

*DSB: Dictionary of Scientific Biography,* 14 vols and 2 suppl. vols, New York (Charles Scribner's Sons) 1970–1980.

Evans, James

1987: "On the Origin of the Ptolemaic Star Catalogue", *Journal for the History of Astronomy* 18, pp. 155–172 and 233–278.

Fomenko, A. T., Kalashnikov, V. V. & Nosovsky, G. V.

1989: "When was Ptolemy's Star Catalogue in "Almagest" Compiled in Reality? Statistical Analysis", *Acta Applicandae Mathematicae* 17, pp. 203–229.

Gauss, Carl Friedrich

1863–1933: *Werke* (eds. Ernst C. J. Schering & Felix Klein), 12 vols, Göttingen (Königliche Gesellschaft der Wissenschaften). Reprinted by Olms, Hildesheim 1973–.

Ginzel, Friedrich Karl

1906–1914: *Handbuch der mathematischen und technischen Chronologie,* 3 vols, Leipzig (Hinrichs).

Glowatzki, Ernst & Göttsche, Helmut

1976: *Die Sehnentafel des Klaudios Ptolemaios. Nach den historischen Formelplänen neu berechnet,* Munich (Oldenbourg).

Grasshoff, Gerd

1990: *The History of Ptolemy's Star Catalogue,* New York (Springer).

Halma, Nicolas

1822–1825: *Commentaire de Théon d'Alexandrie sur le livre III de l'Almageste de Ptolémée; tables manuelles des mouvemens des astres,* 3 vols, Paris (Merlin, Bobée, Eberhart).

Heiberg, Johann Ludvig (ed.)

1898–1903: *Claudii Ptolemaei: Opera quae exstant omnia I, Syntaxis Mathematica,* 2 vols, Leipzig (Teubner).

1907: *Claudii Ptolemaei: Opera quae exstant omnia II, Opera astronomica minora,* Leipzig (Teubner).

Jones, Alexander
1991: "Hipparchus's Computations of Solar Longitudes", *Journal of the History of Astronomy* 22, pp. 101–125.

Kennedy, Edward S.
1988: "Two Medieval Approaches to the Equation of Time", *Centaurus* 31, pp. 1–8.

King, David A.
1973: "Ibn Yūnus' Very Useful Tables for Reckoning Time by the Sun", *Archive for the History of Exact Sciences* 10, pp. 342–394. Reprinted in D. A. King, *Islamic Mathematical Astronomy,* London (Variorum Reprints) 1986 (2nd ed. 1993).

Kunitzsch, Paul
1986–1991: *Claudius Ptolemäus: Der Sternkatalog des Almagest. Die arabisch-mittelalterliche Tradition,* 3 vols, Wiesbaden (Harrassowitz).

Maeyama, Yasukatsu
1984: "Ancient Stellar Observations: Timocharis, Aristyllus, Hipparchus, Ptolemy; the Dates and Accuracies", *Centaurus* 27, pp. 280–310.

Meyier, K. A. de & Hulshoff Pol, E.
1965: *Codices Bibliothecae Publicae Graeci* (vol. 8 of "Bibliotheca Universitatis Leidensis: Codices Manuscripti"), Leiden (University Library).

Moesgaard, Kristian P.
1987: "In Chase of an Origin for the Mean Planetary Motions in Ptolemy's Almagest", *From Ancient Omens to Statistical Mechanics; Essays on the Exact Sciences Presented to Asger Aaboe* (eds. J. L. Berggren & B. R. Goldstein), Copenhagen (University Library), pp. 43–54.

Mogenet, Joseph & Tihon, Anne
1985: *Le Grand Commentaire de Théon d'Alexandrie aux Tables Faciles de Ptolémée. Livre I. Histoire du texte, édition critique, traduction,* Vatican City (Biblioteca Apostolica Vaticana).

Neugebauer, Otto E.
1957: *The Exact Sciences in Antiquity* (2nd ed.), New Haven (Brown University Press).
1958: "The Astronomical Tables P. Lond. 1278", *Osiris* 13, pp. 93–113.
1962: *The Astronomical Tables of al-Khwārizmī. Translation with Commentaries of the Latin Version edited by H. Suter supplemented by Corpus Christi College MS 283,* Det Kongelige Danske Videnskabernes Selskab, historisk-filosofiske Skrifter 4:2, Copenhagen.
1975: *A History of Ancient Mathematical Astronomy,* 3 vols, Berlin (Springer).

Newton, Robert R.
1977: *The Crime of Claudius Ptolemy,* Baltimore (The Johns Hopkins University Press).
1982: *The Origins of Ptolemy's Astronomical Parameters,* Baltimore (The Center for Achaeoastronomy, University of Maryland & The Johns Hopkins University Applied Physics Laboratory).
1985: *The Origins of Ptolemy's Astronomical Tables,* Baltimore (The Center for Archaeoastronomy, University of Maryland & The Johns Hopkins University Applied Physics Laboratory).

North, John D.
1976: *Richard of Wallingford,* 3 vols, Oxford (Clarendon Press).

Pauly: *Paulys Real-Encyclopädie der classischen Altertumswissenschaft* (neue Bearbeitung von Georg Wissowa), 34 vols, Stuttgart (Metzler) 1894–1972.

Pedersen, Olaf
1974: *A Survey of the Almagest,* Odense (University Press).

Petersen, Viggo & Schmidt, Olaf
1967: "The Determination of the Longitude of the Apogee of the Orbit of the Sun According to Hipparchus and Ptolemy", *Centaurus* 12, pp. 73–96.

Rawlins, Dennis
1982: "An Investigation of the Ancient Star Catalogue", *Publications of the Astronomical Society of the Pacific* 94, pp. 359–373.

Rome, Adolphe
1931–1943: *Commentaires de Pappus et de Théon d'Alexandrie sur l'Almageste (Texte établi et annoté),* 3 vols, Rome (Biblioteca Apostolica Vaticana).
1939: "Le problème de l'équation du temps chez Ptolémée", *Annales de la Société Scientifique de Bruxelles* 59, pp. 211–224.

Shevchenko, Michail Y.
1990: "An Analysis of Errors in the Star Catalogues of Ptolemy and Ulugh Beg", *Journal for the History of Astronomy* 21, pp. 187–201.

Smart, William M.
1977: *Textbook on Spherical Astronomy* (6th edition revised by Robin M. Green), Cambridge (University Press).

Stahlman, William D.
1959: *The Astronomical Tables of Codex Vaticanus Graecus 1291* (doctoral thesis), Providence (Brown University). To be published by Garland, New York.

Swerdlow, Noel M.
1992: "The Enigma of Ptolemy's Catalogue of Stars", *Journal for the History of Astronomy* 23, pp. 173–183.

Tihon, Anne
1978: *Le "Petit Commentaire" de Théon d'Alexandrie aux Tables Faciles de Ptolémée (histoire du texte, édition critique, traduction),* Vatican City (Biblioteca Apostolica Vaticana).
1991: *Le "Grand Commentaire" de Théon d'Alexandrie aux Tables Faciles de Ptolémée: Livres II et III. Édition critique, traduction, commentaire,* Vatican City (Biblioteca Apostolica Vaticana).
1992: "Les "Tables faciles" de Ptolémée dans les manuscrits en onciale (IX$^{e}$–X$^{e}$ siècles)", *Revue d'histoire des textes* 23, pp. 47–87.

Tihon, Anne & Mogenet, Joseph
see: Mogenet, Joseph & Tihon, Anne

Toomer, Gerald J.
1984: *Ptolemy's Almagest,* London (Duckworth) and New York (Springer).

Van Brummelen, Glen R.
1993: *Mathematical Tables in Ptolemy's Almagest* (doctoral thesis), Burnaby BC (Simon Fraser University).

Van Dalen, Benno
see: Dalen, Benno van.

Wlodarczyk, Jaroslaw
1990: "Notes on the Compilation of Ptolemy's Catalogue of Stars", *Journal for the History of Astronomy* 21, pp. 283–295.

## NOTES

1. The star catalogue in Ptolemy's *Almagest* was extensively analysed in Newton 1977, Rawlins 1982, Evans 1987, Fomenko, Kalashnikov & Nosovsky 1989, Grasshoff 1990, Shevchenko 1990, and Wlodarczyk 1990. A survey of the secondary literature on the star catalogue can be found in Swerdlow 1992. A complete edition of the versions of the star catalogue in the Arabic and Latin translations of the *Almagest* is presented in Kunizsch 1986–1991. The observations reported and used by Ptolemy were investigated in Petersen & Schmidt 1967, Britton 1969, Newton 1977, Maeyama 1984, Moesgaard 1987, Jones 1991 and Britton 1992. Newton [1982] examined the origin of the astronomical parameter values in the *Almagest*. The method of computation of the mathematically-computed tables was investigated in Glowatzki & Göttsche 1976; Newton 1985 and Van Brummelen 1993.
2. An accurate statistical method for the determination of a *single* unknown parameter in an astronomical table is presented in van Dalen 1989.
3. Gauss calculated the underlying parameter values of Ulugh Beg's table for the equation of time using his newly-invented method of least squares, but his objective was not historical [*Allgemeine geographische Ephemeriden* 3 (1799), pp. 179–185, and Gauss, *Werke*, vol. 12, pp. 64–68]. Rome and Neugebauer investigated the information from primary sources concerning the equation of time as used by Ptolemy, but did not analyse the table in the *Handy Tables* [Rome 1939; Neugebauer 1975, vol. 1, pp. 61–68 and vol. 3, 984–986; see also Neugebauer 1958, pp. 109–111]. North gave a method for determining rough approximations to the underlying parameters of a table for the equation of time based on series expansions [North 1976, vol. 3, pp. 201–205]. Kennedy [1988] recomputed the tables for the equation of time found in the mediaeval astronomical handbooks of Kūshyār ibn Labbān and al-Kāshī.
4. See for example the *Great Commentary on Ptolemy's Handy Tables* by Theon of Alexandria [Mogenet & Tihon 1985, p. 98, lines 10–11 or p. 119, lines 10–11].
5. The equation of time will be explained as far as possible in the terminology of Ptolemy's solar model, which was in use throughout the Middle Ages. A short description of this model can be found in Neugebauer 1957, p. 192; more extensive explanations in Neugebauer 1975, vol. 1, pp. 53–61 or Pedersen 1974, pp. 144–154. Discussions of the equation of time as used by Ptolemy can be found in Neugebauer 1975, vol. 1, pp. 61–68, and vol. 2, pp. 984–986; and in Pedersen 1974, pp. 154–158. For a modern treatment of the equation of time, see Smart 1977, pp. 146–150.
6. The mean suns cannot be found explicitly in Ptolemy's work. They are modern concepts, which we use for the sake of simplicity. Instead of the ecliptical mean sun Ptolemy consistently used the equivalent concept "position of the sun in mean motion". Instead of the equatorial mean sun he applied the concept of simultaneously rising arcs of the equator and the ecliptic.

7. The right ascension of a heavenly body $X$ is defined as the spherical angle between the meridian through the vernal point ♈ and the meridian of $X$, measured in eastward direction. Thus the right ascension of the equatorial mean sun $M$ equals the length of the equatorial arc ♈$M$.
8. The hour angle of a heavenly body $X$ is the spherical angle between the meridian of the observer and the meridian through $X$, measured in westward direction. In this article, it will always be measured in equatorial degrees (as opposed to hours). Note that the hour angle is measured in the opposite direction of the right ascension.
9. At the epoch of Ptolemy's *Almagest* the equation of time was close to its yearly maximum. Thus the resulting equation of time values were practically all negative. If the equation of time at epoch is not close to its minimum or maximum value, the mean solar longitude at epoch may be adjusted in order to obtain an always positive or an always negative equation of time; see Neugebauer 1962, pp. 63–65. Note also footnote 21.
10. An ancient or mediaeval astronomer determined $h(S)$ from an observation of the altitude of the sun or a bright star. Large sets of tables for timekeeping were available, by means of which for instance the number of hours since sunrise could be determined as a function of the solar altitude and longitude [cf. King 1973]. To use these tables, one first had to determine a rough value of the true solar position for the date concerned.
11. Most of the following information was taken from the *DSB*-articles "Ptolemy" and "Theon of Alexandria" by Gerald J. Toomer.
12. In works written before Ptolemy we find no reference to the equation of time as used by Ptolemy; cf. Neugebauer 1975, vol. 1, p. 61; and vol. 2, pp. 584 and 766.
13. Since Theon's *Commentary on the Almagest* is only available in a Greek edition [Rome 1931–1943], I have not been able to study it in detail. The commentary concerning the "inequality of the days" is found in vol. 3, pp. 917–942.
14. Ptolemy's *Introduction to the Handy Tables* is extant in at least 14 manuscripts that are listed in Heiberg 1907, pp. clxxv–clxxix, and was edited in *idem*, pp. 159–185. An edition plus French translation can be found in Halma 1822–1825, vol. 1, pp. 1–26.
15. Book I of the *Great Commentary* is edited, translated and commented upon in Mogenet & Tihon 1985; Book II and III, in Tihon 1991. At present Anne Tihon is preparing an edition and translation of Book IV.
16. The solar declination $\delta$ is the orthogonal distance on the sphere between the sun and the celestial equator. It can be computed as a function of the solar position $\lambda$ according to the formula $\delta(\lambda)=\arcsin(\sin\varepsilon\cdot\sin\lambda)$, where $\varepsilon$ is the obliquity of the ecliptic. Ptolemy and most Islamic astronomers calculated the right ascension according to the formula $\alpha(\lambda)=\arcsin(\tan\delta(\lambda)/\tan\varepsilon)$. Only in exceptional cases an equivalent of the modern formula (2) was used.
17. The solar declination table from the *Almagest* can be found in Heiberg 1898–1903, vol. 1, pp. 80–81 or Toomer 1984, p. 72. The right ascension values for ecliptical arcs of 10 degrees are listed in Heiberg 1898–1903, vol. 1, p. 85 and Toomer 1984, p. 74. The solar declination table from the *Handy Tables* can be found in Leiden BPG 78, f. 97$^r$ or Stahlman 1959, p. 260.
18. Ptolemy's determination of the solar eccentricity can be found in Heiberg 1898–1903, vol. 1, p. 236 or Toomer 1984, p. 155. The worked examples for the determination of

the solar equation are given in Heiberg 1898–1903, vol. 1, p. 240 ff. and Toomer 1984, p. 157 ff. The table for the solar equation can be found in Heiberg 1898–1903, vol. 1 p. 253 or Toomer 1984, p. 167. Assuming that Ptolemy calculated every single tabular entry of his table for the solar equation using the correct formula which he demonstrates in his worked examples, $e=2;29\frac{1}{2}$ fits the table better than $e=2;30$ [van Dalen 1988, pp. 10–12]. See also Van Brummelen 1993, pp. 149–154.

19. The epoch of the Era Nabonassar is 26 Februari 747 B.C. See Toomer 1984, pp. 9–14 for a description of the calendar systems used by Ptolemy. See Ginzel 1906–1914, vol. 1, pp. 143–147 for more information about the Era Nabonassar.
20. The epoch values for the mean and true solar longitudes in the *Almagest* are given in Heiberg 1898–1903, vol. 1, pp. 256–257 and 263 and in Toomer 1984, pp. 168–169 and 172.
21. Like all ancient and mediaeval astronomers, Ptolemy did not make use of negative numbers. Instead he always tabulated the absolute value of quantities like the equation of time and indicated in the table or in the explanatory text in which cases the quantity had to be added or subtracted.
22. The epoch of the Era Philip is 12 November 324 B.C. See Ginzel 1906–1914, vol. 1, p. 147 for more information about the Era Philip.
23. The tables for the solar mean motion, the solar equation and the right ascension in the *Handy Tables* can be found in Leiden BPG 78, ff. 97$^r$, 94$^r$–96$^v$, and 75$^r$–76$^v$ respectively, and in Stahlman 1959, pp. 243, 249–254, and 206–209.
24. Concerning Serapion, see *Pauly,* 2nd series, vol. 2, cols 1666–1667. Neugebauer suggests that the Serapion mentioned by Theon can be identified with the Alexandrian astrologer of the same name listed in *Pauly* as number 1 [Neugebauer 1958, pp. 110–111]. See also Tihon 1992, pp. 74–75.
25. The irregular differences that Rome obtains starting from 5° Capricorn are a result of two errors in the tabular values in Halma 1822–1825, vol. 1, pp. 148–155, namely $T(247)=0;21;41$ (should be 0;21,45) and $T(250)=0;23,16$ (should be 0;23,13). The correct values can be found in the manuscripts Leiden BPG 78 and Vatican gr. 1291.
26. An extensive description of the manuscript BPG 78 can also be found in vol. 8 of the catalogue of manuscript collections in the University Library of Leiden, de Meyier & Hulshoff Pol 1965, pp. 166–171.
27. The complete table is transcribed in Stahlman 1959, pp. 206–209. Since Stahlman did not consult the Leiden manuscript, there may be incidental differences between the values given by Stahlman and those presented in the second and fifth columns of Table 2. However, in general there are so few differences between the copies of the table for the equation of time in the four uncial manuscripts of the *Handy Tables* that the following analysis would be practically identical had we started from one of the other copies. In Section 3.3.6 the results of our final recomputation will be checked against five other manuscripts.
28. The irregular differences for arguments 60 and 61 result from a scribal error: $T(61)=0;6,20$ must be corrected to 0;6,16, which is found in two other manuscripts that I consulted. We will see later that $T(42)=0;8,10$ can be corrected to 0;8,8. This makes the interpolation pattern even more regular. See Section 3.3.6 for an extensive discussion of possible scribal errors.

29. In the displayed part of the table the jumps for multiples of 12° are significantly larger than those for arguments $(6+12n)°$. This is an accidental circumstance which is a result of the method by which the underlying right ascension will be shown to have been computed. In other parts of the table the jumps for multiples of 12° are smaller than those for arguments $(6+12n)°$.
30. Stahlman's argument that the solar equation in the *Almagest* is partly tabulated for multiples of 3 degrees will turn out to be irrelevant, since the solar equation table in the *Almagest* has the *mean* solar anomaly as its independent variable, whereas the table for the equation of time in the *Handy Tables* is a function of the *true* solar longitude.
31. In Leiden BPG 78 the tabular values for the nodes 342 and 360 and for several other arguments are illegible. Therefore I used the values found in Vatican gr. 1291, another early manuscript of the *Handy Tables*, which was used by Stahlman for his edition [1959]. In the Vatican manuscript the table for the right ascension and the equation of time is found on folios $22^r$–$23^v$.
32. I performed a Monte Carlo analysis to show that we obtain reasonably accurate confidence intervals even if the number of tabular values used is only 12. Approximately 91% of the confidence intervals determined in the Monte Carlo analysis contained the actual parameter values.
33. Cf. the remarks concerning Serapion in Section 3.1.2 and about the epoch constant of the equation of time as tabulated by Ptolemy in Section 3.5. The difference of 16′ between the estimated and calculated epoch constants corresponds to the systematic error of 1′4″ in the equation of time values in the *Handy Tables* as noted by Separion.
34. Here I apply the assumption that the tabular errors of a correct tabulation to seconds of the equation of time can be considered to be independent and have approximately a uniform distribution (cf. the Appendix). The errors in the extracted tables for the solar equation and the right ascension are then equal to the sum of two independent, uniformly distributed variables and therefore have a triangular distribution.
35. The solar equation table from the *Almagest* can be found in Heiberg 1898–1903, vol. 1, p. 253 or Toomer 1984, p. 167; the solar equation table from the *Handy Tables* is available in Leiden BPG 78, ff. $94^r$–$96^v$ or Stahlman 1959, pp. 249–254.
36. Inverse linear interpolation in the *Almagest* table leads to 19 differences larger than $7\frac{1}{2}''$ in absolute value; in the *Handy Tables*, to 18. Inverse linear interpolation in a correct solar equation table with values to seconds for every degree of the mean solar anomaly would be accurate enough to produce the extracted solar equation, but there is no reason to believe that Ptolemy had such a table.
37. The algorithm for determining the solar equation as a function of the true solar longitude can be found in Heiberg 1898–1903, vol. 1, pp. 242–243 or Toomer 1984, p. 159. The application of this algorithm to the calculation of the mean solar position at an observed autumnal equinox is described in Heiberg 1898–1903, vol. 1, pp. 254–255 and Toomer 1984, pp. 166 and 168. The table of chords from the *Almagest* can be found in Heiberg 1898–1903, vol. 1, pp. 48–63 or Toomer 1984, pp. 57–60.
38. It can be checked that for the value $\hat{c}$ of the epoch constant the errors in the extracted right ascension values all have the same order of magnitude and show no obvious dependency. Thus the conditions for the least squares estimation as discussed in the Appendix are satisfied.

39. Descriptions of three of these manuscripts can be found in Tihon 1978, pp. 139–141 (Laurentianus gr. 28/26), pp. 103–104 (Laurentianus gr. 28/48) and pp. 88–90 (Ambrosianus gr. H 57 sup). Furthermore, a table of contents of Laurentianus gr. 28/26 is presented in Tihon 1992, pp. 64–66. The manuscript Vatican gr. 1291 is described in Tihon 1992, pp. 61–64. A table of contents of this manuscript can also be found in Neugebauer 1975, vol. 2, pp. 977–978.
40. The only large irregularity in the interpolation pattern occurs between arguments 126 and 132, where the tabular differences are +6, +6, +6, −6, −6, −7″. In the table the local maximum in this interval occurs for argument 129 and amounts to 0;16,21. If we use Ptolemy's method of computation for the nodes to calculate equation of time values for every integer argument, we find a maximum 0;15,49 for true solar longitude 130, different from what we find in the table.
41. I performed least squares estimations on the equation of time values in the *Handy Tables* in Section 3.3.2, but did not give all the results. If all tabular values are used, the least squares estimate for the obliquity is $\hat{\varepsilon}$=23;44,59. If only those arguments are used for which P. London 1278 has tabular values as well, the estimate is $\hat{\varepsilon}$=23;41,50. This is very close to the estimate $\hat{\varepsilon}$=23;41,44 obtained from the table in the papyrus if we leave out the zero values in the sign Aquarius. Below it will be explained that those values were probably set to zero to avoid negative values for the equation of time.

# III

# A table for the true solar longitude in the Jāmi[c] Zīj

In this article I analyse a table that turns out to be related to a table investigated by Prof. E.S. Kennedy in the Festschrift for Willy Hartner (1977). The main part of this article is a reworked version of Section 2.6.3 of my doctoral thesis (van Dalen 1993 in the Bibliography).

Between the eighth and fifteenth centuries Islamic astronomers compiled more than 200 different astronomical handbooks, known by the name of *zīj*. Most of these zījes contained explanatory text and large sets of tables of complicated mathematical functions, by means of which the positions of the sun, moon and planets could be accurately predicted.[1] The Islamic astronomers mostly based their zījes on Ptolemy's planetary models, but they calculated the tables anew, using more accurate methods of computation and more accurate values of the underlying parameters. In some cases the actual parameter values had changed in the course of time; in other cases the determination of the parameters had not been accurate enough to ascertain correct planetary positions over periods of centuries.

Many of the extant manuscripts of zījes contain a mixture of material from various sources. Since these sources are not always explicitly mentioned, it is often difficult to determine the origin of tables in zījes. We can assume that certain mathematical properties of a table, such as the method of computation and the underlying parameter values, are typical for the astronomer who calculated the table. Thus we may be able to identify the origin of a table by investigating its mathematical properties. Usually the tabular headings and the explanatory text in zījes provide little information about the method of computation and the underlying parameter values of the tables. Therefore such information must be extracted from the tabular values themselves. In order to determine the unknown parameter values as reliably as possible and to find important details of the methods of computation, the use of advanced mathematical and statistical methods turns out to be both essential and very effective, as is shown convincingly in van Dalen 1993 and in Van Brummelen 1993.

In this article I will show how mathematical and statistical methods can be used to determine the tabulated function, the underlying parameter values and the author of a table about which no textual information is available. In the analysis of the table I have as much as possible left out the statistical details. A summary of the

1 An overview of all zījes known in 1956 and of the types of tables that occur in zījes can be found in Kennedy 1956.

statistical methods used can be found in the Appendix at the end of this article; for more extensive explanations the reader is referred to my doctoral thesis.

Abu'l-Ḥasan Kūshyār ibn Labbān ibn Bāshahrī al-Jīlī worked as an astronomer in Baghdad around the year 1000. The attribute al-Jīlī indicates that Kūshyār was a native of the region Jīlān in northern Iran. Kūshyār's main achievements were in the fields of arithmetic, trigonometry and astronomy. He wrote a work "The Elements of Hindu Reckoning" about sexagesimal arithmetic and computed extensive trigonometric tables. In his astronomical works Kūshyār made use of the parameters of al-Battānī (c. 900) instead of making his own observations.[2]

It is unclear whether Kūshyār wrote one or two astronomical handbooks. In "The Book of the Astrolabe" he mentions the Jāmiᶜ Zīj ("Comprehensive Astronomical Tables") and the Bāligh Zīj ("Extensive Astronomical Tables") as two different works. Kennedy suggests that the Bāligh Zīj is an abridged version of the Jāmiᶜ.[3] I made a cursory analysis of the tables in four manuscripts of Kūshyār's zīj(es): Istanbul Fatih 3418, Berlin Ahlwardt 5751, Leiden Or. 8 (1054), and Cairo Dār al-Kutub Mīqāt 188/2.[4] The oldest of these manuscripts, Fatih 3418, is entitled "The Book of the Jāmiᶜ Zīj" and is divided into four treatises containing instructions, tables, explanations and proofs respectively. The same division is found in the Berlin and Leiden manuscripts, although the third and fourth treatises are not actually present in the Berlin manuscript. From the given tables of contents and from the coherence of the material in the Istanbul, Berlin and Leiden manuscripts, it can be concluded that, except for the appended tables described below, both explanatory text and tables in the three manuscripts were part of the original zīj written by Kūshyār.[5] The Cairo manuscript contains only a number of Kūshyār's tables.

My analysis revealed that all four manuscripts contain essentially the same set of somewhat more than 50 tables that are listed in the tables of contents referred to in footnote 5. There are, however, small differences between the manuscripts, which may be due to the existence of two different zījes by the hand of Kūshyār ibn Labbān. As far as the date of compilation of the Jāmiᶜ Zīj is concerned, Kūshyār gives his planetary apogee values for the year 962, whence it seems

---

2 More information about Kūshyār ibn Labbān can be found in the article "Kūshyār" in the Dictionary of Scientific Biography (*DSB*).

3 Kennedy 1956, p. 125 (nos 7 and 9).

4 The Istanbul manuscript was copied in the year 1150 and seems to contain the Jāmiᶜ Zīj in its original form. The Berlin manuscript is described in Ahlwardt 1893, pp. 203-206, which also gives an extensive table of contents. The Leiden manuscript is analysed in Kennedy 1956, pp. 156-157. The Cairo manuscript is described in King 1986, p. 45 (no. B70). Sezgin, GAS, vol. 6, pp. 247-248 mentions six more manuscripts that contain fragments of the Jāmiᶜ Zīj, but these do not contain Kūshyār's tables.

5 The table of contents of the explanatory text can be found in Fatih 3418, folios 1ᵛ-2ᵛ; Berlin Ahlwardt 5751, pp. 2-4 (also given in Ahlwardt 1893, pp. 204-205) and Leiden Or. 8 (1054), folios 1ᵛ-2ᵛ. The list of tables can be found in Fatih 3418, folio 37ᵛ; Berlin Ahlwardt 5751, p. 35 (also given in Ahlwardt 1893, p. 205) and Leiden Or. 8 (1054), folio 21ʳ.

plausible that he compiled his zīj(es) shortly after this date.[6] This is confirmed by a reference in Sezgin, GAS, which indicates that from one of the manuscripts of the Jāmiᶜ Zīj it can be concluded that Kūshyār finished his zīj in 964.[7]

At the end of the Berlin and Leiden manuscripts of the Jāmiᶜ Zīj we find a large number of tables that apparently were not part of Kūshyār's original work. In many cases these tables display functions that can also be found in the main set of tables. In the Berlin manuscript, a number of planetary equation tables are attributed to Ibn al-Aᶜlam (c. 960), some other tables to Abū Maᶜshar (Albumasar, c. 850). In the Leiden manuscript, a set of planetary equation tables is taken from the Fākhir Zīj by al-Nasawī (c. 1030), some other tables mention al-Bīrūnī (c. 1000) as their author. However, most of the appended tables in both manuscripts are not attributed.

One of the tables in the manuscript Berlin Ahlwardt 5751 which is not part of the main set of tables, occurs on pages 178-179. The first half of this table is entitled "Table of the Solar Equation", the second half "Table of the Equation of the Mean Solar Position". The argument is the mean solar position and tabular values are displayed in zodiacal signs, degrees, minutes and seconds for every degree of the ecliptic. No further information about the tabulated function or the author of the table is found. In this article we will unravel the mathematical structure of this table and we will determine the values of the underlying parameters. All through this article the table will be referred to as "the true solar longitude table in the Jāmiᶜ Zīj".

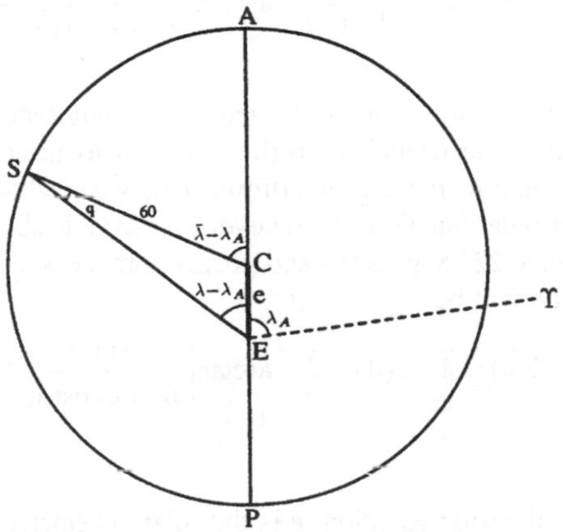


Figure 1: The Ptolemaic solar model

6 See Istanbul Fatih 3418, folio 45ᵛ.
7 See Sezgin, GAS, vol. 5, pp. 343-344.

Islamic astronomers used the solar model described in Ptolemy's Almagest (see Figure 1).[8] According to this model, the sun $S$ moves at uniform speed along a circle with a radius of 60 units. The centre $C$ of this circle is removed from the earth $E$ by a distance $e$, the solar eccentricity. The sun reaches its greatest distance from the earth at the apogee $A$ and its least distance at the perigee $P$. The true solar longitude $\lambda$ is the position of the sun as seen from the earth and is measured by the angle ♈$ES$, where ♈ is the vernal equinoctial point. The angle ♈$EA$ measures the longitude of the apogee and is denoted by $\lambda_A$. The true solar longitude differs from the mean solar longitude $\bar{\lambda}$, a linear function of time, by a variable quantity $q$, called the solar equation. Usually $\bar{\lambda}$ is defined in such a way that the angle $ACS$ is always equal to $\bar{\lambda} - \lambda_A$. Note that we then have $\lambda = \bar{\lambda} = \lambda_A$ when the sun is at its apogee and $\lambda = \bar{\lambda} = \lambda_A + 180^\circ$ when the sun is at its perigee. Furthermore $\lambda < \bar{\lambda}$ when $\lambda_A < \lambda < \lambda_A + 180^\circ$ and $\lambda < \bar{\lambda}$ when $\lambda_A - 180^\circ < \lambda < \lambda_A$. The solar equation can be calculated as a function of the true solar longitude by applying the sine rule to the triangle $CES$:

$$q(\lambda) = \arcsin\left(\frac{e}{60}\sin(\lambda - \bar{\lambda}_A)\right) \tag{1}$$

The solar equation can be calculated as a function of the mean solar longitude by extending the triangle $CES$ to a right-angled triangle $DES$ in which $EDS = 90^\circ$. Then

$$q(\bar{\lambda}) = \arctan\left(\frac{DE}{DS}\right) = \arctan\left(\frac{e\sin(\bar{\lambda} - \lambda_A)}{60 + e\cos(\bar{\lambda} - \lambda_A)}\right). \tag{2}$$

An inspection of the values in the true solar longitude table in the Jāmi$^c$ Zīj reveals that what has been tabulated is the true solar longitude $\lambda$ as a function of the mean solar longitude $\bar{\lambda}$. In fact, the difference between tabular value and argument is roughly a sinusoidal function which never exceeds 2 in absolute value, is negative from 23° Gemini to 22° Sagittarius and positive otherwise. Thus we expect that the tabulated function will be

$$\lambda(\bar{\lambda}) = \bar{\lambda} - q(\bar{\lambda}) = \bar{\lambda} - \arctan\left(\frac{e\sin(\bar{\lambda} - \lambda_A)}{60 + e\cos(\bar{\lambda} - \lambda_A)}\right), \tag{3}$$

where $q$ denotes the solar equation, $e$ is the solar eccentricity and $\lambda_A$ the longitude of the solar apogee.

---

8 For extensive descriptions of the Ptolemaic solar model, see Pedersen 1974, pp. 122-158 or Toomer 1984, pp. 131-172. For exact and approximative methods that were used by mediaeval Islamic astronomers to compute the solar equation, see Kennedy 1977 and Kennedy & Muruwwa 1958.

| $\bar{\lambda}$ | $T(\bar{\lambda})$ | $D^{(1)}(\bar{\lambda})$ | $\bar{\lambda}$ | $T(\bar{\lambda})$ | $D^{(1)}(\bar{\lambda})$ | $\bar{\lambda}$ | $T(\bar{\lambda})$ | $D^{(1)}(\bar{\lambda})$ |
|---|---|---|---|---|---|---|---|---|
| 300 | 301;11,24 | 1; 1,37 | 320 | 321;40, 3 | 1; 1,10 | 340 | 341;56,49 | 1; 0, 4 |
| 301 | 302;13, 1 | 1; 1,37 | 321 | 322;41,13 | 1; 1, 9 | 341 | 342;56,53 | 1; 0,44 |
| 302 | 303;14,38 | 1; 1,37 | 322 | 323;42,22 | 1; 1, 9 | 342 | 343;57,37 | 1; 0,24 |
| 303 | 304;16,15 | 1;1,36 | 323 | 324;43,31 | 1; 1, 9 | 343 | 344;58, 1 | 1; 0,34 |
| 304 | 305;17,51 | 1; 1,36 | 324 | 325;44,40 | 1; 1, 9 | 344 | 345;58,35 | 1; 0,13 |
| 305 | 306;19,27 | 1; 1,32 | 325 | 326;45,49 | 1; 0,51 | 345 | 346;58,48 | 0;59,30 |
| 306 | 307;20,59 | 1; 1,32 | 326 | 327;46,40 | 1; 0,51 | 346 | 347;58,18 | 1; 0,50 |
| 307 | 308;22,31 | 1; 1,32 | 327 | 328;47,31 | 1; 0,50 | 347 | 348;59, 8 | 1; 0,10 |
| 308 | 309;24, 3 | 1; 1,32 | 328 | 329;48,21 | 1; 0,50 | 348 | 349;59,18 | 1; 0,20 |
| 309 | 310;25,35 | 1; 1,32 | 329 | 330;49,11 | 1; 0,50 | 349 | 350;59,38 | 1; 0,10 |
| 310 | 311;27, 7 | 1; 1,18 | 330 | 331;50, 1 | 1; 0,47 | 350 | 351;59,48 | 1; 0, 0 |
| 311 | 312;28,25 | 1; 1,18 | 331 | 332;50,48 | 1; 0,47 | 351 | 352;59,48 | 1; 0, 0 |
| 312 | 313;29,43 | 1; 1,18 | 332 | 333;51,22 | 1; 0,47 | 352 | 353;59,48 | 1; 0, 0 |
| 313 | 314;31,19 | 1; 1,18 | 333 | 334;52,22 | 1; 0,47 | 353 | 354;59,48 | 1; 0, 1 |
| 314 | 315;32,19 | 1; 1,18 | 334 | 335;53, 9 | 1; 0,47 | 354 | 355;59,49 | 1; 0, 0 |
| 315 | 316;33,37 | 1; 1,18 | 335 | 336;53,56 | 1; 0,35 | 355 | 356;59,49 | 0;59,49 |
| 316 | 317;34,55 | 1; 1,17 | 336 | 337;54,31 | 1; 0,35 | 356 | 357;59,38 | 0;59,49 |
| 317 | 318;36,12 | 1; 1,17 | 337 | 338;55, 6 | 1; 0,35 | 357 | 358;59,27 | 0;59,49 |
| 318 | 319;37,29 | 1; 1,17 | 338 | 339;55,41 | 1; 0,35 | 358 | 359;59,16 | 0;59,49 |
| 319 | 320;38,46 | 1; 1,17 | 339 | 340;56,16 | 1; 0,33 | 359 | 360;59, 5 | 0;59,49 |
| | | | | | | 360 | 361;58,54 | |

Table 1: Tabular differences of Kūshyār's true solar longitude table

| $k$ | $\hat{a}_k$ | $\hat{b}_k$ |
|---|---|---|
| 0 | -0.0000501543 | |
| 1 | -1.9684899033 | 0.2530317106 |
| 2 | -0.0002099123 | -0.0004847958 |
| 3 | -0.0131708496 | 0.0049899528 |
| 4 | -0.0001228737 | -0.0003738902 |
| 5 | 0.0004548023 | -0.0002402755 |
| 6 | 0.0001358618 | 0.0002674402 |
| 7 | 0.0006174986 | -0.0000334333 |
| 8 | 0.0002106039 | 0.0005374500 |
| 9 | -0.0001912837 | 0.0002584406 |
| 10 | -0.0000278764 | 0.0003071745 |

Table 2: Fourier coefficients of the reconstructed solar equation

Table 1 displays some of the tabular values $T(\overline{\lambda})$ and their first order differences $D^{(1)}(\overline{\lambda}) \overset{\text{def}}{=} T(\overline{\lambda}+1) - T(\overline{\lambda})$.[9] It can be seen at once that linear interpolation within intervals of 5 degrees of the argument was applied. Moreover, it seems probable that what I will call "distributed linear interpolation" was used: the tabular differences are distributed over the intervals of 5 degrees in such a way that they are increasing or decreasing over as long as possible stretches of the argument.[10] After the obvious scribal errors indicated below have been corrected, 57 out of the 72 intervals of 5 degrees are in agreement with the assumption that distributed linear interpolation was used. In addition to the intervals on which the tabular differences are constant, only one interval is in agreement with the possibility of ordinary interpolation. By means of the irregularities in the first order differences, a number of obvious scribal errors in the tabular values can be corrected. For example, the irregularities in the differences for arguments 340 to 346 can be removed by correcting $T(341)$ to 342;57,13, $T(344)$ to 345;58,25 and $T(346)$ to 347;58,58. The correction $T(339)$=340;56,15, which restores the distributed linear interpolation pattern for arguments 335 to 340, is somewhat less plausible. All corrections made in this way are listed in the Apparatus at the end of this article. For the following analysis of the true solar longitude table we will only make use of the independently calculated tabular values.

To the corrected tabular values for multiples of 5 degrees I applied a least squares estimation as explained in the Appendix, based on the assumption that the tabulated function is given by equation (3). The results were as follows:

| *parameter* | *95 % confidence interval* |
|---|---|
| solar eccentricity | ⟨ 2; 4, 4 , 2; 5 ,11⟩ |
| solar apogee | ⟨82;25, 6 , 82;55,57⟩ |

Even though the 95 % confidence intervals contain historically plausible values of the underlying parameters, we must conclude both from the minimum obtainable standard deviation of the tabular errors (1′38″) and from the sinusoidal error pattern in recomputations for parameter values within the confidence intervals, that the table was not computed according to equation (3).[11] In order to obtain more information about the tabulated function we will make use of Fourier coefficients.

Let $T_q(\overline{\lambda}) \overset{\text{def}}{=} \overline{\lambda} - T(\overline{\lambda})$ denote the solar equation table that can be reconstructed from the true solar longitude values in the Jāmiᶜ Zīj by subtracting them from the

---

9 Practically all astronomical tables in ancient and mediaeval sources display values in sexagesimal notation. In transcribing sexagesimal numbers we will follow the convention that sexagesimal digits are separated by a comma and that the sexagesimal point is indicated by a semicolon. Thus the sexagesimal number 1;59,56 denotes $1 \cdot 60^0 + 59 \cdot 60^{-1} + 56 \cdot 60^{-2}$.

10 In the case of ordinary linear interpolation the differences are distributed evenly between every two independently calculated values. Thus the tabular differences for arguments 300 to 305 would have been 1;1,37, 1;1,36, 1;1,37, 1;1,36 and 1;1,37 respectively (cf. Table 1).

11 For a table with accurate values to seconds, the minimum obtainable standard deviation of the tabular errors is approximately 17‴. This follows from the fact that the distribution of the tabular errors can be shown to be approximately uniform; see van Dalen 1993, Section 1.2.4.

mean solar longitude. $T_q(\overline{\lambda})$, $\overline{\lambda} = 0,5,10,\ldots,355$ are then tabular values for a function $f$ which satisfies $f(x) = g(x - \lambda_A)$ for every $x$, where $g$ is an odd function with period 360°.[12] Thus we can use the approximated Fourier coefficients described in the Appendix.

Table 2 displays the approximated Fourier coefficients

$$\hat{a}_k \stackrel{\text{def}}{=} \frac{2}{n}\sum_{i=1}^{n} T(5i)\cos 5ik$$

for $k = 0,1,2,\ldots,10$ and

$$\hat{b}_k \stackrel{\text{def}}{=} \frac{2}{n}\sum_{i=1}^{n} T(5i)\sin 5ik$$

for $k=1,2,3,\ldots 10$.[13] We can note the following:

- The Fourier coefficients converge rapidly, as can be seen from $\hat{a}_1$, $\hat{a}_3$, $\hat{a}_5$ and $\hat{b}_1$, $\hat{b}_3$, $\hat{b}_5$.
- The coefficients $\hat{a}_2$ and $\hat{a}_4$ are significantly smaller than $\hat{a}_1$ and $\hat{a}_3$. Similarly, $\hat{b}_2$ and $\hat{b}_4$ are significantly smaller than $\hat{b}_1$ and $\hat{b}_3$. In the Appendix it is shown that if the odd, periodic function $g$ as introduced above satisfies the symmetry relation $g(180 - x) = g(x)$ for every $x$, then all Fourier coefficients $\hat{a}_k$ and $\hat{b}_k$ for even $k$ are zero. We conclude that the function $f$ tabulated in Kūshyār's true solar longitude table is such that $g$ contains this symmetry. If the symmetry were present in all tabular values, the approximated Fourier coefficients $\hat{a}_k$ and $\hat{b}_k$ for even $k$ would actually be zero. The fact that they are small but non-zero is a result of scribal and computational errors that will be discovered (and partially corrected) below.
- All approximated Fourier coefficients contain random errors that derive from the rounding errors (and partially from scribal and computational errors) in the tabular values. In the approximated coefficients $\hat{a}_1$, $\hat{a}_3$, $\hat{b}_1$ and $\hat{b}_3$ these errors cannot be recognized, since they are smaller than the Fourier coefficients themselves. In all remaining approximated coefficients the errors overwhelme the actual Fourier coefficients.

12 The solar equation always has this property, regardless of which of the three common methods was used for its computation (cf. Kennedy 1977). Furthermore it can be checked directly that the values $T_q(\overline{\lambda})$ have this property at least approximately.

13 The approximated coefficients for $k>10$ are not displayed, since they show the same behaviour as the coefficients for $k=4$ to 10.

| $\overline{\lambda}$ | $T_q(\overline{\lambda})$ | $T_q(\overline{\lambda})$-$f_3(\overline{\lambda})$ | $\overline{\lambda}$ | $T_q(\overline{\lambda})$ | $T_q(\overline{\lambda})$-$f_3(\overline{\lambda})$ |
|---|---|---|---|---|---|
| 0 | -1;58,54 | +1´´ | 180 | 1;58,54 | +1´´ |
| 5 | −1;56,59 | +3 | 185 | 1;56,59 | −1 |
| 10 | -1;54,10 | +4 | 190 | 1;54,10 | -1 |
| 15 | -1;50,27 | +5 | 195 | 1;50,27 | -2 |
| * 20 | -1;46, 0 | -2 | * 200 | 1;46, 0 | +6 |
| 25 | -1;40,27 | +7 | 205 | 1;40,27 | -4 |
| * 30 | -1;34,59 | -33 | 210 | 1;34,18 | -4 |
| * 35 | -1;27,44 | -9 | * 215 | 1;27,44 | +13 |
| 40 | -1;19,58 | +8 | 220 | 1;19,58 | -4 |
| 45 | -1;11,56 | +6 | 225 | 1;11,54 | -5 |
| 50 | -1; 3,24 | +5 | 230 | 1; 3,24 | -2 |
| 55 | -0;54,24 | +6 | 235 | 0;54,25 | -2 |
| 60 | -0;45, 6 | +2 | 240 | 0;45, 9 | +3 |
| 65 | -0;35,27 | +2 | 245 | 0;35,27 | |
| 70 | -0;25,37 | -1 | 250 | 0;25,37 | +2 |
| 75 | -0;15,33 | +1 | 255 | 0;15,34 | +1 |
| 80 | -0; 5,25 | | 260 | 0; 5,26 | +1 |
| 85 | 0; 4,44 | -1 | 265 | -0; 4,44 | |
| 90 | 0;14,52 | -2 | 270 | -0;14,52 | |
| 95 | 0;24,56 | | 275 | -0;24,55 | |
| 100 | 0;34,48 | -2 | 280 | -0;34,48 | -1 |
| 105 | 0;44,27 | -2 | 285 | -0;44,29 | -3 |
| 110 | 0;53,51 | | 290 | -0;53,49 | -1 |
| 115 | 1; 2,49 | -3 | 295 | -1; 2,49 | -1 |
| 120 | 1;11,23 | -4 | 300 | -1;11,24 | -1 |
| 125 | 1;19,27 | -5 | 305 | -1;19,27 | +1 |
| * 130 | 1;27, 7 | +4 | * 310 | -1;27, 7 | -8 |
| 135 | 1;33,52 | -4 | * 315 | -1;33,37 | +16 |
| 140 | 1;40, 3 | -4 | 320 | -1;40, 3 | +1 |
| * 145 | 1;46, 1 | +27 | * 325 | -1;45,49 | -18 |
| 150 | 1;50,10 | -2 | * 330 | -1;50, 1 | +9 |
| 155 | 1;53,56 | -2 | 335 | -1;53,56 | |
| 160 | 1;56,50 | | 340 | -1;56,49 | +1 |
| 165 | 1;58,47 | | 345 | -1;58,48 | -1 |
| * 170 | 1;59,36 | -11 | 350 | -1;59,48 | -1 |
| 175 | 1;59,49 | | 355 | -1;59,49 | +1 |

**Table 3: Differences between the reconstructed solar equation and an approximation based on approximated Fourier coefficients**

We can now compare tabular values $T_q(5i)$ and $T_q(5i+180)$ to see whether the symmetry $f(x) = -f(x+180)$, which follows from $g(x) = -g(-x)$ combined with $g(180-x) = g(x)$, is in fact present in the reconstructed solar equation table. Furthermore, we can compare the tabular values with approximated functional values $f_K(x)$ obtained from a finite Fourier series based on the approximated coefficients:

$$f_K(x) \stackrel{\text{def}}{=} \tfrac{1}{2}\hat{a}_0 + \sum_{k=1}^{K}(\hat{a}_k \cos kx + \hat{b}_k \sin kx) \tag{4}$$

Since in this case all approximated Fourier coefficients starting from $k=4$ are of the same order of magnitude, we choose $K=3$.

Table 3 shows the reconstructed solar equation values $T_q(5i)$ together with the differences (in seconds) between these values and the approximated functional values $f_3(5i)$. It can be noted that the general error pattern is regular, but that there are many outliers, in particular for arguments 20, 30, 35, 130, 145, 170, 200, 215, 310, 315, 325 and 330 (indicated in Table 3 with an asterisk). The tabular values show the expected symmetry $T_q(5i) = -T_q(5i+180)$ in 20 out of 36 cases. For the pairs of arguments 30/210, 135/315, 145/325, 150/330 and 170/350, $T_q(5i)$ differs by more than 8 seconds from $-T_q(5i+180)$. For eleven other pairs we find a small deviation from the expected symmetry.

It turns out that we can find plausible corrections for four of the outliers. The following list gives the reconstructed solar equation values and the corrected values.[14]

| *reconstructed value* | *corrected value* |
|---|---|
| $T(30) = -1;34,59$ | -1;34,19 |
| $T(170) = 1;59,36$ | 1;59,47 |
| $T(315) = -1;33,37$ | -1;33,52 |
| $T(330) = -1;50,\ 1$ | -1;50,11 |

Note that in all four cases the correction (approximately) restores both the surrounding error pattern and the symmetry $T_q(5i) = -T_q(5i+180)$. Furthermore all four errors are plausible scribal errors. The other eight outliers occur in pairs of tabular values $T_q(5i)/T_q(5i+180)$ and can therefore not be corrected on the basis of the symmetry of the reconstructed solar equation. Since no plausible corrections on the basis of possible scribal mistakes can be given either, we will leave these outliers unchanged for the time being.

Since the solar equation is an almost linear function in the neighbourhood of its zeros, inverse linear interpolation between $T_q(80)$ and $T_q(85)$ or $T_q(260)$ and $T_q(265)$

14 I have also considered the possibility that the original tabular values for the true solar longitude contain more scribal errors than those that we have corrected on the basis of the interpolation pattern. For instance, the correction of $T_q(170)$ indicated in the list corresponds to a correction of the original true solar longitude value 168;0,24 to 168;0,13.

should ordinarily lead to reasonably accurate values for the solar apogee (in this case the results are $\lambda_A \approx 82;40,6$ and $\lambda_A \approx 82;40,20$ respectively). However, because of the large number of errors in the reconstructed solar equation values, it seems advisable to compute a more accurate estimate and a confidence interval for $\lambda_A$ based on all tabular values by means of the Fourier estimator explained in the Appendix. In this way we arrive at an estimate

$$\hat{\lambda}_A \stackrel{\text{def}}{=} \arctan(\hat{a}_1 / \hat{b}_1) = 82;40,1$$

for the apogee. Using

$$\sigma^2 \approx \frac{1}{n-1} \sum_{i=1}^{n} (T_q(5i) - f_3(5i))^2, \qquad (5)$$

with $f_3$ according to equation (4), it follows that the standard deviation $\sigma$ of the tabular errors is approximately equal to 4´´51´´´. Assuming that the distribution of the Fourier estimator is approximately normal we find ⟨82;38,56 , 82;41,6⟩ as an approximate 95 % confidence interval for the solar apogee.

Seeing that none of the remaining outliers occurs for a multiple of 15°, we can obtain a better estimate and a smaller approximate 95 % confidence interval for the solar apogee by applying the Fourier estimator to the set of solar equation values $T_q(15i)$, $i = 1,2,3,\ldots,24$. It turns out that we can then use $K=5$ in our approximation (4) of the functional values, since now $\hat{a}_5$ and $\hat{b}_5$ are significantly larger than the approximated Fourier coefficients with larger indices. The resulting estimate for the solar apogee is $\hat{\lambda}_A = 82;40,6$, the approximated standard deviation of the tabular errors 52´´´47$^{iv}$. An approximate 95 % confidence interval for the apogee is now found as ⟨82;39,46 , 82;40,26⟩. Since the differences $T_q(15i)$ - $f_5(15i)$ in fact do not show any outliers, we conclude that the table for the true solar longitude in the Jāmiᶜ Zīj was very probably computed on the basis of the round solar apogee value $\lambda_A$ = 82°40´. Below I will explain why this value is also historically plausible.

Because of the symmetry $g(180 - x) = g(x)$ discovered above, we know that the solar equation assumes a maximum for $\lambda_A + 90°$, and a minimum for $\lambda_A + 270°$. By means of third order interpolation between the reconstructed values $T_q(165)$, $T_q(170)$, $T_q(175)$ and $T_q(180)$, we find that the maximum is approximately equal to 1;59,55,13. Similarly, we find that the minimum is close to -1;59,55,41. We conclude that the reconstructed solar equation is probably based on a value of $q_{max}$ close to 1°59´55´´ or 1°59´56´´ (or, equivalently, on a value of the solar eccentricity $e$ in the neighbourhood of 2°5´34´´). Of these values only the maximum solar equation value $q_{max}$ = 1°59´56´´ is attested; it occurs in a solar equation table in the Ashrafī Zīj which is attributed to Yaḥyā ibn Abī Manṣūr, one of the astronomers who worked at the court in Baghdad around the year 830.[15] Kennedy found that this solar equation table, which has the mean solar anomaly as its independent variable, was computed according to the so-called "method of declinations", which is probably of Sasanian or early-Islamic origin.[16] We will investigate whether the same holds for the true solar longitude table in the Jāmiᶜ Zīj, i.e. whether the tabulated function is

$$f(x) = q_{\max} \cdot \frac{\arcsin(\sin(x - \lambda_A) \cdot \sin \varepsilon)}{\varepsilon}. \tag{6}$$

Note that this function satisfies both symmetry relations $f(x-\lambda_A) = -f(-x-\lambda_A)$ and $f(180-x-\lambda_A) = f(x-\lambda_A)$ that we have also found in the table in the Jāmiᶜ Zīj.

Disregarding the outliers for arguments 20, 35, 130, 145, 200, 215, 310 and 325, which could not plausibly be corrected, I performed a least squares estimation as explained in the Appendix. Assuming that the true solar longitude table in the Jāmiᶜ Zīj was computed according to the method of declinations, I found that the minimum obtainable standard deviation of the tabular errors is 60´´, and I obtained the following approximate 95 % confidence intervals:

| *parameter* | *95 % confidence interval* |
|---|---|
| maximum solar equation | ⟨ 1;59,55,18, 1;59,55,47⟩ |
| obliquity of the ecliptic | ⟨23;39,15 , 23;48,21 ⟩ |
| solar apogee | ⟨82;39,53 , 82;40,13 ⟩ |

Since the minimum obtainable standard deviation is much higher if we assume any other plausible method of computation,[17] we conclude that the table was very probably computed according to the method of declinations. The confidence interval

15 The Ashrafī Zīj was written in Persian by Muḥammad Sanjar al-Kamālī (Shiraz, south-western Persia, c. 1300). It gives the mean motion parameters and planetary equations from a large number of earlier zījes and is extant in Paris Bibliothèque Nationale Ms. suppl. persan 1488 (288 folios, 1303 A.D.). For more information on the Ashrafī Zīj, see Kennedy 1956, p. 124, no. 4; and Kennedy 1977, p. 183. The solar equation table attributed to Yaḥyā ibn Abī Manṣūr can be found on folio 236ʳ of the Paris manuscript.

16 See Kennedy 1977.

17 We have already seen that the minimum obtainable standard deviation of the tabular errors is 1´38´´ if we assume that the correct formula

$$q(\bar{\lambda}) = \arctan\left(\frac{e\sin(\bar{\lambda} - \lambda_A)}{60 + e\cos(\bar{\lambda} - \lambda_A)}\right)$$

for the solar equation was applied. Assuming the formula

$$q(\lambda) = \arcsin\left(\frac{e}{60}\sin(\lambda - \lambda_A)\right)$$

for the solar equation as a function of the true solar longitude or the "method of sines"

$$q(\bar{\lambda}) = q_{\max} \sin(\bar{\lambda} - \lambda_A),$$

the minimum obtainable standard deviation is 38´´.

for the solar apogee confirms the results that we have found before by means of the Fourier estimator, so indeed $\lambda_A = 82°40'$. Since the method of declinations is only attested for the Ptolemaic obliquity value,[18] we can assume that $\epsilon = 23°51'$. Fixing these two parameter values, we obtain $\langle 1;59,55,27\ ,\ 1;59,56,12 \rangle$ as an approximate 95 % confidence interval for the maximum solar equation $q_{max}$. Thus we see that the true solar longitude table in the Jāmi^c Zīj was probably computed using $q_{max} = 1°59'56$.

**Conclusion:** The table for the true solar longitude which is found on pages 178-179 of the manuscript Berlin Ahlwardt 5751 of Kūshyār ibn Labbān's Jāmi^c Zīj was computed according to the so-called "method of declinations" (formula 4). The underlying parameter values are 1°59′56′′ for the maximum solar equation, 23°51′ for the obliquity of the ecliptic and 82°40′ for the solar apogee.

I will now argue that all three underlying parameter values are historically plausible and that the true solar longitude table in the Jāmi^c Zīj probably derives from Yaḥyā ibn Abī Manṣūr. We have already seen that 1°59′56′′ is the maximum solar equation value underlying Yaḥyā′s solar equation table in the Ashrafī Zīj. Furthermore, we have seen that the method of declinations is only attested with the value 23°51′ of the obliquity of the ecliptic. I conjecture that the solar apogee value 82°40′ is a rounded version of the value 82°39′, which, according to Ibn Yūnus, was observed at Baghdad in the year 214 Hijra by a group of astronomers headed by Yaḥyā ibn Abī Manṣūr.[19] The solar equation tables in Yahyā's Mumtaḥan Zīj and in the contemporary zīj by Ḥabash al-Ḥāsib extant in Istanbul Yeni Cami 784/2, both indicate that the solar apogee is in 82°39′.[20] The two tables are very probably related, since the first 90 values are practically identical. Ḥabash's zīj contains another table based on the same solar equation values, which displays $\lambda_A$ plus the solar equation. Here the apogee is taken equal to 82°40′.[21]

We have seen that the true solar longitude table in the Jāmi^c Zīj was computed by means of what I call "distributed linear interpolation". The extant recension of the Mumtaḥan Zīj contains a table for the normed right ascension, which is based on obliquity 23°51′ and involves the same type of interpolation.[22] Although this table may simply have been copied from Ptolemy's Handy Tables,[23] it seems

---

18 See Kennedy & Muruwwa 1958, p. 118; Kennedy 1977; and Suter 1914, pp. 132-137.

19 See Caussin de Perceval 1804, p. 56 (p. 40 in the separatum).

20 For the table in the Mumtaḥan Zīj, see Escorial Ms. árabe 927, folio 15^r or Yaḥyā ibn Abī Manṣūr, p. 28. For the table in Ḥabash's zīj, see Istanbul Yeni Cami 784/2, folios 90^r-91^r and Debarnot 1987, pp. 41-42. Both tables were analysed in Salam & Kennedy 1967, pp. 494-495. The solar equation table in the Mumtaḥan Zīj is completely different from the table in the Ashrafī Zīj attributed to Yaḥyā.

21 See Istanbul Yeni Cami 784/2, folios 200^v-203^r and Debarnot 1987, p. 58.

22 Escorial Ms. árabe 927, folios 48^v-49^r or Yaḥyā ibn Abī Manṣūr, pp. 93-94. See Neugebauer 1975, vol. 1, pp. 31-32 and 42 for more information about the normed right ascension.

23 The normed right ascension table in the Handy Tables can for instance be found in the manuscript Leiden BPG 78, folios 75^r-76^v. The table is transcribed in Stahlman 1959, pp. 206-209. The normed right ascension in the Mumtaḥan Zīj is practically identical to the table in

probable that it was an original part of Yaḥyā ibn Abī Manṣūr's zīj, and hence that Yaḥyā was familiar with distributed linear interpolation.

Finally, it can be noted that among the appended tables in the manuscript Berlin Ahlwardt 5751 of Kūshyār ibn Labbān's Jāmiᶜ Zīj, we find two tables displaying mean planetary positions at two different epochs according to four astronomers, one of them being Yaḥyā ibn Abī Manṣūr.[24] Thus the compiler of the manuscript apparently had access to Yaḥyā's zīj.

We conclude that there is sufficient reason to believe that the true solar longitude table analysed here, like the solar equation table on folio 236ʳ of the Ashrafī Zīj, originates from Yaḥyā ibn Abī Manṣūr. It seems possible that Yaḥyā's solar equation table as found in the Ashrafī Zīj, was originally contained in the Mumtaḥan Zīj, but was later considered unsatisfactory because of its symmetry (and possibly because of its uncommon value of the maximum equation). Thus we can imagine how in a later recension, like the one that we find in the manuscript Escorial árabe 927, it was replaced, possibly by Ḥabash's table for the solar equation.

Table 4 displays my final recomputation of the solar equation reconstructed from the true solar longitude table in the Jāmiᶜ Zīj. The second and fifth columns contain the reconstructed solar equation values, the third and sixth columns the differences (in seconds) between these values and a recomputation according to formula (6) using the parameter values found above. Apart from the eight outliers (which are again indicated by an asterisk), the number of differences is 40 out of 64 tabular values, the standard deviation of the differences is $1''5'''$.

It seems probable that the true solar longitude table in the Jāmiᶜ Zīj was computed by means of interpolation in a solar equation table like the one in the Ashrafī Zīj. In fact, if linear interpolation were used, the remaining eight outliers in our table could be explained from only two erroneous solar equation values. To see this, we denote the values for the method of declination that were used for the linear interpolation by $q_\delta(\bar{a})$, where $\bar{a} = 1,2,3,\ldots,90$ is the mean solar anomaly. Remembering that $q_\delta(-\bar{a}) = -q_\delta(\bar{a})$ and $q_\delta(180 - \bar{a}) = q_\delta(\bar{a})$, it follows that $T_q(35)$ and hence $-T_q(215)$ were calculated as $\frac{1}{3}q_\delta(47) - \frac{2}{3}q_\delta(48)$, and $T_q(130)$ and $-T_q(310)$ as $\frac{2}{3}q_\delta(47) + \frac{1}{3}q_\delta(48)$. The solar equation values $q_\delta(47) = 1;26,30$ (equal to the value given in the Ashrafī Zīj) and $q_\delta(48) = 1;28,21$ (computational error for 1;27,56?) thus precisely reproduce the four outliers for arguments 35, 130, 215 and 310. In the same way three of the remaining four outliers can be explained if we assume an erroneous value $q_\delta(62) = 1;45,38$ (possible scribal error for the Ashrafī value 1;45,13).

I recomputed the true solar longitude table in the Jāmiᶜ Zīj by using linear interpolation in Yaḥyā ibn Abī Manṣūr' solar equation table in the Ashrafī Zīj. Disregarding the eight outliers, I found 28 differences in 64 values (as compared to 40 differences for the precise recomputation); the standard deviation of the

---

the Handy Tables.

24 See Berlin Ahlwardt 5751, pp. 160-161.

differences was 1´´3´´´. This result is not good enough to conclude that in fact linear interpolation in the Ashrafī table was applied.

| λ | $T_q(\lambda)$ | error | λ | $T_q(\lambda)$ | error |
|---|---|---|---|---|---|
| 0 | −1;58,54 | | 180 | 1;58,54 | |
| 5 | −1;56,59 | +1´´ | 185 | 1;56,59 | −1´´ |
| 10 | −1;54,10 | | 190 | 1;54,10 | |
| 15 | −1;50,27 | | 195 | 1;50,27 | |
| * 20 | −1;46, 0 | −9 | * 200 | 1;46, 0 | +9 |
| 25 | −1;40,27 | +1 | 205 | 1;40,27 | −1 |
| 30 | −1;34,19 | −1 | 210 | 1;34,18 | |
| * 35 | −1;27,44 | −17 | * 215 | 1;27,44 | +17 |
| 40 | −1;19,58 | +1 | 220 | 1;19,58 | −1 |
| 45 | −1;11,56 | | 225 | 1;11,54 | −2 |
| 50 | −1; 3,24 | −1 | 230 | 1; 3,24 | +1 |
| 55 | −0;54,24 | +1 | 235 | 0;54,25 | |
| 60 | −0;45, 6 | −1 | 240 | 0;45, 9 | +4 |
| 65 | −0;35,27 | | 245 | 0;35,27 | |
| 70 | −0;25,37 | −2 | 250 | 0;25,37 | +2 |
| 75 | −0;15,33 | | 255 | 0;15,34 | +1 |
| 80 | −0; 5,25 | | 260 | 0; 5,26 | +1 |
| 85 | 0; 4,44 | −1 | 265 | −0; 4,44 | +1 |
| 90 | 0;14,52 | −1 | 270 | −0;14,52 | +1 |
| 95 | 0;24,56 | +1 | 275 | −0;24,55 | |
| 100 | 0;34,48 | | 280 | −0;34,48 | |
| 105 | 0;44,27 | | 285 | −0;44,29 | −2 |
| 110 | 0;53,51 | +3 | 290 | −0;53,49 | −1 |
| 115 | 1; 2,49 | +1 | 295 | −1; 2,49 | −1 |
| 120 | 1;11,23 | | 300 | −1;11,24 | −1 |
| 125 | 1;19,27 | −1 | 305 | −1;19,27 | +1 |
| * 130 | 1;27, 7 | +8 | * 310 | −1;27, 7 | −8 |
| 135 | 1;33,52 | | 315 | −1;33,52 | |
| 140 | 1;40, 3 | −1 | 320 | −1;40, 3 | +1 |
| * 145 | 1;46, 1 | +30 | * 325 | −1;45,49 | −18 |
| 150 | 1;50,10 | | 330 | −1;50,11 | −1 |
| 155 | 1;53,56 | −1 | 335 | −1;53,56 | +1 |
| 160 | 1;56,50 | | 340 | −1;56,49 | +1 |
| 165 | 1;58,47 | −1 | 345 | −1;58,48 | |
| 170 | 1;59,47 | −1 | 350 | −1;59,48 | |
| 175 | 1;59,49 | −1 | 355 | −1;59,49 | +1 |

Table 4: Final recomputation of the reconstructed solar equation

## Apparatus

Scribal errors in Kūshyār ibn Labbān's table for the true solar longitude corrected on the basis of the interpolation pattern (the corrected digits are given between brackets; possible errors of 1´´ for arguments 204, 206 and 212 were not corrected):

| $\overline{\lambda}$ | $T(\overline{\lambda})$ | | $\overline{\lambda}$ | $T(\overline{\lambda})$ | |
|---|---|---|---|---|---|
| 21 | 22;44,33 | (53´´) | 136 | 134;24,13 | (53´´) |
| 33 | 34;30,18 | (38´´) | 146 | 144;14,59 | (13´) |
| 47 | 48; 8,22 | (32´´) | 147 | 145;12,59 | (19´´) |
| 57 | 57;50,40 | (42´´) | 181 | 179; 1,21 | (29´´) |
| 78 | 78; 9,39 | (29´´) | 279 | 279;32,30 | (50´´) |
| 104 | 103;17,18 | (28´´) | 298 | 299; 7,18 | (58´´) |
| 116 | 116;15,28 | (55´) | 341 | 342;56,53 | (57´13´´) |
| 117 | 115;13,45 | (53´) | 344 | 345;58,35 | (25´´) |
| 130 | 128;32,13 | (53´´) | 346 | 347;58,18 | (58´´) |

## Appendix

This appendix briefly describes the statistical estimators that I use in this article to determine the mathematical structure and unknown parameter values of Kūshyār * ibn Labbān's table for the true solar longitude. More detailed information concerning approximated Fourier coefficients in general and the Fourier estimator in particular can be found in Section 2.3 of my doctoral thesis (van Dalen 1993). A more extensive discussion of least squares estimation can be found in Section 2.4 of my thesis.

The following notations are used throughout this article:
A table for a mathematical function is denoted by $T$, the tabular values of that table by $T(x)$, where $x$ indicates the argument. The tabulated function is always denoted by $f$. The tabular errors $e(x)$ are defined by $e(x) = T(x) - f(x)$ for every argument $x$. Since in this way the tabular error includes the error made by rounding a calculated functional value to the accuracy of the table under consideration, it follows that practically every tabular value contains a non-zero tabular error. An extensive discussion of the distribution of tabular errors can be found in van Dalen 1993, Section 1.2.4. For every parameter that I estimate a so-called 95 % confidence interval is calculated. Such intervals are expected to contain the underlying parameter value in 19 out of 20 cases.

**Approximated Fourier coefficients**
Let $f$ be a $2\pi$-periodic function and assume that the Fourier series of $f$ converges, i.e. that, for every $x \in [0,2\pi]$,

$$f(x) = \tfrac{1}{2}a_0 + \sum_{k=1}^{\infty}(a_k \cos kx + b_k \sin kx),$$

where

$$a_k = \frac{1}{\pi}\int_0^{2\pi} f(x)\cos kx$$

and

$$b_k = \frac{1}{\pi}\int_0^{2\pi} f(x)\sin kx\, dx$$

for every $k$.

Now suppose that we have a table $T$ of the function $f$ with tabular values $T(i\alpha)$ for $i = 1,2,3,\ldots,n$, where $n$ is a multiple of 4 and $\alpha = 2\pi/n$. Since for every $2\pi$-periodic, twice continuously differentiable function $h$ we have

$$\int_0^{2\pi} h(x)dx = \frac{2\pi}{n}\sum_{i=1}^{n} h(i\alpha) + \frac{\pi^3}{3n^3}h''(\xi),$$

where $\xi \in [0,2\pi]$, it follows that the Fourier coefficients $a_k$ and $b_k$ of the function $f$ can be approximated by

$$\tilde{a}_k \stackrel{\text{def}}{=} \frac{2}{n}\sum_{i=1}^{n} f(i\alpha)\cos ik\alpha$$

and

$$\tilde{b}_k \stackrel{\text{def}}{=} \frac{2}{n}\sum_{i=1}^{n} f(i\alpha)\sin ik\alpha$$

respectively. I found that in practice the errors in these approximations can be neglected if the number of tabular values $n$ is at least equal to twelve.

Since we do not know the functional values $f(i\alpha)$ themselves, we have to approximate them by the given tabular values $T(i\alpha)$. Thus we will estimate the Fourier coefficients $a_k$ and $b_k$ by

$$\hat{a}_k \stackrel{\text{def}}{=} \frac{2}{n}\sum_{i=1}^{n} T(i\alpha)\cos ik\alpha$$

and

$$\hat{b}_k \stackrel{\text{def}}{=} \frac{2}{n}\sum_{i=1}^{n} T(i\alpha)\sin ik\alpha$$

respectively. If the tabular errors $T(i\alpha)$ - $f(i\alpha)$ are denoted by $e(i\alpha)$, then we have

$$\hat{a}_k - \tilde{a}_k = \frac{2}{n}\sum_{i=1}^{n} e(i\alpha)\cos ik\alpha$$

and

$$\hat{b}_k - \tilde{b}_k = \frac{2}{n}\sum_{i=1}^{n} e(i\alpha)\sin ik\alpha.$$

Assuming that the tabular errors are mutually independent and have a common mean 0 and variance $\sigma^2$, we find that $\hat{a}_k$ and $\hat{b}_k$ have a negligible bias (namely the error made in the approximation of the integral in the definition of the Fourier coefficients by a finite sum) and a variance Var $\hat{a}_k$ = Var $\hat{b}_k = 2\sigma^2/n$. Furthermore $\hat{a}_k$ and $\hat{b}_k$ have a zero covariance. Using the Central Limit Theorem we find that both $\hat{a}_k$ and $\hat{b}_k$ have a distribution which is approximately normal.

In Islamic astronomical handbooks we find tables of $2\pi$-periodic functions $f$ that, for some odd function $g$ and a constant $\lambda_A$, satisfy the relation $f(x) = g(x-\lambda_A)$ for every $x$.[25] If the Fourier series of $f$ converges, the Fourier series of $g$ converges as well, and we can derive the Fourier coefficients of $g$ from those of $f$. For every $x \in [0,2\pi]$ we have

$$\begin{aligned} g(x) &= f(x+\lambda_A) \\ &= \tfrac{1}{2}a_0 \\ &\quad + \sum_{k=1}^{\infty} a_k(\cos kx \cdot \cos k\lambda_A - \sin kx \cdot \sin k\lambda_A) \\ &\quad + \sum_{k=1}^{\infty} b_k(\sin kx \cdot \cos k\lambda_A + \cos kx \cdot \sin k\lambda_A) \\ &= \tfrac{1}{2}a_0 \\ &\quad + \sum_{k=1}^{\infty} (a_k \cos k\lambda_A + b_k \sin k\lambda_A)\cos kx \\ &\quad + \sum_{k=1}^{\infty} (b_k \cos k\lambda_A - a_k \sin k\lambda_A)\sin kx. \end{aligned}$$

Since g is an odd function, it follows that for every $k$ we have $a_k\cos k\lambda_A + b_k\sin k\lambda_A = 0$. In cases where $g$ also satisfies the symmetry relation $g(180-x) = g(x)$ for every $x$, it can be shown that $f(x+180) = -f(x)$ for every $x$ and that $a_k = b_k = 0$ when $k$ is even. If the tabular values satisfy the symmetry relation $T(x+180) = -T(x)$ for every $x$, then $\hat{a}_k = \hat{b}_k = 0$ for even $k$.

---

25 The notation $\lambda_A$ is chosen since in practice the constant will often be the solar apogee.

**Fourier estimator**

As above, let $f$ be a $2\pi$-periodic function with convergent Fourier series. Assume that $f(x) = g(x-\lambda_A)$ for every $x$, where $g$ is an odd function and $\lambda_A$ an unknown constant. Let $T$ be a table for $f$ with tabular values $T(i\alpha)$, $i = 1,2,3,\ldots,n$, where $n$ is a multiple of 4 and $\alpha = 2\pi/n$. Again let $e(i\alpha)$ denote the tabular errors $T(i\alpha) - f(i\alpha)$ and $\sigma^2$ the common variance of these errors. Let $a_k$ and $b_k$ denote the Fourier coefficients of the function $f$ and let $\hat{a}_k$ and $\hat{b}_k$ be the estimators for these coefficients introduced above.

We have seen that $a_k$ and $b_k$ satisfy the relation $a_k \cos k\lambda_A + b_k \sin k\lambda_A = 0$ for every $k$. Provided that $b_k$ is not equal to zero, this implies that $\tan k\lambda_A = -a_k / b_k$ for every $k$ and that $\hat{\lambda}_A(k)$ defined by $\tan \hat{\lambda}_A(k)$ is an estimator for $\lambda_A$ for every $k$. The calculation of the accuracy of this estimator is straightforward and can be found in my doctoral thesis.[26] It turns out that the bias of $\hat{\lambda}_A$ is negligible (it is of the order of $\sigma^3$ for $\sigma \to 0$) and that

$$\operatorname{Var} \hat{\lambda}_A = \frac{180^2}{\pi^2} \frac{2\sigma^2}{nk^2(a_k^2 + b_k^2)} + O(\sigma^3).$$

Since the Fourier coefficients converge rapidly for most functions $f$ of which we find tables in Islamic astronomical handbooks, the estimator with the smallest variance is obtained for $k=1$. I call this estimator the "Fourier estimator". The distribution of the Fourier estimator is in most cases very close to normal.

**Least Squares estimation**

Let $T$ be a table with tabular values $T(x)$, $x \in X$, for the function $f_\theta$ which depends on the parameter vector $\theta$. Let the objective function $\Phi(\theta)$ be defined as the sum of the squares of the tabular errors:

$$\Phi(\theta) = \sum_{x \in x} (t(x) - f_0(x))^2.$$

A least squares estimate for the parameter vector $\theta$ is a vector $\hat{\theta}$ that minimizes $\Phi(\theta)$.

Whenever $f_\theta$ is a non-linear function, we need an iterative optimization procedure in order to calculate a least squares estimate for the parameter vector underlying a particular table. It turns out that for most types of tables in Islamic astronomical handbooks the so-called Gauss-Newton procedure converges rapidly to a least squares estimate. Once the estimate $\hat{\theta}$ has been obtained, we can approximate the standard deviation $\sigma$ of the tabular errors from $\sigma^2 \approx \Phi(\hat{\theta})/n$. I refer to this approximation as the "minimum obtainable standard deviation" of the tabular errors. If the minimum obtainable standard deviation is much larger than what we expect on the basis of the number of sexagesimal digits of the tabular values, then

26 See van Dalen 1993, pp. 49-50.

it is probable that $f_\theta$ is not the tabulated function. If the minimum obtainable standard deviation is small enough and the tabular errors can be assumed to be independent and to have zero means and equal variances, then separate confidence intervals for all underlying parameters of the table under consideration can be computed from the found least squares estimate. More information about least squares estimation can for instance be found in van Dalen 1993, Section 2.4 or in Draper & Smith 1981, Chapter 10.

## Bibliography

AHLWARDT, Wilhelm: Verzeichniss der arabischen Handschriften, Band 5, Die Handschriften-Verzeichnisse der Königlichen Bibliothek zu Berlin, vol. 17, Berlin (Asher) 1893

CAUSSIN DE PERCEVAL, G.: Le livre de la grande table Hakémite, Notices et Extraits des Manuscrits de la Bibliothèque Nationale 7 1804, pp. 16-240 (1-224 in the separatum)

DALEN, Benno van: Ancient and Mediaeval Astronomical Tables: mathematical structure and parameter values (unpublished doctoral thesis), Utrecht (University of Utrecht) 1993

DEBARNOT, Marie-Thérèse: The Zīj of Ḥabash al Ḥāsib: A Survey of MS Istanbul Yeni Cami 784/2, From Deferent to Equant: a Volume of Studies in the History of Science in the Ancient and Medieval Near East in Honor of E.S. Kennedy (eds. David A. King & George Saliba), New York (New York Academy of Sciences) 1987, pp. 35-69

DRAPER, Norman R. & SMITH, Harry: Applied Regression Analysis (2nd ed.), New York (Wiley) 1981

DSB: Dictionary of Scientific Biography, 14 vols and 2 suppl. vols, New York (Charles Scribner's Sons) 1970-1980

KENNEDY, Edward S.: A Survey of Islamic Astronomical Tables, Transactions of the American Philosophical Society 46-2 1956, pp. 123-177. Reprinted by the American Philosophical Society, Philadelphia 1989

KENNEDY, Edward S.: The Solar Equation in the Zīj of Yaḥyā ibn Abī Manṣūr, Prismata: Festschrift für Willy Hartner (eds. Yasukatsu Maeyama & Walter G. Saltzer), Wiesbaden (Franz Steiner) 1977, pp. 183-186, Reprinted in Kennedy et. al. 1983, pp. 136-139

KENNEDY, Edward S. et. al.: Studies in the Islamic Exact Sciences, Beirut (American University of Beirut) 1983

KENNEDY, Edward S. & Muruwwa, Ahmad: Bīrūnī on the Solar Equation, Journal of Near Eastern Studies 17 1958, pp. 112-121, Reprinted in Kennedy et. al. 1983, pp. 603-612

KING, David A.: Survey of the Scientific Manuscripts in the Egyptian National Library, Winona Lake IN (The American Research Center in Egypt) 1986

NEUGEBAUER, Otto E.: A History of Ancient Mathematical Astronomy, 3 vols, Berlin (Springer) 1975

PEDERSEN, Olaf: A Survey of the Almagest, Odense (University Press) 1974

SALAM, Hala & Kennedy, Edward S.: Solar and Lunar Tables in Early Islamic Astronomy, Journal of the American Oriental Society 87 1967, pp. 492-497. Reprinted in Kennedy et. al. 1983, pp. 108-113

SEZGIN, GAS: Fuat Sezgin (ed.), Geschichte des arabischen Schrifttums, 9 vols, Leiden (Brill) 1971-

STAHLMAN, William D.: The Astronomical Tables of Codex Vaticanus Graecus 1291 (doctoral thesis), Providence (Brown University) 1959. To be published by Garland, New York

SUTER, Heinrich: Die astronomischen Tafeln des Muḥammed ibn Mūsā al Khwārizmī in der Bearbeitung des Maslama ibn Aḥmed al-Madjrīṭī und der lateinischen Übersetzung des Adelard von Bath, Det Kongelige Danske Videnskabernes Selskab hist.-fil. Skrifter 3:1, Copenhagen 1914. Reprinted in Suter 1986, vol. 1, pp. 473-751

SUTER, Heinrich: Beiträge zur Geschichte der Mathematik und Astronomie im Islam, 2 vols., Frankfurt am Main (Institut für Geschichte der Arabisch-Islamischen Wissenschaften) 1986

TOOMER, Gerald J.: Ptolemy's Almagest, London (Duckworth) and New York (Springer) 1984

VAN BRUMMELEN, Glen R.: Mathematical Tables in Ptolemy's Almagest (unpublished doctoral thesis), Burnaby BC (Simon Fraser University) 1993

YAḤYĀ ibn Abī Manṣūr: The Verified Astronomical Tables for the Caliph al-Ma'mun (Al-Zīj al-Ma'mūnī al-mumtaḥan) (facsimile edition), Frankfurt am Main (Institut für Geschichte der Arabisch-Islamischen Wissenschaften) 1986

## Correction

III 185, lines 2–3 of the Appendix: 'Kūshyār ibn Labbān's table for the true solar longitude' → 'the table for the true solar longitude in the Berlin manuscript of Kūshyār ibn Labbān's Jāmiᶜ Zīj'.

# IV

## AL-KHWĀRIZMĪ'S ASTRONOMICAL TABLES REVISITED: ANALYSIS OF THE EQUATION OF TIME

*Contents*

1. *Introduction*

Al-Khwārizmī was an influential Muslim mathematician, astronomer and geographer who lived in Baghdad in the first half of the 9th century A.D. His main astronomical work was a *zīj*, an astronomical handbook with tables and explanatory text, called the *Sindhind Zīj*. This work was largely based on Indian methods, as opposed to most later Islamic astronomical handbooks which utilized the Greek planetary models laid out in Ptolemy's *Almagest*. The *Sindhind Zīj* is only extant in a Latin translation of a recension by Maslama al-Majrīṭī (Cordoba, c. 980). Through this translation and the so-called *Toledan Tables* some of the Indian methods used by al-Khwārizmī made their way to Western Europe.

The mathematical structure and underlying parameter values of practically all tables in the Latin version of al-Khwārizmī's *Sindhind Zīj* have been determined. Using the mathematical information obtained in this way the origin of most of the tables could be ascertained. One of the very few tables of which the mathematical structure has not yet been established, is the table for the equation of time. In this article I will present a full analysis of this table and will show that it is based on two Ptolemaic parameter values plus a value found by the group of astronomers who compiled the *Mumtaḥan Zīj* (Baghdad, c. 830).

The main mathematical tool used for the analysis of al-Khwārizmī's table for the equation of time is the method of least squares. The application of this method will be described step by step. In this way it is hoped to enable the reader to perform similar determinations of the unknown parameters of an astronomical table by means of the computer program TA (Table-Analysis), available from the author.

In Section 2 of this article information is presented concerning al-Khwārizmī's life and works. Section 3 gives an overview of the available primary and secondary sources related to the *Sindhind Zīj*. In Section 4 I present a detailed survey of previous results about al-Khwārizmī's astronomical tables, including the most important technical details and ample references. After the explanation of the equation of time in Section 5, al-Khwārizmī's table for the equation of time will be extensively analysed in Section 6. A summary of the results of this analysis can be found in Section 7.

## 2. *Al-Khwārizmī's life and works*

Abū Ja<sup></sup>cfar Muḥammad ibn Mūsā al-Khwārizmī lived in the first half of the 9th century A.D.[2] His name indicates that his ancestors came from Khorezm, a region south of the Aral sea. According to the historian al-Ṭabarī (Baghdad, 839-923 A.D.), al-Khwārizmī himself came from Quṭrubbul, a suburb of Baghdad.

Al-Khwārizmī was active in Baghdad as a mathematician, astronomer and geographer during the reigns of the Abbasid caliphs al-Ma'mūn (813-833), al-Mu<sup></sup>ctaṣim (833-842) and al-Wāthiq (842-847). During the reign of al-Ma'mūn he became a member of the "House of Wisdom", a scientific institution strongly supported by the caliph (cf. the *EI*² article "Bayt al-ḥikma"). Al-Khwārizmī's works on algebra and astronomy were dedicated to al-Ma'mūn and were hence probably written before 833. His treatise on Hindu numerals refers to the work on algebra and must therefore be later; his treatise on the Jewish calendar gives an example for the year 823 / 824. The dating of al-Khwārizmī's remaining known works, a treatise on geography, a chronicle, a treatise on the sundial and two treatises on the astrolabe is problematic.

Al-Khwārizmī's works were influential both in the Arab world and in medieval Europe. His work on algebra *al-Kitāb al-mukhtaṣar fī ḥisāb al-jabr wa'l-muqābala* (*The Compendium on Calculation by Completing and Balancing*) was in use as a textbook for several centuries and served as an archetype for treatises on algebra by later authors. The Latin translation of this work stood at the basis of the development of European algebra, to which it gave its name.

The Latin translation of al-Khwārizmī's work on arithmetic with Hindu numerals, of which the original Arabic text is no longer extant, initiated a number of 12th and 13th-century European works on arithmetic. The titles of many of these works contained the Latinized version "Algorismus" of al-Khwārizmī's name, from which our word "algorithm" derives.

---

[2] Most of the following information was taken from the *DSB* article "al-Khwārizmī" by Gerald J. Toomer. For more extensive references and further biographical and bibliographical information the reader is also referred to the *EI*² article "al-Khwārazmī" by Juan Vernet, and to Sezgin 1971-1984, vol. 5, pp. 228-241 and vol. 6, pp. 140-143.

Al-Khwārizmī's main astronomical work was called the *Sindhind Zīj*.[3] It was largely based on Indian methods and parameter values taken from the *Sindhind*, an Arabic translation (made around the year 770 by al-Fazārī) of the *Brāhmasphuṭasiddhānta* by the 7th-century Indian astronomer Brahmagupta. Other elements were taken from the *Shāh Zīj*, a non-extant Persian work of the 6th century, and from the *Khaṇḍakhādyaka*, another work by Brahmagupta. The *Sindhind Zīj* existed in a larger version, which included explanations of the models used, and a smaller version containing only tables and instructions for their use. Neither version is extant in the original Arabic. The smaller version became known in Spain in the 9th century and a recension of it was made by the 10th-century Muslim mathematician and astronomer Abū'l-Qāsim Maslama ibn Aḥmad al-Faraḍī al-Majrīṭī, who worked in Cordoba.[4] According to the 11th-century historian and astronomer Ṣāᶜid al-Andalusī, al-Majrīṭī converted the planetary tables in al-Khwārizmī's *zīj* from the Persian to the Arabic calendar and adapted some of the tables to the geographical longitude of Cordoba. Al-Majrīṭī's recension is only extant in a 12th-century Latin translation by Adelard of Bath, which is the main source for research on al-Khwārizmī's astronomical tables (cf. the following section).

## 3. *Sources for the study of al-Khwārizmī's astronomical tables*

For the study of al-Khwārizmī's *Sindhind Zīj* the following

---

[3] The Arabic word *zīj* derives from the middle Persian *zīg*. It indicates an astronomical handbook with tables and explanatory text.

[4] Al-Majrīṭī is known to have written a work on commercial arithmetic (*Muᶜāmalāt*) and was the first Andalusian astronomer who made astronomical observations of his own. His disciples, among whom were Ibn al-Ṣaffār, Ibn al-Samḥ, ᶜAmr ibn ᶜAbd al-Raḥmān al-Kirmānī and Ibn Barghūth, were influential mathematicians and astronomers throughout Spain. For further information, see the *DSB* article "al-Majrīṭī" and the *EI*[2] article "al-Madjrīṭī" by Juan Vernet. See also Vernet & Catalá 1965 and Samsó 1992, pp. 80-110.

primary sources are available:[5]

1) *The Latin translation by Adelard of Bath of al-Majrīṭī's recension of the smaller version of al-Khwārizmī's zīj.* This translation is available in nine manuscripts, some of which contain fragments only. The relationship between the manuscripts was discussed extensively in Mercier 1987, which lists the manuscripts in footnote 9 on page 89. The manuscripts Chartres Bibliothèque publique No. 214 (173), Madrid Biblioteca Nacional No. 10016, Oxford Bodleian Library Cod. Auct. F.I. 9 (Bernard No. 4137) and Paris Bibliothèque Mazarine No. 3642 (1258) were used by Suter for his edition and commentary published in 1914. Neugebauer (1962) translated the Latin version of al-Khwārizmī's *zīj* into English and provided a new commentary giving many new insights into the mathematical structure and the origin of the tables. He included a complete edition and translation of the manuscript Oxford Corpus Christi College Ms. 283. Recently Pedersen (1992) established that a set of astronomical rules in the Latin manuscript Oxford Merton College 259 is close to al-Khwārizmī's original *zīj*.
2) *The commentary on the larger version of al-Khwārizmī's zīj by Ibn al-Muthannā.* This 10th-century work is lost in the Arabic original. A Latin translation by Hugo Sanctallensis is available in the manuscripts Oxford Bodleian Library Arch. Selden B 34, Oxford Bodleian Library Savile 15, and Cambridge Gonville and Caius College 456. Two Hebrew translations, one of which by Ibn Ezra, can be found in the manuscripts Parma Biblioteca Palatina 2636 (De Rossi 212) and Oxford Bodleian Library Ms. Michael 400. The Latin translation was edited in Millás Vendrell 1963; the Hebrew versions were edited and translated in Goldstein 1967.
3) *The commentary on al-Khwārizmī's zīj by Ibn Masrūr.* This 10th-century commentary, entitled *Kitāb ʿilal al-zījāt* (*Book of the reasons of the zījes*) and available as Cairo Taymūr Math. 99 (see King 1986, no. B37, p. 38), has not been published. Kennedy and Ukashah consulted the manuscript for their investigation of al-Khwārizmī's tables for planetary latitude (1969), King for his research about lunar

[5] Detailed information about the manuscripts listed below can be found in the secondary sources indicated.

crescent visibility tables (1987).

4) *The Toledan Tables*. The *Toledan Tables* were written by the Andalusian astronomer al-Zarqālī (or Azarquiel) in the 11th century. The original Arabic has been lost, but various Latin versions of both the tables and the explanatory text are extant in more than 100 manuscripts scattered all over Western Europe. These manuscripts contain several tables from al-Khwārizmī's original *zīj*, some of which are not found in al-Majrīṭī's recension. The *Toledan Tables* were described in Zinner 1935 and Millás Vallicrosa 1943-1950, pp. 22-71, and were extensively analysed by Toomer (1968). The explanatory text from a group of manuscripts was published by F.S. Pedersen (1987), who is currently preparing a complete edition of the tables.

A commentary on al-Khwārizmī's *zīj* by al-Farghānī, mentioned by al-Bīrūnī and Ibn al-Muthannā, is non-extant. Some of the tables in the only extant manuscript of al-Battānī's *Ṣābi' Zīj* (Escorial árabe 908) are explicitly attributed to Maslama al-Majrīṭī and can thus be used to identify additions by al-Majrīṭī in the Latin translation of al-Khwārizmī's *zīj* (cf. Nallino 1899-1907, vol. 2, pp. 300 ff.).

Valuable information concerning the transmission of Indian and Persian astronomical knowledge to Baghdad in the 8th century can be found in *Kitāb ᶜilal al-zījāt* (*The book of the reasons behind astronomical tables*) by ᶜAlī ibn Sulaymān al-Hāshimī ("al-Hāshimī" in the bibliography). This information was explored by Pingree in his publications 1968a, 1968b and 1970.

The most important secondary sources dealing with al-Khwārizmī's *Sindhind Zīj* have been mentioned above. Many articles have appeared about particular tables in the *zīj*; these are listed in the bibliography and will be referred to in my survey of previous results about the mathematical structure and origin of the tables in al-Majrīṭī's recension in the following section.

## 4. *Survey of previous results about al-Khwārizmī's astronomical tables*

In this Section I summarize the most important previous results concerning the mathematical structure and origin of the tables in al-Majrīṭī's recension of al-Khwārizmī's *Sindhind Zīj*. For every table or group of tables references are given to the table numbers in Suter's edition published in 1914 (for tables displaying multiple functions, the respective columns are indicated by 1°, 2°, etc.); furthermore, to the relevant page numbers in Neugebauer's translation and commentary (1962) and in Goldstein's edition of Ibn al-Muthannā's commentary (1967). In these publications the reader can also find complete technical descriptions of the functions tabulated in al-Khwārizmī's *zīj*. References to other secondary sources will only be given for results that cannot be found in one of the three above-mentioned works. The tables are listed in the order in which they occur in Suter's edition. Note that tables 57b (multiplication of sexagesimal fractional digits) and 116 ("houses, judges and decans") were left out, because they were not mathematically computed. Al-Khwārizmī's tables for timekeeping, the *qibla* and the construction of sundials and astrolabes, which were not part of his *zīj*, are described in King 1983.

### CHRONOLOGICAL TABLES

(Suter 1-3a, Neugebauer pp. 82-89, Goldstein pp. 16-25)

Like most Islamic astronomical handbooks, al-Khwārizmī's original *zīj* contained a set of chronological tables similar to the set in the Latin translation of al-Majrīṭī's recension. Al-Majrīṭī made some small modifications to the tables for the *notae* (days of the week of year and month beginnings; cf. Goldstein p. 88 and al-Hāshimī, pp. 231-234). Furthermore, although he maintained the epoch and the year beginning (1 October) of the Byzantine calendar, he moved the intercalary day from the end of February to the end of December (Table 3a).

### MEAN MOTIONS

(Suter 4-20, Neugebauer pp. 90-95, Goldstein pp. 26-28 and 190-191)

Al-Khwārizmī's original mean motion tables were calculated for the Persian calendar and the Era Yazdigird. They were based on the Indian mean motion theory, which assumes that at the time of the creation all planets and their apogees and nodes had a mean position equal to 0° Aries. Al-Khwārizmī's original mean motion values were probably given to an accuracy of sexagesimal thirds (cf. Goldstein pp. 28 and 152), and were calculated for the meridian of Ujjain in Central India (in Arabic sources called Arīn).

According to Ṣāᶜid al-Andalusī, al-Majrīṭī adapted al-Khwā-rizmī's mean motion tables to the Arabic calendar. The mean motion tables preserved in the Latin translation of al-Khwārizmī's *zīj* are indeed based on the Arabic calendar and are intended for the meridian of Arīn. It can be shown that most of the tables are in agreement with the Indian period relations occurring in Brahmagupta's works (Burckhardt 1961 and Toomer 1964, pp. 207-208; see also Mercier 1987, pp. 90-92).

SOLAR EQUATION
(Suter 21-26 3°, Neugebauer pp. 19-21 and 95-96)

Ibn al-Muthannā gives hardly any information concerning the solar equation table in al-Khwārizmī's original *zīj*. However, there is little doubt that the table in al-Majrīṭī's recension stems from al-Khwārizmī. This table was computed according to the so-called *method of declinations* described by al-Bīrūnī (Kennedy & Muruwwa 1958, p. 118), whereas Indian astronomers used the *method of sines*.[6] Since Ibn al-Qifṭī states that al-Khwārizmī took his planetary equations from "the Persians", it seems plausible that the method of declinations derives from the *Shāh Zīj*. Al-Majrīṭī's maximum solar equation 2°14′ occurs both in the *Khaṇḍakhādyaka* (Neugebauer p. 96) and in the *Shāh Zīj* (Kennedy & Van der Waerden 1963, p. 326). His value 77°55′ for the longitude of the solar apogee is in agreement with the mean motion system used by Brahmagupta (Pingree 1965). The same values for eccentricity and longitude of the apogee were found by Neugebauer (pp. 90-91) to underlie the small table for the mean solar position at the entry of the sun in the zodiacal signs (Suter 4). The solar equation table in al-Majrīṭī's recension was *not* computed by means of linear interpolation between values for multiples of 3¾°, as suggested by Ibn al-Muthannā (Goldstein pp. 42-43).

LUNAR EQUATION
(Suter 21-26 4°, Neugebauer pp. 21 and 96)

Al-Majrīṭī's recension tabulates only a single lunar equation. Like the solar equation, this table was computed according to the *method of declinations* and has the same maximum value (4°56′) as the *Khaṇḍa-khādyaka* (which employs the *method of sines*) and the *Shāh Zīj*. No traces of linear interpolation can be recognized. The equation is probably of

---

[6] A planetary equation $q$ computed according to the *method of sines* is given by $q(x) = q_{max} \cdot \sin x$ , where $q_{max}$ is the maximum equation. An equation computed by the *method of declinations* is given by $q(x) = q_{max} \cdot \delta(x) / \varepsilon$, where $\delta$ represents the solar declination for obliquity of the ecliptic $\varepsilon$.

Persian origin.[7]

SOLAR DECLINATION

(Suter 21-26 5°, Neugebauer pp. 96-97, Goldstein pp. 49 and 64-66)

Al-Khwārizmī's original *zīj* contained two tables for the solar declination. In one of these tables al-Khwārizmī followed Ptolemy, although he replaced the obliquity value 23°51'20" used both in the *Almagest* and in the *Handy Tables* by 23°51'0". In the other table he followed the Indian tradition by displaying differences between declination and "versed declination" values for multiples of 15° based on the obliquity value 24°. Al-Majrīṭī's recension only contains the Ptolemaic table; the *Toledan Tables* include both the Ptolemaic table (Toomer 1968, pp. 27-28) and, as part of the explanatory text, the Indian values (Millás Vallicrosa 1943-1950, pp. 43-45).

LUNAR LATITUDE

(Suter 21-26 6°, Neugebauer pp. 97-98, Goldstein pp. 89-92 and 211-213)

The lunar latitude table in al-Majrīṭī's recension was computed according to the *method of sines* and has a maximum value of 4°30'. This is in agreement with the commentaries of both Ibn al-Muthannā and Ibn Masrūr (Kennedy & Ukashah 1969, pp. 95-96). The same maximum lunar latitude can be found in Indian sources such as the *Sūryasiddhānta* and the *Khaṇḍakhādyaka* (Sengupta 1934, p. 32) and, according to Ibn Yūnus, in the *Shāh Zīj* (Delambre 1819, pp. 138-139).

PLANETARY EQUATIONS

(Suter 27-56 3°-5°, Neugebauer pp. 22-30 and 98-101, Goldstein pp. 30-45 and 192-198)

Al-Khwārizmī's calculation of the true planetary positions as described by Ibn al-Muthannā is based on Indian methods which were fully explained by Neugebauer (1956, pp. 12-26). The tables and instructions in al-Majrīṭī's recension are in agreement with these methods. The maximum equations agree very well with those from the *Shāh Zīj* as reported by Ibn Hibintā and al-Bīrūnī (Kennedy 1956a, pp. 170-172). The equations of centre were computed according to the *method of sines* using

[7] Note that, for instance, Brahmagupta applies a second correction to the lunar mean motion, which is derived from the solar equation (Sengupta 1934, pp. 21-22).

linear interpolation within intervals of 15°.[8] The equations of anomaly correspond fairly well to the simple eccentric model and are thus approximately given by $\tan q(x) = e \cdot \sin x / (60 + e \cdot \cos x)$, where $q$ is the equation and $e$ the eccentricity. The constant longitudes of the planetary apogees implicit in the column for the *modified apogee* and confirmed by the explanatory text and by Ibn al-Muthannā agree with values calculated from the *Khaṇḍakhādyaka* (Toomer 1964, p. 207).

## PLANETARY STATIONS

**(Suter 27-56 6°, Neugebauer pp. 30-31 and 101, Goldstein pp. 45-49 and 198)**

Both the theory and the tables for the planetary stations in al-Majrīṭī's recension are Ptolemaic. Ibn al-Muthannā confirms the presence of the tables among the planetary equation tables in al-Khwārizmī's original *zīj*. The tabular values are close to those in the *Handy Tables*, but not always identical with them. The same tables for the planetary stations occur in the *Toledan Tables* (Toomer 1968, p. 60).

## PLANETARY LATITUDES

**(Suter 27-56 7°-8°, Neugebauer pp. 34-41 and 101-103, Goldstein pp. 92-94 and 213-215)**

Al-Khwārizmī's rules for the determination of the planetary latitudes given in the commentaries of Ibn Masrūr and Ibn al-Muthannā and in al-Majrīṭī's recension are of Indian origin. The maximum latitudes mentioned in the commentaries are the same as those in al-Majrīṭī's tables and occur in Indian sources like the *Sūryasiddhānta* and the *Khaṇḍakhādyaka*. The second latitude tables (column 8) were computed according to the *method of sines* and are accurate to seconds. The first latitude tables (column 7) are not in full agreement with the Indian rules. Toomer (1964, pp. 205-206) suggested that this could be the result of an error made by al-Majrīṭī when he replaced the value 150 for the radius of the base circle by 60 (cf. the section about the sine below). However, Kennedy & Ukashah (1969) showed that the tables agree with the incorrect explanation of the Indian rules presented in the commentaries of Ibn Masrūr and Ibn al-Muthannā. The constant longitudes of the planetary nodes, given in the tabular headings, agree with calculations based on the *Khaṇḍakhādyaka* (Toomer 1964, p. 207). Al-Majrīṭī's planetary latitude

---

[8] In the equation of centre for Mars the use of two additional independently calculated values for arguments 82½° and 97½° can be recognized.

tables are also present in the *Toledan Tables* (Toomer 1968, pp. 69-70).

LUNAR VISIBILITY
(Suter 57a, Neugebauer pp. 42-44 and 103, Goldstein pp. 96-104 and 218-225)

The presence of a table for lunar crescent visibility in al-Khwārizmī's original *zīj* cannot be ascertained from the commentaries by Ibn al-Muthannā and Ibn Masrūr. However, a table attributed to al-Khwārizmī can be found in various sources (King 1987, pp. 189-192). This table can be shown to be based on the Indian visibility criterion with obliquity of the ecliptic 23°51′ and geographical latitude 33°. The different table in al-Majrīṭī's recension was studied by Kennedy & Janjanian (1965) and by King (1987, pp. 192-197). A systematic analysis by Hogendijk (1988, pp. 32-35) led to the conclusion that the table was based on the Indian visibility criterion and either obliquity 23°35′ and latitude 41°35′ or obliquity 23°51′ and latitude 41°10′.

SINE
(Suter 58-58a, Neugebauer p. 104, Goldstein pp. 49-62)

Al-Khwārizmī's original *zīj* contained sine and versed sine values for so-called *kardajas* ("sections", multiples of 15 degrees), which were computed for a radius of the base circle equal to 150′. Such values derive from Indian sources (see, for example, the *Khaṇḍakhādyaka*, Sengupta 1934, p. 32) and are also present in the explanatory text of the *Toledan Tables* (Millás Vallicrosa 1943-1950, pp. 43-44). According to Ibn al-Muthannā's commentary, the intermediate values for integer degrees had to be filled in by means of interpolation. One possible way in which al-Khwārizmī could have done this was discovered by Hogendijk (1991). He found that a table for a function called *sine of the hours*, which follows al-Khwārizmī's treatise on the astrolabe in a manuscript in Berlin, is based on the Indian sine values for *kardajas* and a special type of linear interpolation.

The sine table in al-Majrīṭī's recension is based on radius 60 and must therefore be a later addition. Bjørnbo noted that it was computed by halving Ptolemy's chord values and truncating the result after the second sexagesimal fractional digit (1909, pp. 12-13).[9]

RIGHT ASCENSION
(Suter 59-59b, Neugebauer pp. 46-48 and 104-105, Goldstein pp. 69-76 and 202-204)

Al-Khwārizmī's original *zīj* contained a table of the right ascension for every degree of the ecliptic starting with Capricorn, thus following Ptolemy's *Handy Tables*. The table in al-Majrīṭī's recension also starts from Capricorn and, like the solar declination, is based on the obliquity value 23°51′0″. We conclude that it is very probably al-Khwārizmī's original table.

OBLIQUE ASCENSION
(Neugebauer pp. 48-55, Goldstein pp. 76-81 and 204-206)

Neither al-Khwārizmī's original *zīj* nor al-Majrīṭī's recension contain a table for the oblique ascension. Instead it is explained both in Ibn al-Muthannā's commentary and in al-Majrīṭī's recension how to calculate the rising times by means of a right ascension table, a shadow length table for gnomon length $G=12$ units, a table for the *diminutions of the rising times for the entire earth* displaying $R \cdot \tan \delta / G$ (where $R$ is the radius of the base circle and $\delta$ the solar declination), and inverse interpolation in a sine table for radius $R$. These rules are of Indian origin and can also be found in the *Toledan Tables*. Of the required tables al-Majrīṭī's recension contains the right ascension (for al-Khwārizmī's obliquity value 23°51′0″), the shadow length (for $G=12$), and the sine

---

[9] In my opinion the sine table in al-Majrīṭī's recension of the *Sindhind Zīj* is different from the sine table for radius 60 in the *Toledan Tables*: the number of differences between the two tables which cannot be explained as scribal errors is large enough to make it plausible that the tables were calculated independently (cf. Toomer 1968, p. 29). The *Toledan Tables* also contain a sine table for radius 150 (Toomer 1968, p. 27), which is practically identical to the table in a Latin manuscript with tables for Newminster (England), which was published in Neugebauer & Schmidt 1952, pp. 226-227. From the fact that nearly all values in this table end in 0, 2, 5 or 7, it can be concluded that it was computed from a sine table for radius 60 by multiplying by 2½, possibly in order to construct a set of tables for determining oblique ascensions based on al-Khwārizmī's parameter values (see below). The underlying sine table for radius 60 is different from the table in al-Majrīṭī's recension.

(for $R=60$ instead of al-Khwārizmī's $R=150$), but omits the table for the diminutions. In the *Toledan Tables* we find a table for $R \cdot \tan \delta / G$, which was shown by Lesley (1957, p. 125-127) to be based on $R=150$, $G=12$ and obliquity 23°51′0″.[10] In Ibn al-Muthannā's commentary three values of al-Khwārizmī's table for $R \cdot \tan \delta / G$ are mentioned (Goldstein p. 80; Millás Vendrell 1963, p. 145). Since the table in the *Toledan Tables* displays the same values (disregarding a couple of scribal mistakes), it is probably al-Khwārizmī's original table.

SHADOW LENGTH (cotangent)
(Suter 60, Neugebauer p. 105, Goldstein pp. 87-89)

From Ibn al-Muthannā's commentary it becomes clear that the calculation of the length of the shadow cast by a gnomon is extensively described in al-Khwārizmī's original *zīj*. However, no mention is made of a table for this function. Ibn al-Muthannā states that al-Khwārizmī took the gnomon length equal to 12 units, in agreement with the cotangent table in al-Majrīṭī's recension. Since many Islamic *zīj*es contained a cotangent table for gnomon length 12, it is nevertheless possible that the table is a later addition. In my opinion, the cotangent values in al-Majrīṭī's recension were calculated independently from those in al-Battānī's *zīj* and the *Toledan Tables*.

TRUE SOLAR AND LUNAR MOTION
(Suter 61-66, Neugebauer pp. 57-63 and 105-107, Goldstein pp. 94-96, 104-109, 216-217 and 226-230.)

Suter (p. 90) showed that al-Majrīṭī's table for the true solar and lunar motion and the apparent radii of the sun, moon and shadow agree with the rules given in the *Khaṇḍakhādyaka* and Ibn al-Muthannā's commentary.

EQUATION OF TIME
(Suter 67 68, Neugebauer pp. 63-65 and 107-108)

In Ibn al-Muthannā's commentary no mention is made of the equation of time. Al-Hāshimī presents a Ptolemaic description of the calculation of the equation of time and states that the same method is used

---

[10] The same table is found in the Latin manuscript with tables for Newminster mentioned in footnote 9; see Neugebauer & Schmidt 1952, p. 226.

in the *Shāh Zīj* and in the *zīj*es by al-Khwārizmī and Abū Maʿshar (al-Hāshimī, pp. 156-157 and 279). He does not give parameter values or other details of the method of computation and does not mention tables for the equation of time in the above-mentioned works.

Al-Majrīṭī's recension of al-Khwārizmī's *zīj* contains a table for the equation of time with values to seconds of an hour for every degree of solar longitude. From the instructions for the use of this table (Suter p. 25; Neugebauer pp. 61-62) it follows that the argument of the table is the true solar longitude and that the equation of time values must always be added to mean solar time to obtain true solar time. The equation of time as tabulated in al-Majrīṭī's recension is typically Ptolemaic; Indian astronomers only corrected for the solar velocity component (cf. Section 5) and thus obtained a sine-wave instead of a function with four local extreme values (cf. Figure 3). In Section 6 of this article the mathematical structure and the underlying parameter values of the table for the equation of time in al-Majrīṭī's recension will be determined.

MEAN OPPOSITIONS AND CONJUNCTIONS
(Suter 69-72, Neugebauer pp. 108-115, Goldstein pp. 94 and 216)

The tables for mean conjunctions and oppositions in al-Majrīṭī's recension were computed for a length of the mean synodic month very close to an Indian value reported by al-Bīrūnī. Since the tables are based on the Arabic calendar and are said to be for the geographical longitude of Cordoba, they were probably modified by al-Majrīṭī. The difference in geographical longitude between the tables for mean oppositions and conjunctions and those for mean motions is approximately 63°. This signifies the first (implicit) occurrence of the so-called "meridian of water", which was used, in particular, by Andalusian and Western-Maghribian geographers and astronomers (Comes 1992-1994, pp. 43-44). The tables for mean oppositions and conjunctions in the *Toledan Tables* are based on parameter values different from those used in al-Majrīṭī's recension (Toomer 1968, pp. 78-81).

LUNAR ECLIPSES
(Suter 73-76, Neugebauer pp. 66-69 and 116-120, Goldstein pp. 109-120 and 231-235)

The organization of the eclipse tables in al-Majrīṭī's recension is purely Ptolemaic. However, Neugebauer found that only the table for lunar eclipses at apogee could be based on Ptolemaic parameter values; the tables for the remaining three cases are based on the Indian value

4°30′ of the maximum lunar latitude. The lunar eclipse tables in al-Majrīṭī's recension are identical to those in the *Toledan Tables* (Toomer 1968, pp. 91-93).

PARALLAX

(Suter 77-77a, Neugebauer pp. 69-76 and 121-126, Goldstein pp. 121-130 and 236-238)

The parallax tables and explanatory text in al-Majrīṭī's recension derive from al-Khwārizmī's original *zīj*. Kennedy (1956b) showed that the latitude component is in complete agreement with the parallax theory in the *Sūryasiddhānta*. The longitude component contains Indian elements as well (in particular the value 24° for the obliquity of the ecliptic), but was computed using an iterative procedure described by Ḥabash al-Ḥāsib (Baghdad, c. 830).

SOLAR ECLIPSES

(Suter 78, Neugebauer pp. 73-76 and 126-128, Goldstein pp. 120-142 and 236-241)

See above under Lunar Eclipses.

EQUATION OF THE HOUSES

(Suter 79-90, Neugebauer pp. 78 and 128-129, Goldstein pp. 84-86 and 209-210)

Ibn al-Muthannā's commentary describes the method by which the equation of the houses can be computed, but does not mention the presence in al-Khwārizmī's original *zīj* of a table for that purpose. The theory underlying the table in al-Majrīṭī's recension is Ptolemaic (Suter pp. 96-98). The underlying parameter values are 23°35′ for the obliquity of the ecliptic and approximately 38°43′ for the geographical latitude (Toomer 1968, pp. 140-143). The table is thus probably an addition by al-Majrīṭī. The same table can be found in the *Toledan Tables*.

PROJECTION OF THE RAYS

(Suter 91-114, Neugebauer pp. 78-81 and 129-131)

The table for the projection of the rays from al-Khwārizmī's original *zīj* can be found in an astrological work by Ibn Hibintā (Kennedy & Krikorian-Preisler 1972) and in the *Toledan Tables* (Toomer 1968, pp. 147-151). Toomer found that the table was computed for obliquity 23°51′ and the latitude of Baghdad (close to 33°). The table for the projection of the rays in al-Majrīṭī's recension is indicated to be for geographical latitude 38°30′, i.e. probably for Cordoba. We can thus conclude that it is an addition by al-Majrīṭī. Hogendijk (1989) discussed the mathematical structure of both tables for the projection of the rays. He found that al-Majrīṭī's table is based on al-Khwārizmī's obliquity value 23°51′, but that it presents a significant improvement of al-Khwārizmī's method of computation.

EXCESS OF REVOLUTION
(Suter 115, Neugebauer pp. 131-132, Goldstein pp. 143-144 and 242)

The table for the excess of revolution in al-Majrīṭī's recension is based on a sidereal year of 365;15,30,22,30 days.[11] This value occurs in various Indian sources, for instance in the *Brāhmasphuṭasiddhānta*, and is confirmed by Ibn al-Muthannā to have been used by al-Khwārizmī. Note that the *Shāh Zīj* uses the value 365;15,32,30 (Kennedy 1956a, p. 147).

*Summary*

With regard to their origin the tables in the Latin translation of al-Majrīṭī's recension of al-Khwārizmī's *Sindhind Zīj* can be divided into five groups (the numbers mentioned are the table numbers in Suter's edition published in 1914):

I. Tables deriving from al-Khwārizmī's original *zīj*:
A) *based on Indian methods and / or parameter values:*
1) Mean motion tables: motions and positions (4-20)
2) Lunar latitude (21-26 6°)
3) Planetary equations: structure (27-56 3°-5°)
4) Planetary latitudes (27-56 7°-8°)
5) True solar and lunar motion (61-66)
6) Lunar eclipses: parameter values for eclipses at apogee (73-76)
7) Parallax (77-77a)
8) Solar eclipses: parameter values (78)
9) Excess of revolution (115)

B) *based on Persian methods and / or parameter values:*
1) Solar equation (21-26 3°)
2) Lunar equation (21-26 4°)

---

[11] Sexagesimal numbers will be written in the conventional way: sexagesimal digits are separated by a comma and the sexagesimal point is represented by a semicolon. For example: 365;15,30 or 6,5;15,30 denotes $365 + 15/60 + 30/60^2$.

3) Planetary equations: parameter values (27-56 3°-5°)

C) *based on Ptolemaic methods and / or parameter values:*
1) Solar declination (21-26 5°)
2) Planetary stations (27-56 6°)
3) Right ascension (59-59b)
4) Lunar eclipses: organization and parameter values for eclipses at perigee (73-76)
5) Solar eclipses: organization (78)

II: Tables modified by al-Majrīṭī:
1) Chronological tables (1-3a)
2) Mean motion tables: epoch (4-20)
3) Mean conjunctions and oppositions (69-72)

III: Tables added or replaced by al-Majrīṭī:
1) Lunar crescent visibility (57a)
2) Sine (58-58a)
3) Cotangent (60)
4) Equation of the houses (79-90)
5) Projection of the rays (91-114)

Al-Khwārizmī's original table for lunar crescent visibility is extant in various sources (see above for references). His sine values for *kardajas* based on radius of the base circle 150 can be found in the *Toledan Tables*; a sine table with arguments 1,2,3,...,90 based on these values was reconstructed by Hogendijk. Al-Khwārizmī's original table for the projection of the rays has come down to us in a work by Ibn Hibintā and in the *Toledan Tables*.

The table for the equation of time (67-68) belongs to one of the groups I-C, II or III. The analysis in Section 6 below will enable us to make a more detailed statement about its origin.

## 5. *The equation of time*

If we want to measure local time by the solar position (for example, by means of a sundial), we define noon as the time of the daily culmination of the sun. The period of time between two consecutive culminations can then be divided into 24 equal hours. In case the sun were

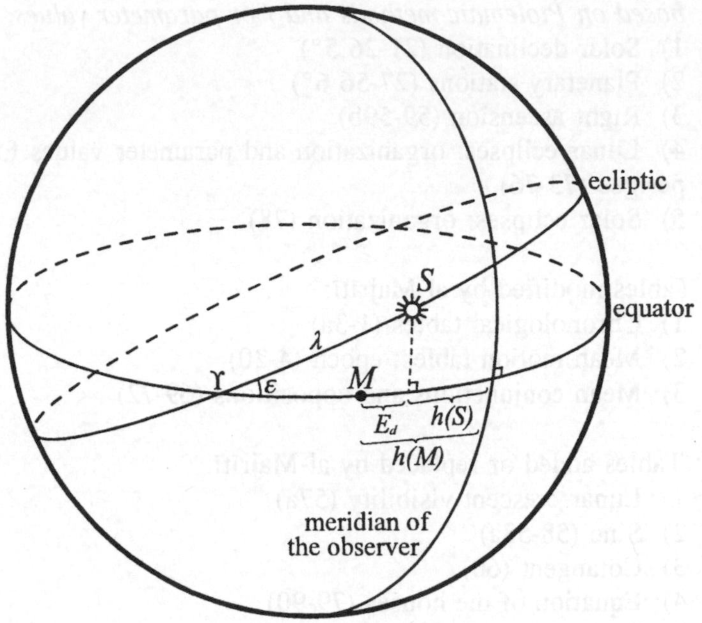


*Figure 1* *Graphical explanation of the equation of time.*

positioned in the plane of the equator and moved with a uniform apparent velocity, the arc of the equator which crosses the meridian of the observer between each two consecutive culminations would be the same all through the year, namely 360° plus the daily solar motion. Consequently, each day and each hour would have precisely the same length. The time obtained on the basis of the assumption that the sun moves at a uniform speed in the plane of the equator is called *local mean solar time*; it differs at most a constant amount from the time that we use today. Ancient and medieval astronomers used mean solar time for the calculation of planetary longitudes: they applied corrections to the so-called mean planetary longitudes, which were linear functions of time and could therefore be determined by multiplying the mean solar time elapsed by the average planetary motion.

Since in reality the sun moves in the plane of the ecliptic with a non-uniform velocity, *local true solar time*, which is defined by the daily culmination of the *true* sun, differs a variable quantity from mean solar time. The difference between true and mean solar time is called the *equation of time* (in Arabic: *taᶜdīl al-ayyām bi-layālī-hā*; in Latin: *equatio dierum cum noctibus suis*). It is determined by two factors: the non-uniformity of the solar motion, and the fact that normally an arc of the ecliptic does not cross the meridian of the observer in the same period of time as an equatorial arc of the same length.

We will now define mean and true solar time mathematically and will derive a formula for the equation of time as a function of the solar position.[12] First note that the *hour angle* of a heavenly body $X$ is the spherical angle between the meridian of the observer and the meridian of $X$, measured in *westward* direction. By $S$ I will denote the true sun; by $M$ the virtual equatorial mean sun, which moves on the equator at a constant speed with the same period as the true sun.

Now mean solar time can be defined as the hour angle $h(M)$ of the equatorial mean sun, true solar time as the hour angle $h(S)$ of the true sun. The equation of time $E_d$ expressed in equatorial degrees is the difference between true and mean solar time (cf. Figure 1, which depicts the heavenly sphere; the earth, located in the centre of the sphere, has not been indicated):

$$E_d = h(S) - h(M). \qquad (1)$$

In order to express $E_d$ as a function of the solar position, we now consider the Ptolemaic eccentric solar model, which was used by most medieval astronomers (see Figure 2).[13] In this model the true sun $S$ moves at a constant speed on a circle with radius 60, which lies in the plane of the ecliptic. The centre $C$ of this circle is a distance $e$, called the *solar eccentricity*, removed from the Earth $E$. The sun reaches its largest distance from the Earth at the apogee $A$, its smallest distance at the

---

[12] Extensive descriptions of the equation of time as used by Ptolemy can be found in Neugebauer 1975, vol. 1, pp. 61-68 and Pedersen 1974, pp. 154-158. The methods by which the Islamic astronomers Kūshyār ibn Labbān (10th century) and al-Kāshī (15th century) computed their tables for the equation of time were explained in Kennedy 1988.

[13] More extensive explanations of the Ptolemaic solar model can be found in Neugebauer 1975, vol. 1, pp. 53-61 or Pedersen 1974, pp. 144-154.

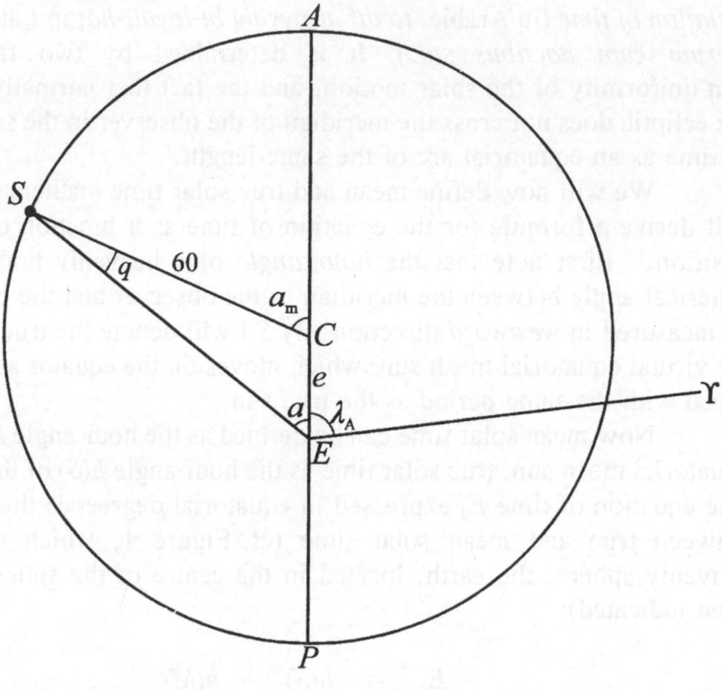


*Figure 2* *The Ptolemaic solar model.*

perigee $P$. Now the *true solar longitude* $\lambda$ is defined as the angle $\measuredangle\, \Upsilon ES$, measured in eastward direction from the vernal point $\Upsilon$, under which the sun is seen from the Earth (in order to keep Figure 2 as clear as possible I have not indicated $\lambda$ in it). In the following calculations we will also make use of the *true solar anomaly* $a$, which is defined by the angle $\measuredangle\, AES$ between the apogee $A$ and the sun $S$. We have $\lambda = a + \lambda_A$, where $\lambda_A$ is the longitude of the solar apogee, given by the angle $\measuredangle\, \Upsilon EA$.

In order to calculate the true solar position $\lambda$, we will apply a trigonometrically computed correction to a linear function of time. For this function we can, for instance, take the *mean solar anomaly* $a_m$, which is the angle $\measuredangle\, ACS$ between the apogee and the sun as seen from the centre $C$ of the eccentric solar orbit. Since the sun moves at a uniform speed around $C$, this function is in fact linear. However, ancient and medieval astronomers usually tabulated a different function, namely the *mean solar longitude* $\lambda_m$, which is defined by $\lambda_m = a_m + \lambda_A$ (for the same reason as above, $\lambda_m$ has not been indicated in Figure 2).

Since the sum of the angles in triangle *ECS* is now $a + (180° - a_m) + \measuredangle ESC$, it follows that the difference between the true solar anomaly $a$ and the mean solar anomaly $a_m$ (and hence the difference between the true solar longitude $\lambda$ and the mean solar longitude $\lambda_m$) equals the angle $\measuredangle ESC$. This angle is called the *solar equation* and will be denoted by $q$. It can be determined geometrically as a function of $a_m$ by extending the triangle *SCE* to a right-angled triangle *SXE* (not indicated in the figure) and then expressing the sine or the tangent of the angle $q$ in terms of the sides of the extended triangle. This yields:

$$q^m (a^m) = \arctan (e \cdot \sin a^m / ( 60 + e \cdot \cos a^m ) ) \qquad (2)$$

where $q_m$ denotes the solar equation as a function of the mean solar anomaly (the equivalent formula based on an expression for $\sin q_m$ is somewhat more complicated). For $a_m$ between 0° and 180° the solar equation must be subtracted from the mean solar anomaly (or longitude) in order to obtain the true solar anomaly (or longitude); for $a_m$ between 180° and 360° medieval astronomers added the absolute value of the solar equation to the mean solar anomaly (or longitude). Since our formula yields a negative equation for values of $a_m$ between 180° and 360°, we can write in general $a = a_m - q_m(a_m)$ or $\lambda = \lambda_m - q_m(\lambda_m - \lambda_A)$.

The solar equation can also be expressed as a function of the true solar anomaly and will then be denoted by $q$. By applying the sine law to triangle *SCE* we find

$$q(a) = \arcsin ( e \cdot \sin a / 60 ), \qquad (3)$$

and we have $a = a_m - q(a)$ or $\lambda = \lambda_m - q(\lambda - \lambda_A)$ for all values of $a$ and $\lambda$.

Now we can return to figure 1. We first note that both the mean solar longitude $\lambda_m$ and the position of the equatorial mean sun $M$ on the equator are linear functions of time. Since the equatorial mean sun has the same period as the true sun (namely a solar year), it follows that at any time the position of the equatorial mean sun $M$ on the equator can be expressed as $\lambda_m + c$ for a certain constant $c$. Because the right ascension of a heavenly body $X$ is the spherical angle between the meridian through the vernal point ♈ and the meridian of $X$ measured in *eastward* direction, it follows from formula 1 that at any time the equation of time equals the difference in the right ascension of the equatorial mean sun and the true sun. Thus we have:

$$E_d = \lambda_m - \alpha(\lambda) + c,$$

where $\alpha$ denotes the right ascension. For values of $\lambda$ between 0° and 90° $\alpha$ can be calculated from $\alpha(\lambda) = \arctan(\cos \varepsilon \cdot \tan \lambda)$, where $\varepsilon$ denotes the obliquity of the ecliptic. For values between 90° and 360° $\alpha$ follows from the symmetry relations $\alpha(180-\lambda) = 180 - \alpha(\lambda)$ and $\alpha(180+\lambda) = 180 + \alpha(\lambda)$. The constant $c$ determines the synchronization of the true sun and the virtual equatorial mean sun. Since ancient and medieval astronomers often determined this constant in such a way that it depended on the epoch (i.e. the starting point) of their planetary tables, I will call it the epoch constant.

In the above formula the equation of time $E_d$ is expressed in equatorial degrees. However, in Ptolemy's *Handy Tables* and in most medieval astronomical handbooks (including al-Majrīṭī's recension of al-Khwārizmī's *zīj*) the equation of time was tabulated in hours, minutes and possibly seconds. Since 24 hours correspond to one daily rotation of 360° plus the daily solar motion of approximately 0°59′8″, an accurate factor for the conversion from degrees to hours would be (360° + 0°59′8″) / 24 ≈ 15°2′28″ per hour. However, Ptolemy and many medieval astronomers neglected the daily solar motion and used the factor 15°/ hour instead. Thus if $E_h$ denotes the equation of time expressed in hours, we often have $E_h = 1/15 \cdot E_d$. From now on I will write $E$ instead of $E_h$, since we will only be dealing with the equation of time expressed in hours.

Like the solar equation, the equation of time can be expressed both as a function of the true solar longitude (denoted by $E(\lambda)$ ) and as a function of the mean solar longitude (denoted by $E_m(\lambda_m)$ ). From the above we find

$$E(\lambda) = 1/15 \cdot (\lambda + q(\lambda - \lambda_A) - \alpha(\lambda) + c) \quad (4)$$

and

$$E_m(\lambda_m) = 1/15 \cdot (\lambda_m - \alpha(\lambda_m - q_m(\lambda_m - \lambda_A)) + c) \quad (5)$$

respectively.

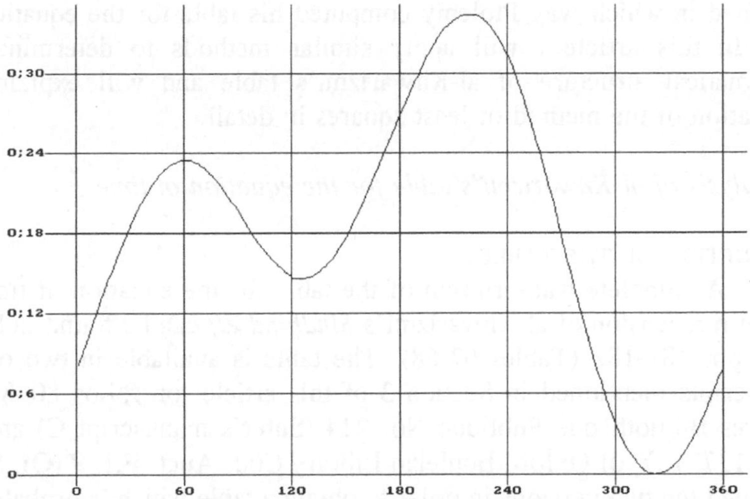


*Figure 3 al-Khwārizmī's values for the equation of time (horizontally: the solar longitude in ecliptical degrees; vertically: the equation of time in hours).*

We have thus seen that the equation of time is based on five different parameters: the obliquity of the ecliptic, the solar eccentricity, the longitude of the solar apogee, the epoch constant and the conversion factor. Many medieval astronomers included in their handbooks a table for the equation of time, very often without an indication of the underlying parameter values. Furthermore it was not always clear whether the equation of time was tabulated as a function of the mean or the true solar longitude (i.e. whether the *argument* or *independent variable* of the table was the mean or true solar longitude); in both cases a plot of the tabular values would have the general shape shown in Figure 3 (here the solar longitude has been plotted horizontally, the equation of time in hours vertically).

We can conclude that the analysis of a table for the equation of time is often a difficult matter. Until now only very few tables for the equation of time in ancient and medieval sources have been mathematically explained. Kennedy (1988) recomputed the tables for the equation of time found in the *zīj*es of Kūshyār ibn Labbān and al-Kāshī. In a recent article (van Dalen 1994) I have described a number of methods which can be used to analyse tables for the equation of time and have explained in which way Ptolemy computed his table for the equation of time. In this article I will apply similar methods to determine the mathematical structure of al-Khwārizmī's table and will explain the application of the method of least squares in detail.

## 6. *Analysis of al-Khwārizmī's table for the equation of time*

### Description of the table.

A complete transcription of the table for the equation of time in the Latin recension of al-Khwārizmī's *Sindhind Zīj* can be found in Suter 1914, pp. 181-182 (Tables 67-68). The table is available in two of the manuscripts mentioned in Section 3 of this article: on folios $80^r$-$80^v$ of Chartres Bibliothèque Publique No. 214 (Suter's manuscript C) and on folios $137^r$-$137^v$ of Oxford Bodleian Library Cod. Auct. F.I. 9 (O). Suter combined the two versions in order to obtain a table which is probably as close as possible to al-Khwārizmī's original table. He found that the two versions have only very few scribal errors in common.

The values of al-Khwārizmī's table for the equation of time are given in minutes and seconds of an hour for every degree of solar longitude starting from Aries. The minimum value is assumed when the solar position is 22° Aquarius and amounts to $0^h0^m0^s$. This implies that the user of the table did not have to distinguish the cases where the tabular values are additive ("positive") and where they are subtractive ("negative"). Apparently the author of the table chose his epoch constant *c* especially in order to obtain this property (cf. the technical explanation in the previous section).

For the following analysis I have used the values given by Suter (see Tables 4a to 4c at the end of this section), a plot of which is shown in Figure 3. The scarce information about the equation of time in al-Khwārizmī's original *zīj* which can be found in other primary sources has been summarized in the paragraph "Equation of time" of Section 4.

### Conversion factor.

During a first inspection of al-Khwārizmī's table for the equation of time it can be noted that practically all tabular values are multiples of four seconds. The only exceptions occur in the neighbourhood of the (local and global) minimum and maximum values of the table. It seems reasonable to assume that also in those regions the values were originally multiples of four seconds, but that they were adjusted in order to avoid obvious jumps in the tabular values.

The presence of mere multiples of 4 seconds can be explained from formulas 4 and 5 for the equation of time: the last step of the calculation is in both cases the division by the conversion factor, which was usually taken equal to 15°/ hour, sometimes to 15°2′28″/ hour. Apparently the author of the table in al-Majrīṭī's recension calculated the equation of time to an accuracy of equatorial minutes, i.e. his values for

$$\lambda + q(\lambda - \lambda_A) - \alpha(\lambda) + c$$

(or $\lambda_m - \alpha(\lambda_m - q_m(\lambda_m - \lambda_A)) + c$ in case the argument of the table was the mean solar longitude) were all multiples of 60 seconds. By dividing by the conversion factor 15 he then obtained tabular values for the equation of time expressed in hours which were all multiples of 4 seconds.

INDEPENDENT VARIABLE.

From the explanatory text in al-Majrīṭī's recension of al-Khwārizmī's *zīj* (Suter 1914, p. 25; Neugebauer 1962, pp. 61-62) it follows that the independent variable of the table for the equation of time is the *true* solar position. Furthermore, the equation found must always be *subtracted* from true solar time to obtain mean solar time; thus the tabular values equal true solar time minus mean solar time as in formula 1.

The second fact can easily be verified from the tabular values: whenever we compute the equation of time by subtracting mean solar time from true solar time, the resulting function has, for a solar position running from 0° to 360°, a local maximum, a local minimum, a global maximum, and a global minimum respectively. As can be seen from Figure 3 (or from Figure 32 in Neugebauer 1962, p. 108) this is in fact the case for the table for the equation of time in al-Majrīṭī's recension.[14]

---

[14] Ptolemy's table for the equation of time in the *Handy Tables* is an example of a table of which the tabular values must always be *added* to true solar time in order to obtain mean solar time (cf. Neugebauer 1975, vol. 2, pp. 984-986).

There is no easy way to verify the independent variable of a table for the equation of time: both as a function of the true solar longitude and as a function of the mean solar longitude the equation of time has the form shown in Figure 3. We can, however, derive a number of properties of a table for the equation of time as a function of the true solar longitude which do not hold if the independent variable is the mean solar longitude and which could thus be used to investigate which independent variable was used in al-Majrīṭī's table. In particular, using symmetry relations satisfied by the right ascension and the solar equation we can reconstruct these functions from a given table for the equation of time as a function of the true solar longitude.

RECONSTRUCTION OF THE UNDERLYING RIGHT ASCENSION AND SOLAR EQUATION.

As we have seen in the technical explanation in Section 5, the equation of time is built up from two components, the right ascension and the solar equation. Both components satisfy a number of symmetry relations. In particular, we have for the right ascension $\alpha$

$$\alpha(180-\lambda) = 180 - \alpha(\lambda) \quad \text{and} \quad \alpha(180+\lambda) = 180 + \alpha(\lambda)$$

for every value of $\lambda$. These formulae state mathematically that, for instance, the rising time at sphaera recta of Aries, which can be calculated as $\alpha(30) - \alpha(0)$, is equal to that of Virgo ( $\alpha(180) - \alpha(150)$ ) and to that of Libra ( $\alpha(210) - \alpha(180)$ ).

For the solar equation as a function of the true solar longitude we have

$$q(\lambda_A+a) = -\,q(\lambda_A-a) \quad \text{and} \quad q(\lambda_A+180+a) = -\,q(\lambda_A+a)$$

for every value of the true solar anomaly $a$. Here $\lambda_A$ is the longitude of the solar apogee and $a = \lambda - \lambda_A$, as in the description of the solar model in Section 5. These formulae state mathematically that the absolute value of the solar equation only depends on the distance of the sun from either apogee or perigee, the sign of the equation being different on the two sides of the line connecting apogee and perigee.[15]

From the symmetry relations satisfied by the right ascension and the solar equation we can derive relations between certain values for the equation of time as a function of the true solar longitude. First note that, for every value of $\lambda$, $E(\lambda) = 1/15 \cdot (\lambda + q(\lambda - \lambda_A) - \alpha(\lambda) + c)$ (formula 4) and

$$\begin{aligned} E(180+\lambda) &= 1/15 \cdot (180 + \lambda + q(180+\lambda-\lambda_A) - \alpha(180+\lambda) + c) \\ &= 1/15 \cdot (180 + \lambda - q(\lambda-\lambda_A) - 180 - \alpha(\lambda) + c) \\ &= 1/15 \cdot (\lambda - q(\lambda-\lambda_A) - \alpha(\lambda) + c). \qquad (6) \end{aligned}$$

Therefore, by *adding* two values for the equation of time for arguments 180° apart, we obtain:

$$\begin{aligned} E(\lambda) + E(180+\lambda) &= 1/15 \cdot (\lambda + q(\lambda-\lambda_A) - \alpha(\lambda) + c + \\ &\qquad \lambda - q(\lambda-\lambda_A) - \alpha(\lambda) + c) \\ &= 1/15 \cdot (2\lambda - 2\alpha(\lambda) + 2c), \end{aligned}$$

from which we find:

$$\alpha(\lambda) = \lambda + c - 7\tfrac{1}{2} \cdot (E(\lambda) + E(180+\lambda)). \qquad (7)$$

Thus we can reconstruct the right ascension underlying a given table for the equation of time as a function of the true solar longitude, provided that we know the value of $c$ (or have a good approximation for it).

By *subtracting* two values for the equation of time for arguments 180° apart, we can in a similar way reconstruct the underlying solar equation. We have:

$$\begin{aligned} E(\lambda) - E(180+\lambda) &= 1/15 \cdot (\lambda + q(\lambda-\lambda_A) - \alpha(\lambda) + c \\ &\qquad - \lambda + q(\lambda-\lambda_A) + \alpha(\lambda) - c) \\ &= 1/15 \cdot (2 \cdot q(\lambda-\lambda_A)), \end{aligned}$$

[15] The solar equation as a function of the mean solar longitude only satisfies a symmetry relation similar to the first of the relations above, namely $q_m(\lambda_A + a_m) = -q_m(\lambda_A - a_m)$, where $a_m$ is the mean solar anomaly ($a_m = \lambda_m - \lambda_A$).

leading to

$$q(\lambda-\lambda_A) \;=\; 7\tfrac{1}{2}\cdot(\,E(\lambda) - E(180+\lambda)\,). \qquad (8)$$

Thus we can reconstruct the solar equation underlying a table for the equation of time as a function of the true solar longitude even if no value of the epoch constant $c$ is available.

Neither the formulae derived here nor similar formulae hold for the equation of time as a function of the mean solar longitude. Since in formula 5 the solar equation $q_m$ occurs "within" the right ascension (in the term $\alpha(\lambda_m - q_m(\lambda_m-\lambda_A))$ ), the terms will not cancel out so nicely regardless which equation of time values we add or subtract.

Assuming that al-Khwārizmī's table presents the equation of time as a function of the true solar longitude, we can now reconstruct the underlying solar equation using formula 8. I found that the resulting values are close to solar equation values computed for Ptolemy's solar eccentricity 2;30 and a longitude of the apogee in the neighbourhood of 84°40′, which, as far as I know, is not attested. If the independent variable of al-Khwārizmī's table were the mean solar longitude, the reconstructed table would have shown systematic divergences from solar equation tables computed for any values of the eccentricity and the longitude of the apogee. Therefore we can conclude that the independent variable is *not* the mean solar longitude. More evidence for this conclusion could be found by reconstructing the right ascension underlying al-Khwārizmī's table for the equation of time using formula 7. In order to do that, we first have to find an approximation for the epoch constant $c$.

### APPROXIMATION OF THE EPOCH CONSTANT *C*.

The epoch constant $c$ can be approximated from the values of a table for the equation of time as a function of the true solar longitude by once again applying the symmetry relations satisfied by the right ascension and the solar equation. For every value of $\lambda$ we have

$$\begin{aligned} E(180-\lambda) &= 1/15\cdot(\,180 - \lambda + q(180-\lambda-\lambda_A) - \alpha(180-\lambda) + c\,) \\ &= 1/15\cdot(\,180 - \lambda - q(-\lambda-\lambda_A) - 180 + \alpha(\lambda) + c\,) \\ &= 1/15\cdot(\,-\lambda - q(-\lambda-\lambda_A) + \alpha(\lambda) + c\,) \qquad (9) \end{aligned}$$

and

$$\begin{aligned}
E(360-\lambda) &= 1/15 \cdot (360 - \lambda + q(360-\lambda-\lambda_A) - \alpha(360-\lambda) + c) \\
&= 1/15 \cdot (360 - \lambda + q(180+(180-\lambda)-\lambda_A) \\
&\qquad - \alpha(180+(180-\lambda)) + c) \\
&= 1/15 \cdot (360 - \lambda - q(180-\lambda-\lambda_A) \\
&\qquad - 180 - \alpha(180-\lambda) + c) \\
&= 1/15 \cdot (360 - \lambda + q(-\lambda-\lambda_A) - 360 + \alpha(\lambda) + c) \\
&= 1/15 \cdot (-\lambda + q(-\lambda-\lambda_A) + \alpha(\lambda) + c). \qquad (10)
\end{aligned}$$

Using formulae 4, 6, 9 and 10, we can now show that for any value of the true solar longitude $\lambda$ the sum of the four equation of time values for arguments $\lambda$, $180-\lambda$, $180+\lambda$ and $360-\lambda$ is constant:

$$\begin{aligned}
E(\lambda) + E(180-\lambda) + E(180+\lambda) + E(360-\lambda) &= \\
= \; & 1/15 \cdot (\lambda + q(\lambda-\lambda_A) - \alpha(\lambda) + c) + \\
& 1/15 \cdot (-\lambda + q(180-\lambda-\lambda_A) + \alpha(\lambda) + c) + \\
& 1/15 \cdot (\lambda - q(\lambda-\lambda_A) - \alpha(\lambda) + c) + \\
& 1/15 \cdot (-\lambda - q(180-\lambda-\lambda_A) + \alpha(\lambda) + c) \\
= \; & 1/15 \cdot (4c) \\
= \; & 4/15 \cdot c.
\end{aligned}$$

If we have a total of $n$ values for the equation of time, where $n$ is a multiple of 4 and the corresponding arguments are $360°/n$, $2 \cdot 360°/n$, ..., $360°$, we can build $n/4$ groups of four values of which the sum equals $4/15 \cdot c$.[16] As a result, the sum of all available values for the equation of time equals $n/4 \cdot (4/15 \cdot c)$. This implies that

[16] For $\lambda=0°$ and $\lambda=90°$ we obtain only two values. However, we have

$$\begin{aligned}
& E(0) + E(90) + E(180) + E(270) = \\
&\quad = 1/15 \cdot (q(0-\lambda_A) + c) + 1/15 \cdot (90 + q(90-\lambda_A) - 90 + c) + \\
&\quad\quad 1/15 \cdot (180 - q(0-\lambda_A) + c) + 1/15 \cdot (270 - q(90-\lambda_A) - 270 + c) \\
&\quad = 4/15 \cdot c.
\end{aligned}$$

$$c = (15/n) \cdot \sum_{i=1}^{n} E(i \cdot 360° / n) \qquad (11)$$

Again, this formula does not hold for the equation of time as a function of the mean solar longitude. However, with a large computational effort it can be shown that for the equation of time as a function of the mean solar longitude formula 11 holds at least approximately, i.e. we have

$$c \approx (15/n) \cdot \sum_{i=1}^{n} E_m(i \cdot 360° / n)$$

where $n$ is again the total number of tabular values.

Neither for the reconstruction of the underlying right ascension and solar equation tables (formulae 7 and 8) nor for the approximation of the epoch constant $c$ (formula 11) will we in practice be able to use the exact values $E(\lambda)$ for the equation of time. Instead we will have to use tabular values $T(\lambda)$ which were rounded to a fixed number of sexagesimal digits. These tabular values contain at least rounding errors, which are relatively small. (For instance, the differences between exact functional values and values rounded to seconds are half a second at most). Furthermore, they could contain relatively large errors like computational errors or scribal mistakes. Nevertheless, in most cases formula 11 (with $E$ replaced by $T$) will give us a good approximation for $c$.

In the case of al-Khwārizmī's table for the equation of time the application of formula 11 leads to $c \approx 4;30,3$. As was shown above, al-Khwārizmī computed $\lambda + q(\lambda - \lambda_A) - \alpha(\lambda) + c$ to an accuracy of minutes. Furthermore, he apparently chose the epoch constant in such a way that the minimum value of his equation of time became zero. It therefore seems natural to assume that his epoch constant also had an accuracy of minutes. In that case, the value he used was probably $c = 4;30$.

| λ | reconstructed right ascension | differences | λ | reconstructed right ascension | differences |
|---|---|---|---|---|---|
| 0 | 0;10, 0 | 0;10, 0 | 90 | 89;48,30 | −0;11,30 |
| 10 | 9;20, 0 | 0;10,20 | 100 | 100;44,30 | −0;10,14 |
| 20 | 18;33,30 | 0; 8,47 | 110 | 111;33, 0 | −0; 9, 1 |
| 30 | 27;56,30 | 0; 6,20 | 120 | 122;11, 0 | −0; 4,45 |
| 40 | 37;33,30 | 0; 3,14 | 130 | 132;30,30 | −0; 1,35 |
| 50 | 47;27, 0 | −0; 0,55 | 140 | 142;31,45 | 0; 2, 1 |
| 60 | 57;38,30 | −0; 5,45 | 150 | 152;15,30 | 0; 5,40 |
| 70 | 68; 9,30 | −0; 8,29 | 160 | 161;43, 0 | 0; 7,43 |
| 80 | 78;55,30 | 0; 9,46 | 170 | 171; 0, 0 | 0; 9,40 |

Table 1 *Right ascension reconstructed from al-Khwārizmī's table for the equation of time under the assumption that the independent variable is the true solar longitude. The 3rd and 6th columns display the differences between the reconstructed values and accurate right ascension values for obliquity 23°51′.*

Assuming that the argument of al-Khwārizmī's table for the equation of time is the true solar longitude and that the epoch constant used was 4;30, we can now reconstruct the underlying right ascension using formula 7. A selection of the resulting values is shown in Table 1 together with the differences between these values and values recomputed for al-Khwārizmī's obliquity value 23°51′. It turns out that the agreement is very bad indeed. Elementary properties of the right ascension such as $\alpha(0)=0$ and $\alpha(90)=90$ are not satisfied by the reconstructed values and we find differences up to 12′ which display an obvious pattern[17] and hence are probably caused by a systematic error in our reconstruction (cf. the explanation of the method of least squares below). It can be checked that this error does not lie in our values of the epoch constant and the obliquity of the ecliptic: for no values of these parameters will the pattern in the differences in Table 1 disappear. We must therefore conclude that the argument of al-Khwārizmī's table is not the true solar longitude.

Since we concluded from the good agreement of the reconstructed solar equation with recomputed values that the argument of

[17]The differences show a clear sine-wave pattern. They are practically zero for $\lambda \approx 45°$ and $\lambda \approx 135°$, reach a maximum of approximately 11′ in the neighbourhood of 0° and 180° and a minimum of approximately −12′ around 90°.

al-Khwārizmī's table for the equation of time is not the mean solar longitude either, we have to consider the possibility that the equation of time was tabulated by methods different from those presented in Section 5. A powerful mathematical tool which can be used to determine multiple unknown parameter values from an astronomical table and to find more information about the tabulated function, is the *method of least squares*. In the following pages this method will be explained in detail.

### METHOD OF LEAST SQUARES.

The use of the method of least squares for the determination of the parameter values underlying a given astronomical table will be illustrated by means of Table 2. The 1st column of this table contains arguments $\lambda$ of a table for the equation of time, the 2nd column tabular values $T(\lambda)$, in this case taken from al-Majrīṭī's recension of al-Khwārizmī's *zīj*. The 3rd column contains equation of time values $E(\lambda)$ computed for a historically plausible set of parameter values, namely obliquity 23°51′, solar eccentricity 2;20 (corresponding to al-Khwārizmī's maximum equation 2°14′), solar apogee 77°55′ (as given by al-Majrīṭī), epoch constant 4°30′ (found above) and conversion factor 15, under the assumption that the independent variable is the true solar longitude. The 4th column contains the differences $T(\lambda)-E(\lambda)$ between al-Majrīṭī's tabular values and our computation, the 5th column the squares of these differences. The sum of the squares (taken over all tabular values present in al-Majrīṭī's recension) is indicated at the end of the table.

From the 4th column of Table 2 it can be seen as follows that either our assumption that al-Khwārizmī's table has the true solar longitude as its independent variable or the chosen parameter values are incorrect. Normally, when we recompute a medieval astronomical table using the correct formula and parameter values, we find differences which have more or less random values and a maximum size of at most a couple of units of the final sexagesimal position. An example of such randomly distributed differences is shown in Figure 4, where the solar position has been plotted horizontally and the differences (each indicated by a dot) vertically. In the 4th column of Table 2 we not only find differences up to 100 units (namely 1′44″), but in a plot of these differences (Figure 5) we can also clearly recognize a non-random pattern, which has more or less the shape of the equation of time itself (cf. Figure 3). The presence of such patterns in the differences usually points to the use of an

| arg. λ | T(λ) (al-Majrīṭī) | E(λ) (computed) | differences T(λ) − E(λ) | squares of the differences |
|---|---|---|---|---|
| 10 | 0;11,28 | 0;13, 5,40,24, 1 | −0; 1,37,40,24, 1 | 0; 0, 2,39, 0, 5 |
| 20 | 0;15, 8 | 0;16,47,55,48,54 | −0; 1,39,55,48,54 | 0; 0, 2,46,26, 3 |
| 30 | 0;18,28 | 0;20, 2,21,52,44 | −0; 1,34,21,52,44 | 0; 0, 2,28,24,41 |
| 40 | 0;21, 4 | 0;22,30,18,57, 8 | −0; 1,26,18,57, 8 | 0; 0, 2, 4,10,26 |
| 50 | 0;22,48 | 0;23,57,55,18,24 | −0; 1, 9,55,18,24 | 0; 0, 1,21,29, 3 |
| 60 | 0;23,28 | 0;24,18,27,17,28 | −0; 0,50,27,17,28 | 0; 0, 0,42,25,42 |
| 70 | 0;23, 0 | 0;23,34,23,43,45 | −0; 0,34,23,43,45 | 0; 0, 0,19,43, 3 |
| 80 | 0;21,32 | 0;21,58,21,35,23 | −0; 0,26,21,35,23 | 0; 0, 0,11,34,50 |
| 90 | 0;19,40 | 0;19,51,56,41,35 | −0; 0,11,56,41,35 | 0; 0, 0, 2,22,41 |
| 100 | 0;17,32 | 0;17,42, 7,59,49 | −0; 0,10, 7,59,49 | 0; 0, 0, 1,42,41 |
| 110 | 0;15,52 | 0;15,56, 0,34,33 | −0; 0, 4, 0,34,33 | 0; 0, 0, 0,16, 5 |
| 120 | 0;14,44 | 0;14,55,28,26, 2 | −0; 0,11,28,26, 2 | 0; 0, 0, 2,11,39 |
| 130 | 0;14,44 | 0;14,53,38,17,34 | −0; 0, 9,38,17,34 | 0; 0, 0, 1,32,54 |
| 140 | 0;15,44 | 0;15,53,39,28,19 | −0; 0, 9,39,28,19 | 0; 0, 0, 1,33,16 |
| 150 | 0;17,36 | 0;17,49,38,24,14 | −0; 0,13,38,24,14 | 0; 0, 0, 3, 6, 3 |
| 160 | 0;20,24 | 0;20,28,41,33,54 | −0; 0, 4,41,33,54 | 0; 0, 0, 0,22, 1 |
| 170 | 0;23,32 | 0;23,33,13,53,43 | −0; 0, 1,13,53,43 | 0; 0, 0, 0, 1,31 |
| 180 | 0;26,52 | 0;26,43, 2,19, 9 | 0; 0, 8,57,40,51 | 0; 0, 0, 1,20,18 |
| 190 | 0;29,52 | 0;29,36,56,49,11 | 0; 0,15, 3,10,49 | 0; 0, 0, 3,46,36 |
| 200 | 0;32,24 | 0;31,54,16,28, 2 | 0; 0,29,43,31,58 | 0; 0, 0,14,43,36 |
| 210 | 0;34, 0 | 0;33,16,14,32,26 | 0; 0,43,45,27,34 | 0; 0, 0,31,54,44 |
| 220 | 0;34,28 | 0;33,27,36,53,53 | 0; 1, 0,23, 6, 7 | 0; 0, 1, 0,46,21 |
| 230 | 0;33,36 | 0;32,18,41, 6,52 | 0; 1,17,18,53, 8 | 0; 0, 1,39,37,34 |
| 240 | 0;31,24 | 0;29,47,28,59,23 | 0; 1,36,31, 0,37 | 0; 0, 2,35,15,30 |
| 250 | 0;27,44 | 0;26, 1,42,15, 8 | 0; 1,42,17,44,52 | 0; 0, 2,54,24,26 |
| 260 | 0;23, 4 | 0;21,19,28,46,11 | 0; 1,44,31,13,49 | 0; 0, 3, 2, 4,32 |
| 270 | 0;17,52 | 0;16, 8, 3,18,25 | 0; 1,43,56,41,35 | 0; 0, 3, 0, 4,32 |
| 280 | 0;12,32 | 0;11, 0, 1,38,37 | 0; 1,31,58,21,23 | 0; 0, 2,20,58,58 |
| 290 | 0; 7,44 | 0; 6,27,53,26,34 | 0; 1,16, 6,33,26 | 0; 0, 1,36,32,37 |
| 300 | 0; 3,48 | 0; 2,58,35,17, 8 | 0; 0,49,24,42,52 | 0; 0, 0,40,41,32 |
| 310 | 0; 1,12 | 0; 0,49,45,17,10 | 0; 0,22,14,42,50 | 0; 0, 0, 8,14,51 |
| 320 | 0; 0, 2 | 0; 0, 8,24,40,40 | −0; 0, 6,24,40,40 | 0; 0, 0, 0,41, 6 |
| 330 | 0; 0,20 | 0; 0,51,45,10,37 | −0; 0,31,45,10,37 | 0; 0, 0,16,48,15 |
| 340 | 0; 1,52 | 0; 2,49, 6, 9,10 | −0; 0,57, 6, 9,10 | 0; 0, 0,54,20,42 |
| 350 | 0; 4,28 | 0; 5,44, 8,53, 6 | −0; 1,16, 8,53, 6 | 0; 0, 1,36,38,32 |
| 360 | 0; 7,48 | 0; 9,16,57,40,51 | −0; 1,28,57,40,51 | 0; 0, 2,11,54, 7 |
| SUM OF THE SQUARES OF THE DIFFERENCES: | | | | 0; 6,18,43,18, 0 |

Table 2 *Illustration of the method of least squares*

incorrect formula or incorrect parameter values for the computation.

In order to find the values of the underlying parameters which yield the best agreement with the given table for the equation of time, we can now use the *method of least squares*. The 5th column of table 2 contains the squares of the differences in the 4th column; in the bottom row we find the sum of these squares over all tabular values present in al-Majrīṭī's recension (of these values only every tenth is displayed in Table 2). If we use different sets of parameter values for the computation in the 3rd column, then the differences in the 4th column, the squares of the differences in the 5th column, and the sum of the squares will all be different. *According to the method of least squares, the parameter values are determined in such a way that the sum of the squares of the differences between al-Majrīṭī's table and the computed table is as small as possible.* Expressed mathematically, the parameter values are determined by minimizing the sum

$$\sum_{\lambda} \left( T(\lambda) - E(\lambda) \right)^2,$$

which is taken over all values of λ for which tabular values are available. Since squares are positive, the sum of the squares of the differences can only be small if the absolute value of all differences is small, i.e. if all computed values are close to the given tabular values.

Instead of the sum of the squares of the differences we will mostly use the so-called *standard deviation* of the differences. The standard deviation is calculated by dividing the sum of the squares of the differences by the total number of tabular values and taking the square root of the quotient, i.e. the standard deviation is the square root of the average squared difference.[18] The standard deviation is a popular measure for the size of the differences between any two sets of comparable values. In our example in Table 2 the mean square of the differences is approximately 0;0,1,3,7,13 (namely 0;6,18,43,18 / 360), and the standard deviation 0;1,1,32,25. We will see below that if we recompute a table with values to seconds using the correct formula and parameter values, the standard deviation of the differences between tabular values and computed values is approximately 0;0,0,17. Thus for our recomputation of al-Majrīṭī's table the standard deviation of the

---

[18]For statistical purposes the standard deviation is usually calculated by dividing the sum of the squares of the differences by $n-1$, where $n$ is the total number of tabular values, and taking the square root. In this way the standard deviation yields a better approximation to one of the parameters describing the statistical properties of the differences.

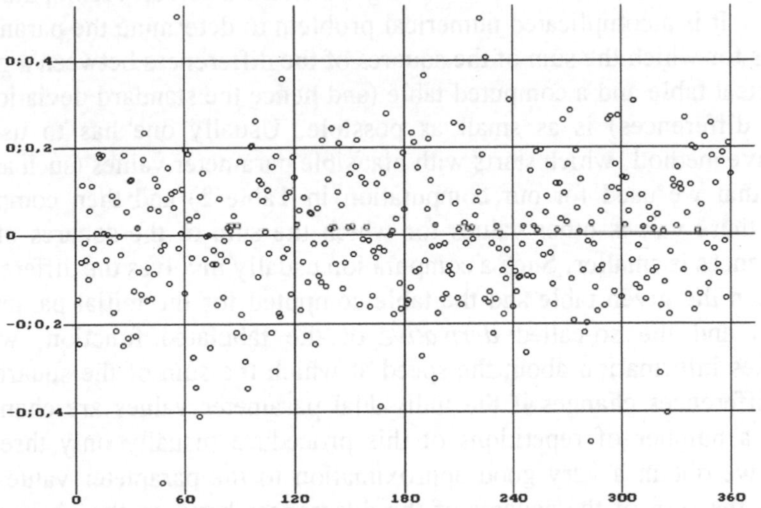


***Figure 4*** ***Random differences between equation of time values to seconds and recomputed values (horizontally: the solar longitude; vertically: the differences in hours).***

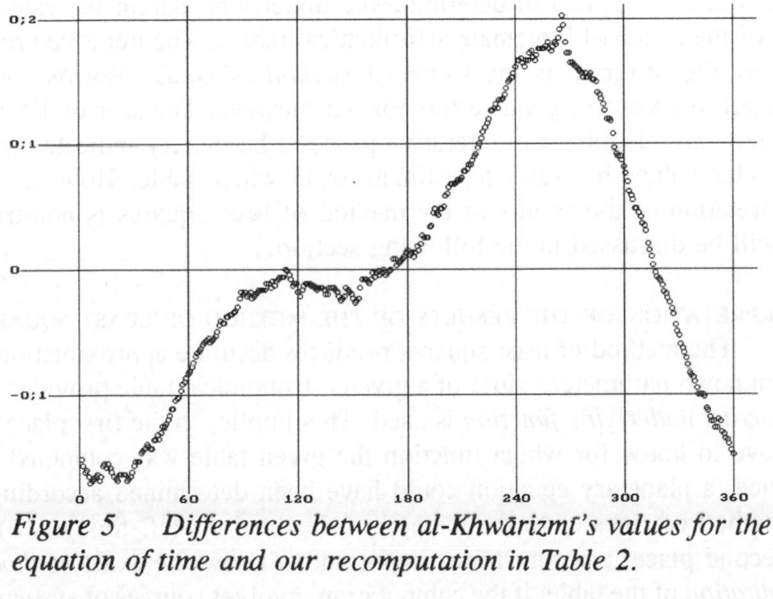


***Figure 5*** ***Differences between al-Khwārizmī's values for the equation of time and our recomputation in Table 2.***

differences is more than 200 times larger than for a correct recomputation.

It is a complicated numerical problem to determine the parameter values for which the sum of the squares of the differences between a given historical table and a computed table (and hence the standard deviation of these differences) is as small as possible. Usually one has to use an iterative method, which starts with plausible parameter values (such as the ones that we used for our computation in Table 2) and then computes from these values other values for which the sum of the squares of the differences is smaller. Such a computation usually involves the differences between the given table and the table computed for the initial parameter values and the so-called *derivative* of the tabulated function, which supplies information about the speed at which the sum of the squares of the differences changes if the individual parameter values are changed. After a number of repetitions of this procedure (usually only three or four) we obtain a very good approximation to the parameter values for which the sum of the squares of the differences between the given table and a computed table is as small as possible. The values found will be called the *least squares estimates* of the parameters underlying the given table.

In my computer program TA (Table-Analysis) the method of least squares can be applied to determine the underlying parameter values of most of the standard Ptolemaic astronomical tables. The iterative process used by the program is the so-called *method of Gauss-Newton*, which turns out to give very good results for our purpose. The user of TA need not know any details of the iterative process; he merely indicates which parameter values he wants to estimate from which table. However, the interpretation of the results of the method of least squares is non-trivial and will be discussed in the following section.

INTERPRETATION OF THE RESULTS OF THE METHOD OF LEAST SQUARES.

The method of least squares produces accurate approximations to the unknown parameter values of a given astronomical table provided that the *correct underlying function* is used. This implies in the first place that we have to know for which function the given table was computed; for instance, a planetary equation could have been determined according to the method of sines or to the method of declinations (cf. footnote 6). In the second place it is sometimes important to know the exact *method of computation* of the table: if the computation involves sources of systematic error, such as linear interpolation, severe truncation of intermediate results and the use of inaccurate auxiliary tables, the results of the method of least squares could be invalid. In order to decide whether the results are valid the following three criterions should be applied:

1) *The standard deviation of the differences between the given historical table and a table computed for the least squares estimates of the underlying parameters should be reasonably small.*[19] First note that we cannot expect this standard deviation to be equal to zero. Even if we had chosen the correct underlying function and parameter values for our recomputation of al-Khwārizmī's table for the equation of time in Table 2, the tabular values in the 2nd column would have been rounded versions of the exact values in the 3rd column and the differences in the 4th column would have been between −0;0,0,30 and +0;0,0,30. It can be shown statistically that in that case the standard deviation of the differences between the exact and the rounded values is approximately 0;0,0,17.[20]

As a result the standard deviation of the differences between a given historical table with values to seconds and a table computed for the least squares estimates of the underlying parameters will normally not be smaller than 0;0,0,17. If the standard deviation is much larger than 0;0,0,17, we have to consider the possibility that we have chosen an incorrect underlying function.

2) *The differences between the given historical table and a computation based on the least squares estimates of the underlying parameters should be random and not display obvious patterns.* If the differences display sine-wave or other regular patterns, we can be certain

---

[19]Bear in mind that this standard deviation is the smallest possible for all sets of parameter values.

[20]The differences between correctly rounded tabular values and exactly calculated functional values can be assumed to have a so-called uniform probability distribution. If the tabular values were calculated to seconds, this implies that all possible digits occur approximately equally often in the third sexagesimal fractional position of the calculated functional values. The expected standard deviation of such uniformly distributed differences can be calculated as approximately 0;0,0,17. If the tabular values were calculated to minutes, the expected standard deviation would be approximately 0;0,17, etc.

that we have used an incorrect underlying function or that we have neglected aspects of the computation of the table which caused systematic errors in the tabular values.[21] If the differences seem to be random, we have probably chosen the correct underlying function even if the standard deviation of the differences is large. Examples of differences with obvious patterns can be found in Figures 5, 6 and 7; an example of random differences is shown in Figure 4.

3) *The least squares estimates should be (close to) historically plausible values for the parameters.* In practice there is only a limited number of possibilities for the values of the underlying parameters of a given historical table. These are either values attested in historical sources (like Ptolemy's value 23°51′20″ for the obliquity of the ecliptic and al-Battānī's 2;4,45 for the solar eccentricity) or round numbers (like al-Khwārizmī's value 4°30′ for the epoch constant, which we found above). If the least squares estimates are far removed from historically plausible parameter values, we have probably chosen an incorrect underlying function.

CONFIDENCE INTERVALS.

Even if we have chosen the correct underlying function, the least squares estimates of the parameters of a given astronomical table are normally not identical with the parameter values actually used for the computation. Those values are usually round numbers (see above), whereas the least squares estimates are numerically determined quantities which could in principle have any value; for example, 23;34,59,45,18,6 for the obliquity or 2;4,45,17,23,15 for the solar eccentricity.[22] After

---

[21]Note that before we applied the method of least squares in our example in Table 2 obvious patterns in the differences had two possible causes: an incorrect underlying function or incorrect values of the underlying parameters. Since we now consider the differences between the given table and a computation based on the least squares estimates of the parameters, which were determined in such a way that the differences are minimized, we can be certain that the cause of the patterns is an incorrect underlying function.

[22]The reason that the least squares estimates are not normally equal to the actual historical parameter values is that the tabular values contain rounding and possibly other types of errors. Even if we would use exact functional values for the

applying the method of least squares we thus have to find historically plausible, round numbers in the neighbourhood of the estimates in such a way that the standard deviation of the differences between the given table and a recomputation for the historical values is only slightly larger than the standard deviation for the least squares estimates. The decision how far the historically plausible values can be removed from the least squares estimates can be made on the basis of so-called *95 % confidence intervals* for the underlying parameters. These are statistically determined intervals around the least squares estimates which are expected to contain the parameter values used for the computation in 19 out of 20 cases.

For example, if we find a 95 % confidence interval ⟨ 23;34,57 , 23;35,6 ⟩ for the obliquity of the ecliptic, we can safely conclude that the underlying parameter value is 23°35′, since this is the only historically plausible value in the neighbourhood of the confidence interval. However, if we find a 95 % confidence interval ⟨ 2;4,27 , 2;4,56 ⟩ for the solar eccentricity, our table could be based on either of the attested values 2;4,35,30 (corresponding to a maximum solar equation of 1°59′) and 2;4,45 (corresponding to 1°59′10″).

## Application of the Method of Least Squares to al-Khwārizmī's Table.

We have already found that the conversion factor underlying al-Khwārizmī's table for the equation of time is 15°/ hour. Furthermore, we expect that the argument of the table is the true solar longitude. Under these assumptions the results of the application of the method of least squares (as displayed by my program TA) are as follows:

Equation of Time, al-Khwārizmī (Suter Tables 67-68)
LEAST SQUARES ESTIMATION from the values for arguments 1, 2, ..., 360.

FINAL RESULT (after 3 iterations)

| Parameter | Estimate | 95 % Confidence Interval |
|---|---|---|
| OBLIQUITY | 23;50, 6,30,45, 1 | ⟨ 23;44,58, 0,34,57 , 23;55,13,58,27,47 ⟩ |
| ECCENTRICITY | 2;29,50,28,18,53 | ⟨ 2;28,39,20,50,30 , 2;31, 1,35,47,15 ⟩ |
| APOGEE | 84;40,33,21,39,30 | ⟨ 84;13,20,52,13, 6 , 85; 7,45,51, 5,54 ⟩ |
| EPOCH CONSTANT | 4;30, 3, 0, 0, 0 | ⟨ 4;29,14,56,34, 9 , 4;30,51, 3,25,51 ⟩ |

Standard Deviation of the Differences: 0;0,31,0,51,32

---

application of the method of least squares, the estimates need not be equal to the actual parameter values because of the internal rounding in our computer.

Although we have found historically plausible values for the obliquity of the ecliptic and the solar eccentricity (Ptolemy's and al-Khwārizmī's value 23;51 for the obliquity and Ptolemy's value 2;30 for the eccentricity lie in the middle of the respective 95 % confidence intervals), we cannot be satisfied with the result. We have seen that all tabular values are multiples of four seconds. Therefore the standard deviation of the differences between al-Khwārizmī's table and a table computed for the least squares estimates would approximately be 4 · 0;0,0,17 = 0;0,1,28 if we had used the correct function and method of computation. The standard deviation found is more than 20 times as large. Since furthermore the differences display a clear sine wave pattern with an amplitude of approximately 45 seconds (Figure 6), we must conclude that we have not used the correct underlying function, i.e. that al-Khwārizmī's table is not an ordinary table for the equation of time as a function of the true solar longitude.

We will therefore perform the least squares estimation for other possibilities of the underlying function. If we assume that the argument of the table is the *mean* solar longitude, the results are as follows:

EQUATION OF TIME, AL-KHWĀRIZMĪ (SUTER TABLES 67-68)
LEAST SQUARES ESTIMATION FROM THE VALUES FOR ARGUMENTS 1, 2, ..., 360.

FINAL RESULT (AFTER 3 ITERATIONS)

| PARAMETER | ESTIMATE | 95 % CONFIDENCE INTERVAL |
|---|---|---|
| OBLIQUITY | 23;35,31,17,18,32 | ⟨ 23;29,12,34,43,30 , 23;41,48,24,45,48 ⟩ |
| ECCENTRICITY | 2;36,11,51,25, 0 | ⟨ 2;34,41,48,17,40 , 2;37,41,54,32,20 ⟩ |
| APOGEE | 85;17,30,12,50,45 | ⟨ 84;47,11,46, 4, 3 , 85;47,48,39,37,27 ⟩ |
| EPOCH CONSTANT | 4;30, 3, 0, 0, 0 | ⟨ 4;29, 4,42,13,34 , 4;31, 1,17,46,26 ⟩ |

STANDARD DEVIATION OF THE DIFFERENCES: 0;0,37,37,19,59

We now find a completely different plausible value for the obliquity of the ecliptic (the common Islamic value 23;35), but a practically impossible value for the solar eccentricity. Furthermore, the minimum possible standard deviation is again much larger than the value 0;0,1,28 which we expect for the correct underlying function, and the differences between al-Khwārizmī's table and a table computed on the basis of the estimates again display an obvious pattern (this time more complicated than an ordinary sine-wave; see Figure 7). We conclude that the tabulated function is not the equation of time as a function of the mean solar longitude either.

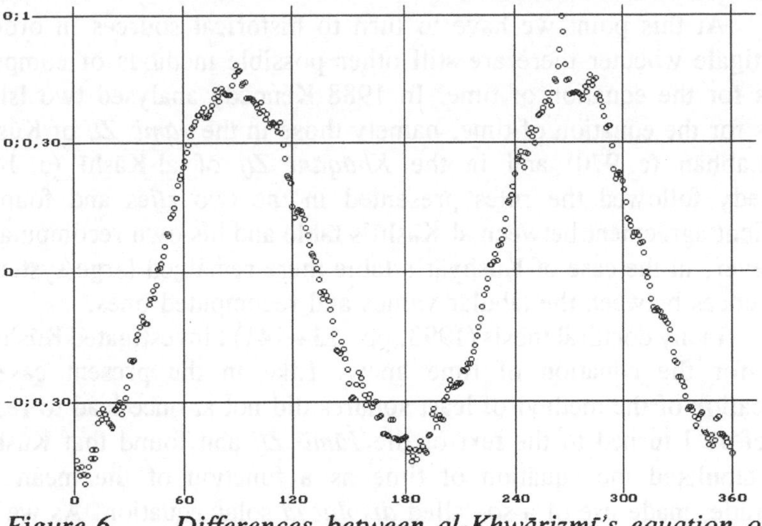


***Figure 6*** ***Differences between al-Khwārizmī's equation of time and the best possible recomputation under the assumption that the argument is the* true *solar longitude.***

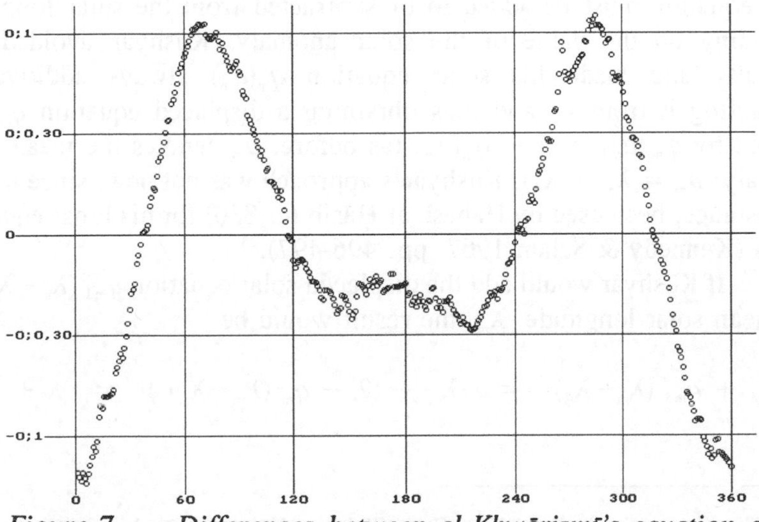


***Figure 7*** ***Differences between al-Khwārizmī's equation of time and the best possible recomputation under the assumption that the argument is the* mean *solar longitude.***

DISPLACED SOLAR EQUATION.

At this point we have to turn to historical sources in order to investigate whether there are still other possible methods of computing tables for the equation of time. In 1988 Kennedy analysed two Islamic tables for the equation of time, namely those in the *Jāmiᶜ Zīj* of Kūshyār ibn Labbān (c. 970) and in the *Khāqānī Zīj* of al-Kāshī (c. 1420). Kennedy followed the rules presented in the two *zīj*es and found an excellent agreement between al-Kāshī's table and his own recomputation. However, in the case of Kūshyār's table there remained large systematic differences between the tabular values and recomputed ones.

In my doctoral thesis (1993, pp. 134-141) I investigated Kūshyār's table for the equation of time anew. Like in the present case, an application of the method of least squares did not at once lead to results. Therefore I turned to the text of the *Jāmiᶜ Zīj* and found that Kūshyār, who tabulated the equation of time as a function of the mean solar longitude, made use of a so-called *displaced* solar equation. As we have seen in Section 5, the solar equation as determined by Ptolemy and most Islamic astronomers is sometimes additive and sometimes subtractive. This implies that the user of a solar equation table had to decide whether the solar equation must be added to or subtracted from the solar longitude depending on the value of the solar anomaly. Kūshyār avoided this difficulty and made his solar equation $q_m(a_m)$ always additive by subtracting it from 2° and thus obtaining a displaced equation $q_{md}(a_m)$ defined by $q_{md}(a_m) = 2 - q_m(a_m)$ (as before, $a_m$ denotes the mean solar anomaly: $a_m = \lambda_m - \lambda_A$). Kūshyār's approach was not new, since it had, for instance, been used by Ḥabash al-Ḥāsib (c. 830) for his lunar equation tables (Kennedy & Salam 1967, pp. 496-497).[23]

If Kūshyār would add the displaced solar equation $q_{md}(\lambda_m - \lambda_A)$ to the mean solar longitude $\lambda_m$, the result would be

$$\lambda_m + q_{md}(\lambda_m - \lambda_A) = \lambda_m + (2 - q_m(\lambda_m - \lambda_A)) = \lambda + 2$$

[23]The manuscript Istanbul Yeni Cami 784/2 of Ḥabash's *zīj* also contains a table for $\lambda_A + q_m(a_m)$, where $\lambda_A$ is the longitude of the solar apogee and $q_m(a_m)$ the solar equation as a function of the mean solar anomaly (cf. Debarnot 1987, p. 58). From this table the true solar position can be calculated by taking the tabular value for the desired mean solar anomaly and adding it to that anomaly.

| $\lambda_m$ | "ordinary" solar equation | $\lambda_m$ | displaced solar equation | $\lambda_{ms}$ | shifted displaced solar equation |
|---|---|---|---|---|---|
| 356 | −0; 8, 2 | 356 | 2; 8, 2 | 356 | 2; 4, 1 |
| 357 | −0; 6, 1 | 357 | 2; 6, 1 | 357 | 2; 2, 1 |
| 358 | −0; 4, 1 | 358 | 2; 4, 1 | 358 | 2, 0, 0 |
| 359 | −0; 2, 1 | 359 | 2; 2, 1 | 359 | 1;57,59 |
| 0 | 0; 0, 0 | 0 | 2, 0, 0 | 0 | 1;55,59 |
| 1 | 0; 2, 1 | 1 | 1;57,59 | 1 | 1;53,59 |
| 2 | 0; 4, 1 | 2 | 1;55,59 | 2 | 1;51,58 |
| 3 | 0; 6, 1 | 3 | 1;53,59 | 3 | 1;49,58 |
| 4 | 0; 8, 2 | 4 | 1;51,58 | 4 | 1;47,58 |
| 86 | 1;58,30 | 86 | 0; 1,30 | 86 | 0; 1,10 |
| 87 | 1;58,41 | 87 | 0; 1,19 | 87 | 0; 1, 2 |
| 88 | 1;58,50 | 88 | 0; 1,10 | 88 | 0; 0,56 |
| 89 | 1;58,58 | 89 | 0; 1, 2 | 89 | 0; 0,52 |
| 90 | 1;59, 4 | 90 | 0; 0,56 | 90 | 0; 0,50 |
| 91 | 1;59, 8 | 91 | 0; 0,52 | 91 | 0; 0,52 |
| 92 | 1;59,10 | 92 | 0; 0,50 | 92 | 0; 0,57 |
| 93 | 1;59, 8 | 93 | 0; 0,52 | 93 | 0; 1, 4 |
| 94 | 1;59, 3 | 94 | 0; 0,57 | 94 | 0; 1,12 |

Table 3 *The displacement and the shift of Kūshyār's solar equation table*

instead of the desired true solar longitude $\lambda$ itself (cf. Section 5). Kūshyār therefore replaced $\lambda_m$ by a *shifted* mean solar longitude $\lambda_{ms}$ defined by $\lambda_{ms} = \lambda_m - 2$. The addition of the displaced solar equation to the corresponding shifted mean solar longitude then yielded the true solar longitude $\lambda$. In order to tabulate the displaced solar equation as a function of the shifted mean solar longitude Kūshyār had to shift all values two degrees backwards, thus tabulating

$$q_{md}\,(\lambda_{ms} - \lambda_A) \;=\; 2 - q_m\,(\lambda_{ms} - \lambda_A + 2)$$

(cf. Table 3).[24] In that way he could calculate the true solar position

[24]Thus the zeros $q_m(0°)=0;0,0$ and $q_m(180°)=0;0,0$ of Kūshyār's original solar equation lead to displaced solar equation values $q_{md}(-2°) = (q_{md}(358°)=)\ 2;0,0$ and $q_{md}(178°) = 2;0,0$. The maximum value $q_m(92°) = 1;59,10$ leads to a minimum $q_{md}(90°) = 0;0,50$ and the minimum value $q_m(268°) = -1;59,10$ to a maximum $q_{md}(266°) = 3;59,10$ (in each case the argument of $q_{md}$ is the shifted mean solar

corresponding to a given shifted mean solar longitude $\lambda_{ms}$ by adding $q_{md}(\lambda_{ms}-\lambda_A)$ to $\lambda_{ms}$:

$$\begin{aligned} \lambda_{ms} + q_{md}(\lambda_{ms} - \lambda_A) &= (\lambda_m - 2) + (2 - q_m(\lambda_{ms} + 2 - \lambda_A)) \\ &= \lambda_m - q_m(\lambda_m - \lambda_A) \\ &= \lambda \end{aligned}$$

It now seems natural that also the argument of Kūshyār's table for the equation of time was the shifted mean solar longitude $\lambda_{ms}$ instead of the ordinary mean solar longitude $\lambda_m$. Thus we expect that the tabulated function is:

$$\begin{aligned} E_{ms}(\lambda_{ms}) &= E_m(\lambda_{ms}+2) \\ &= 1/15 \cdot (\lambda_{ms} + 2 - \alpha(\lambda_{ms} + 2 - q_m(\lambda_{ms}+2)) + c) \end{aligned}$$

(cf. formula 5 and note that the shifted equation of time for argument $\lambda_{ms}$ corresponds to the ordinary equation of time for argument $\lambda_m$, which is equal to $\lambda_{ms} + 2°$). We thus see that the equation of time as a function of the shifted mean solar longitude can be obtained from the "ordinary" equation of time by shifting all values two degrees backwards.

If we take into account that what Kūshyār calls the "mean solar longitude" is in fact a *shifted* mean solar longitude, we find a good agreement between his table for the equation of time and a recomputation following the rules presented in his *zīj* (van Dalen 1993, pp. 138-139).

THE SHIFT IN AL-KHWĀRIZMĪ'S TABLE FOR THE EQUATION OF TIME.

Different from Kūshyār's table for the equation of time, al-Khwārizmī's table is expected to have the true solar longitude as its independent variable. Although al-Khwārizmī's solar equation is not of the displaced type described above, it could nevertheless be worth while to investigate whether his equation of time values were shifted. For a given shift $\Delta$ we define the shifted true solar longitude $\lambda_s$ by $\lambda_s = \lambda - \Delta$. The shifted equation of time $E_s$ as a function of $\lambda_s$ is then given by:

$$E_s(\lambda_s) = E(\lambda_s + \Delta)$$

---

longitude).

$$= 1/15 \cdot (\lambda_s + \Delta + q\,(\lambda_s + \Delta) - \alpha(\lambda_s + \Delta) + c).$$

Again the resulting function is obtained from the "ordinary" equation of time by shifting all tabular values Δ degrees backwards. However, because of the shift some of the properties of the equation of time as a function of the true solar longitude which we derived from the symmetry relations satisfied by the right ascension and the solar equation (formulas 7, 8 and 11 above) are no longer valid. Formula 11 now holds only approximately, formula 8 yields a shifted solar equation $q(\lambda_s+\Delta) = 7\frac{1}{2} \cdot (E_s(\lambda_s) - E_s(180+\lambda_s))$, and instead of formula 7 we obtain:

$$\alpha\,(\lambda_s+\Delta) - (\lambda_s+\Delta) \;=\; c - 7\tfrac{1}{2} \cdot (\,E_s(\lambda_s) \;+\; E_s(\lambda_s + 180°)\,) \qquad (12)$$

for every value of $\lambda_s$. Because $\lambda_s+\Delta$ equals $\lambda$ and $\alpha(\lambda) - \lambda = 0$ whenever $\lambda$ is a multiple of 90°, we expect that the right hand side of formula 12 equals 0 whenever $\lambda_s$ equals a multiple of 90° minus Δ. Since we usually do not have an exact value for $c$ and since the values of $\lambda_s$ for which the right hand side of formula 12 is precisely equal to zero need not be among the arguments of our table, this property allows us only in exceptional cases to determine the shift.

A more effective method for determining the shift is to regard it as a fifth parameter of the equation of time and to approximate it together with the other underlying parameters using the method of least squares. If we assume that the independent variable of al-Majrīṭī's table is a shifted true solar longitude, the results are as follows:

EQUATION OF TIME AL-KHWĀRIZMĪ (SUTER TABLES 67-68)
LEAST SQUARES ESTIMATION FROM THE VALUES FOR ARGUMENTS 1, 2, ..., 360.

FINAL RESULT (AFTER 3 ITERATIONS)

| PARAMETER | ESTIMATE | 95 % CONFIDENCE INTERVAL |
|---|---|---|
| OBLIQUITY | 23;51,51, 2,41,32 | ⟨23;51,21, 8,10,36 , 23;52,20,56,37,11⟩ |
| ECCENTRICITY | 2;29,50,28,18,53 | ⟨ 2;29,43,33,23,37 , 2;29,57,23,14, 8⟩ |
| APOGEE | 82;39, 3,53,30,19 | ⟨82;36, 8,48, 1, 3 , 82;41,58,58,59,35⟩ |
| EPOCH CONSTANT | 4;30, 3, 0, 0, 0 | ⟨ 4;29,58,19,38,52 , 4;30, 7,40,21, 8⟩ |
| SHIFT | −2; 1,29,28, 9,11 | ⟨−2; 2,43,23, 2,39 , −2; 0,15,33,15,43⟩ |

STANDARD DEVIATION OF THE DIFFERENCES: 0;0,3,0,55,44

We first note that the minimum possible standard deviation of the differences between al-Khwārizmī's table and computed values based on the assumption of a shift is much smaller than the standard deviations we obtained before. In fact, the standard deviation found is only twice as large as the value 0;0,1,28 which we expect if we have chosen the correct underlying function for a table of which all values are multiples of four seconds.

Secondly we note that the least squares estimates for the shifted equation of time are close to historically plausible values for all underlying parameters: Ptolemy's and al-Khwārizmī's value 23°51′ for the obliquity of the ecliptic, Ptolemy's value 2;30 for the solar eccentricity, and the value 82°39′ (or possibly 82°40′) for the longitude of the solar apogee. This value was determined from observations made by order of the caliph al-Ma'mūn (c. 830) and was used in the *zījes* of al-Khwārizmī's contemporaries Yaḥyā ibn Abī Manṣūr and Ḥabash al-Ḥāsib. The value 4;30 for the epoch constant is in agreement with what we have found before using formula 11, and the estimated shift is close to −2° (i.e. 2 degrees forwards). The fact that some of the plausible parameter values lie just outside of their 95 % confidence intervals, could point to small systematic errors in the computation of the table. As was mentioned before, possible causes of such errors are linear interpolation in the equation of time itself or in the underlying tables, systematic truncation of intermediate results, etc.

If we recompute al-Khwārizmī's table for the equation of time for the historically plausible parameter values mentioned above, we find that the differences between table and recomputation are generally smaller than 7 seconds and display no obvious global pattern (see Tables 4a to 4c and Figure 8). There are some local patterns in the differences (for example, the small mountains around arguments 6, 65 and 216, and the somewhat larger one around 266°). These could be indications of the small systematic errors indicated above. However, the general pattern of the differences is random enough to conclude that the historically plausible parameter values found above were in fact used for the computation of al-Khwārizmī's table for the equation of time.

The use of the method of least squares has (indirectly) confirmed that al-Khwārizmī's table displays the equation of time as a function of the true solar longitude and that the conversion factor used is 15°/ hour. If we apply the method of least squares for the equation of time as a function of a shifted *mean* solar longitude, we find a minimum possible standard deviation of 19 seconds and differences between table and recomputation

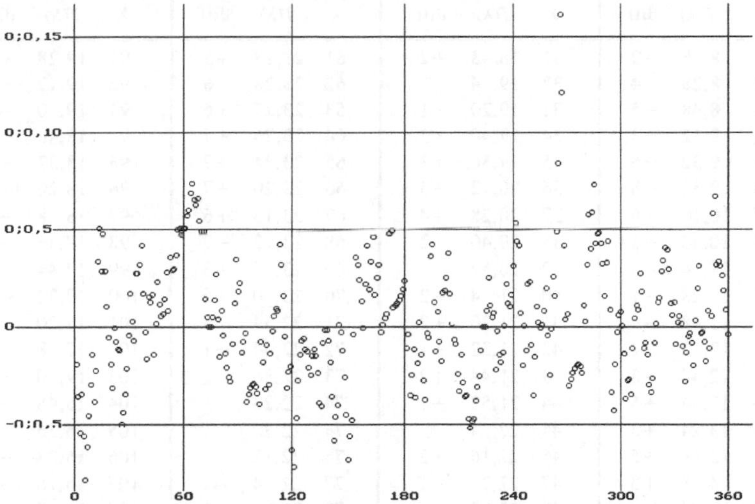


*Figure 8 Differences between al-Khwārizmī's values for the equation of time and our final recomputation under the assumption that the tabular values were shifted.*

showing obvious sine-wave patterns. If we assume that the conversion factor is 15;2,28 instead of 15, the minimum possible standard deviation of the differences is again 3 seconds and the differences are just as random as in the case of conversion factor 15, but the least squares estimates are further away from historically plausible values.

Assuming that the shift of al-Khwārizmī's table for the equation of time is −2° precisely, we can easily reconstruct the underlying "ordinary" table for the equation of time. From this table we can reconstruct the right ascension and the solar equation according to formulas 7 and 8. It turns out that both underlying tables contain large groups of small errors of the same sign pointing to the presence of some source of small systematic error. I have not been able to determine this source, but it is probably the same one that also caused the above-mentioned local patterns in the differences in Tables 4a to 4c and Figure 8.

| λ | T(λ) | diff | λ | T(λ) | diff | λ | T(λ) | diff | λ | T(λ) | diff |
|---|---|---|---|---|---|---|---|---|---|---|---|
| 1 | 8, 8 | −2 | 31 | 18,48 | +2 | 61 | 23,28 | +5 | 91 | 19,28 | +2 |
| 2 | 8,28 | −4 | 32 | 19, 4 | | 62 | 23,28 | +6 | 92 | 19,12 | −1 |
| 3 | 8,48 | −5 | 33 | 19,20 | −1 | 63 | 23,27 | +6 | 93 | 19, 0 | −1 |
| 4 | 9,12 | −3 | 34 | 19,40 | +3 | 64 | 23,26 | +7 | 94 | 18,48 | |
| 5 | 9,32 | −6 | 35 | 19,56 | +3 | 65 | 23,24 | +7 | 95 | 18,32 | −4 |
| 6 | 9,52 | −8 | 36 | 20,12 | +3 | 66 | 23,20 | +7 | 96 | 18,20 | −3 |
| 7 | 10,16 | −6 | 37 | 20,28 | +4 | 67 | 23,16 | +6 | 97 | 18, 8 | −3 |
| 8 | 10,40 | −5 | 38 | 20,40 | +2 | 68 | 23,12 | +7 | 98 | 17,56 | −3 |
| 9 | 11, 4 | −3 | 39 | 20,52 | | 69 | 23, 6 | +5 | 99 | 17,44 | −3 |
| 10 | 11,28 | −1 | 40 | 21, 4 | −2 | 70 | 23, 0 | +5 | 100 | 17,32 | −3 |
| 11 | 11,48 | −4 | 41 | 21,20 | +2 | 71 | 22,54 | +5 | 101 | 17,20 | −3 |
| 12 | 12,12 | −2 | 42 | 21,32 | +1 | 72 | 22,44 | +1 | 102 | 17, 8 | −4 |
| 13 | 12,40 | +3 | 43 | 21,44 | +2 | 73 | 22,36 | | 103 | 17, 0 | −1 |
| 14 | 13, 4 | +5 | 44 | 21,52 | −1 | 74 | 22,28 | | 104 | 16,48 | −2 |
| 15 | 13,24 | +3 | 45 | 22, 4 | | 75 | 22,20 | | 105 | 16,36 | −3 |
| 16 | 13,48 | +5 | 46 | 22,16 | +2 | 76 | 22,12 | | 106 | 16,28 | −1 |
| 17 | 14, 8 | +3 | 47 | 22,24 | +1 | 77 | 22, 4 | +1 | 107 | 16,16 | −3 |
| 18 | 14,28 | +1 | 48 | 22,32 | +1 | 78 | 21,56 | +3 | 108 | 16, 8 | −1 |
| 19 | 14,48 | −1 | 49 | 22,40 | +1 | 79 | 21,44 | | 109 | 16, 0 | |
| 20 | 15, 8 | −2 | 50 | 22,48 | +1 | 80 | 21,32 | −2 | 110 | 15,52 | +1 |
| 21 | 15,28 | −3 | 51 | 22,56 | +3 | 81 | 21,24 | +1 | 111 | 15,44 | +1 |
| 22 | 15,52 | | 52 | 23, 0 | +1 | 82 | 21,12 | | 112 | 15,36 | +1 |
| 23 | 16,12 | −1 | 53 | 23, 8 | +4 | 83 | 21, 0 | −1 | 113 | 15,32 | +4 |
| 24 | 16,32 | −1 | 54 | 23,12 | +3 | 84 | 20,48 | −2 | 114 | 15,24 | +3 |
| 25 | 16,52 | −1 | 55 | 23,16 | +3 | 85 | 20,36 | −3 | 115 | 15,16 | +2 |
| 26 | 17, 8 | −5 | 56 | 23,20 | +4 | 86 | 20,24 | −3 | 116 | 15, 8 | −1 |
| 27 | 17,28 | −4 | 57 | 23,24 | +5 | 87 | 20,14 | −1 | 117 | 15, 0 | −3 |
| 28 | 17,48 | −3 | 58 | 23,26 | +5 | 88 | 20, 4 | +1 | 118 | 14,56 | −2 |
| 29 | 18, 8 | −2 | 59 | 23,27 | +5 | 89 | 19,54 | +3 | 119 | 14,52 | −2 |
| 30 | 18,28 | | 60 | 23,28 | +5 | 90 | 19,40 | +2 | 120 | 14,44 | −6 |

Table 4a *al-Khwārizmī's values for the equation of time ( T(λ) ) and the differences between al-Khwārizmī's values and our final recomputation (1st part)*

| $\lambda$ | $T(\lambda)$ | diff | $\lambda$ | $T(\lambda)$ | diff | $\lambda$ | $T(\lambda)$ | diff | $\lambda$ | $T(\lambda)$ | diff |
|---|---|---|---|---|---|---|---|---|---|---|---|
| 121 | 14,40 | −7 | 151 | 17,48 | −7 | 181 | 27, 8 | −1 | 211 | 34, 8 | −1 |
| 122 | 14,40 | −5 | 152 | 18, 4 | −6 | 182 | 27,28 | | 212 | 34,12 | −3 |
| 123 | 14,40 | −3 | 153 | 18,20 | −5 | 183 | 27,44 | −4 | 213 | 34,16 | −4 |
| 124 | 14,40 | −1 | 154 | 18,40 | | 184 | 28, 4 | −2 | 214 | 34,20 | −4 |
| 125 | 14,40 | −1 | 155 | 19, 0 | +4 | 185 | 28,24 | −1 | 215 | 34,22 | −5 |
| 126 | 14,40 | −1 | 156 | 19,16 | +3 | 186 | 28,40 | −4 | 216 | 34,24 | −5 |
| 127 | 14, 0 | −1 | 157 | 19,32 | +2 | 187 | 29, 0 | −2 | 217 | 34,26 | −5 |
| 128 | 14,41 | −1 | 158 | 19,48 | +1 | 188 | 29,20 | | 218 | 34,27 | −5 |
| 129 | 14,42 | −2 | 159 | 20, 4 | | 189 | 29,36 | −1 | 219 | 34,28 | −4 |
| 130 | 14,44 | −2 | 160 | 20,24 | +2 | 190 | 29,52 | −3 | 220 | 34,28 | −3 |
| 131 | 14,48 | −2 | 161 | 20,44 | +4 | 191 | 30, 8 | −4 | 221 | 34,27 | −3 |
| 132 | 14,52 | −1 | 162 | 21, 0 | +2 | 192 | 30,28 | | 222 | 34,26 | −1 |
| 133 | 14,56 | −1 | 163 | 21,20 | +3 | 193 | 30,44 | −1 | 223 | 34,24 | |
| 134 | 15, 0 | −2 | 164 | 21,40 | +5 | 194 | 31, 4 | +4 | 224 | 34,20 | |
| 135 | 15, 4 | −4 | 165 | 21,56 | +2 | 195 | 31,20 | +4 | 225 | 34,16 | +1 |
| 136 | 15,10 | −4 | 166 | 22,16 | +3 | 196 | 31,32 | +1 | 226 | 34,12 | +3 |
| 137 | 15,20 | −1 | 167 | 22,36 | +3 | 197 | 31,44 | −1 | 227 | 34, 4 | +1 |
| 138 | 15,28 | | 168 | 22,52 | | 198 | 32, 0 | +1 | 228 | 33,56 | |
| 139 | 15,36 | | 169 | 23,12 | | 199 | 32,12 | −1 | 229 | 33,48 | +1 |
| 140 | 15,44 | −1 | 170 | 23,32 | | 200 | 32,24 | −2 | 230 | 33,36 | −2 |
| 141 | 15,52 | −2 | 171 | 23,52 | +1 | 201 | 32,40 | +2 | 231 | 33,28 | |
| 142 | 16, 0 | −4 | 172 | 24,16 | +5 | 202 | 32,52 | +2 | 232 | 33,16 | −1 |
| 143 | 16, 8 | −6 | 173 | 24,36 | +5 | 203 | 33, 4 | +3 | 233 | 33, 4 | −2 |
| 144 | 16,20 | −5 | 174 | 24,52 | +1 | 204 | 33,16 | +4 | 234 | 32,52 | −1 |
| 145 | 16,32 | −4 | 175 | 25,12 | +1 | 205 | 33,24 | +2 | 235 | 32,40 | |
| 146 | 16,48 | | 176 | 25,32 | +1 | 206 | 33,32 | | 236 | 32,28 | +2 |
| 147 | 17, 0 | | 177 | 25,52 | +1 | 207 | 33,40 | −1 | 237 | 32,12 | +1 |
| 148 | 17,12 | −1 | 178 | 26,12 | +2 | 208 | 33,48 | −1 | 238 | 31,56 | +1 |
| 149 | 17,24 | −3 | 179 | 26,32 | +2 | 209 | 33,54 | −2 | 239 | 31,40 | +2 |
| 150 | 17,36 | −4 | 180 | 26,52 | +2 | 210 | 34, 0 | −3 | 240 | 31,24 | +3 |

Table 4b *al-Khwārizmī's values for the equation of time ( $T(\lambda)$ ) and the differences between al-Khwārizmī's values and our final recomputation (2nd part)*

| λ | T(λ) | diff | λ | T(λ) | diff | λ | T(λ) | diff | λ | T(λ) | diff |
|---|---|---|---|---|---|---|---|---|---|---|---|
| 241 | 31, 8 | +5 | 271 | 17,16 | −3 | 301 | 3,28 | | 331 | 0,24 | −1 |
| 242 | 30,48 | +4 | 272 | 16,44 | −3 | 302 | 3, 8 | −1 | 332 | 0,32 | |
| 243 | 30,28 | +4 | 273 | 16,12 | −2 | 303 | 2,48 | −4 | 333 | 0,40 | +1 |
| 244 | 30, 4 | +1 | 274 | 15,40 | −2 | 304 | 2,32 | −3 | 334 | 0,48 | |
| 245 | 29,40 | −2 | 275 | 15, 8 | −2 | 305 | 2,16 | −3 | 335 | 0,56 | −1 |
| 246 | 29,20 | | 276 | 14,36 | −2 | 306 | 2, 0 | −3 | 336 | 1, 4 | −3 |
| 247 | 28,56 | −1 | 277 | 14, 4 | −2 | 307 | 1,48 | −1 | 337 | 1,16 | −1 |
| 248 | 28,36 | +2 | 278 | 13,32 | −2 | 308 | 1,36 | | 338 | 1,28 | −1 |
| 249 | 28,12 | +2 | 279 | 13, 4 | +1 | 309 | 1,24 | +1 | 339 | 1,40 | |
| 250 | 27,44 | −1 | 280 | 12,32 | | 310 | 1,12 | +1 | 340 | 1,52 | −1 |
| 251 | 27,16 | −4 | 281 | 12, 4 | +3 | 311 | 1, 3 | +3 | 341 | 2, 4 | −2 |
| 252 | 26,52 | −2 | 282 | 11,36 | +6 | 312 | 0,52 | +2 | 342 | 2,20 | |
| 253 | 26,28 | +1 | 283 | 11, 4 | +4 | 313 | 0,40 | −1 | 343 | 2,36 | +2 |
| 254 | 26, 0 | | 284 | 10,36 | +6 | 314 | 0,32 | −1 | 344 | 2,52 | +3 |
| 255 | 25,32 | | 285 | 10, 8 | +7 | 315 | 0,24 | −1 | 345 | 3, 4 | |
| 256 | 25, 4 | | 286 | 9,36 | +4 | 316 | 0,16 | −3 | 346 | 3,20 | |
| 257 | 24,36 | +1 | 287 | 9, 8 | +5 | 317 | 0,10 | −3 | 347 | 3,36 | −1 |
| 258 | 24, 8 | +2 | 288 | 8,40 | +5 | 318 | 0, 6 | −3 | 348 | 3,52 | −2 |
| 259 | 23,36 | | 289 | 8,12 | +4 | 319 | 0, 4 | −1 | 349 | 4,12 | +1 |
| 260 | 23, 4 | −2 | 290 | 7,44 | +3 | 320 | 0, 2 | | 350 | 4,28 | −1 |
| 261 | 22,36 | | 291 | 7,16 | +2 | 321 | 0, 1 | +1 | 351 | 4,48 | +1 |
| 262 | 22, 8 | +3 | 292 | 6,52 | +3 | 322 | 0, 0 | +1 | 352 | 5,12 | +6 |
| 263 | 21,36 | +2 | 293 | 6,28 | +4 | 323 | 0, 1 | +3 | 353 | 5,32 | +7 |
| 264 | 21, 8 | +5 | 294 | 6, 0 | +1 | 324 | 0, 2 | +3 | 354 | 5,48 | +3 |
| 265 | 20,40 | +8 | 295 | 5,32 | −3 | 325 | 0, 4 | +4 | 355 | 6, 8 | +3 |
| 266 | 20,16 | +16 | 296 | 5, 8 | −4 | 326 | 0, 6 | +4 | 356 | 6,28 | +3 |
| 267 | 19,40 | +12 | 297 | 4,48 | −2 | 327 | 0, 8 | +3 | 357 | 6,48 | +3 |
| 268 | 19, 0 | +4 | 298 | 4,28 | | 328 | 0,10 | +1 | 358 | 7, 8 | +2 |
| 269 | 18,24 | | 299 | 4, 8 | +1 | 329 | 0,14 | +1 | 359 | 7,28 | +1 |
| 270 | 17,52 | +1 | 300 | 3,48 | +1 | 330 | 0,20 | +1 | 360 | 7,48 | |

Table 4c *al-Khwārizmī's values for the equation of time ( T(λ) ) and the differences between al-Khwārizmī's values and our final recomputation (3rd part)*

7. *Conclusions*

The mathematical analysis of the table for the equation of time in the Latin translation of al-Majrīṭī's recension of al-Khwārizmī's *Sindhind Zīj* has led to the following results:

A. The independent variable of the table is the true solar longitude, in agreement with the explanatory text in the Latin translation of al-Majrīṭī's recension.
B. The factor used for the conversion from equatorial degrees to hours is 15°/ hour. This can be concluded from the fact that practically all tabular values are multiples of 4″ and is confirmed by an application of the method of least squares.
C. The underlying value of the obliquity of the ecliptic is 23°51′. This value also underlies the tables for the solar declination and the right ascension in al-Majrīṭī's recension. It is a rounded version of the value 23°51′20″ used by Ptolemy in the *Almagest* and the *Handy Tables*.
D. The underlying solar equation was computed on the basis of the Ptolemaic solar theory. The value of the eccentricity is Ptolemy's 2;30, which corresponds to a maximum equation of 2°23′. The solar equation table in al-Majrīṭī's recension is of Indian / Persian origin and is based on a maximum equation of 2°14′.
E. The underlying longitude of the solar apogee is 82°39′, the value determined by the group of astronomers working for the caliph al-Ma'mūn (c. 830).[25] Note that neither the Indian value 77°55′ given in al-Majrīṭī's instructions for calculating the true solar longitude, nor the Ptolemaic value 65°30′ were used. It seems natural that Ptolemy's outdated longitude of the solar apogee was replaced with the result of recent observations, but then the same should have been done with the solar eccentricity (the maximum solar equation determined by al-Ma'mūn's astronomers was 1°59′).
F. The underlying value of the epoch constant is 4°30′. As we have seen, the epoch constant was determined in such a way that the

---

[25]On the basis of the 95 % confidence intervals given above we cannot choose between the value 82°39′, which was used in the *zīj*es of Yaḥyā ibn Abī Manṣūr and Ḥabash al-Ḥāsib, and the rounded value 82°40′, which occurs in a table in Ḥabash's *zīj* (see Debarnot 1987, p. 58).

minimum equation of time became zero. Since the minimum occurs for argument 322° (22° Aquarius), corresponding to argument 320° for the unshifted table, we expect $c \approx \alpha(320) - 320 + q(320-\lambda_A)$ (cf. formula 4). For the parameter values found above this yields c ≈ 4;30,22, which is rounded to 4;30.

G. The values for the equation of time in al-Majrīṭī's recension were shifted forwards 2 degrees, i.e. the actual equation of time value for 0° occurs for argument 2°, the one for 1° for argument 3°, etc. I have not been able to find a satisfactory explanation for this shift. Neugebauer explains how a small shift in the solar longitude is required to make the minimum equation of time equal to 0 (1962, pp. 64-65); however, this shift is smaller than 1′. Furthermore, there is no reason to believe that al-Khwārizmī's table for the equation of time belonged to a set of solar tables based on a displaced equation, such as Kūshyār's. In fact, al-Khwārizmī should have chosen his displacement larger than 2°, since his maximum solar equation is 2°23′.

From the above we can conclude that the table for the equation of time in the Latin translation of al-Majrīṭī's recension of al-Khwārizmī's *Sindhind Zīj* fits into the group of Ptolemaic tables in that work which probably stem from al-Khwārizmī (group I-C in Section 4): it is based on the Ptolemaic values for the obliquity and the solar eccentricity and has a minimum value equal to zero following the table for the equation of time in the *Handy Tables*. The longitude of the solar apogee stems from the astronomers employed by the caliph al-Ma'mūn and was used in the earliest Islamic astronomical handbooks which were mainly based on the Ptolemaic planetary models. Nevertheless, we cannot be certain that the table was computed by al-Khwārizmī, since none of the sources listed in Section 3 mentions a table for the equation of time in al-Khwārizmī's original *zīj*. In any case we can conclude that either the whole table or the underlying parameter values were transmitted from Eastern to Western Islam.

*Acknowledgement*

It is a great pleasure to thank Professor Juan Vernet and the other members of the Departamento de Árabe for their warm hospitality during my three visits to Barcelona.

I would like to thank Dr. Fritz Saaby Pedersen (Copenhagen) for his useful information concerning al-Khwārizmī and the *Toledan Tables* and for our pleasant discussions through electronic mail. The constructive comments of Professor David A. King and Silke Ackermann (both in Frankfurt am Main) enabled me to make drastic improvements to several sections of this article.

I first presented the results found in this article at the XIXth International Congress of History of Science in Zaragoza (Spain), August 1993. My stay in Zaragoza was made possible by the Netherlands Organization for Scientific Research (NWO) and the Stichting Mathematisch Centrum (Amsterdam). The article was finished during a stay in Frankfurt am Main sponsored by the Alexander von Humboldt Foundation.

8. *Bibliography*

Bjørnbo, Axel Anthon

1909 Al-Chwarizmi's trigonometriske Tavler, in: *Festskrift til H. G. Zeuthen*, Copenhagen (Høst), pp. 1-17.

Burckhardt, Johann Jakob

1961 Die mittleren Bewegungen der Planeten im Tafelwerk des Khwārizmī, *Vierteljahrsschrift der naturforschenden Gesellschaft in Zürich* 106, pp. 213-231.

Comes, Mercè

1992-1994 The "Meridian of Water" in the Tables of Geographical Coordinates of al-Andalus and North Africa, *Journal for the History of Arabic Science* 10, pp. 41-52.

Dalen, Benno van

1993 *Ancient and Mediaeval Astronomical Tables: mathematical structure and parameter values*, doctoral thesis, Utrecht University.

1994 On Ptolemy's Table for the Equation of Time, *Centaurus* 37, pp. 97-153.

Debarnot, Marie-Thérèse
1987 The *Zīj* of Ḥabash al-Ḥāsib: A Survey of MS Istanbul Yeni Cami 784/2, in: *From Deferent to Equant: A Volume of Studies in the History of Science in the Ancient and Medieval Near East in Honor of E.S. Kennedy* (David A. King & George A. Saliba, eds.), New York (New York Academy of Sciences), pp. 35-69.
Delambre, Jean Baptiste Joseph
1819 *Histoire de l'astronomie du moyen âge*, Paris (Courcier). Reprint: New York and London (Johnson Reprint Corporation) 1965.
DSB *Dictionary of Scientific Biography*, 14 vols and 2 suppl. vols, New York (Charles Scribner's Sons) 1970-1980.
EI$^2$ *The Encyclopaedia of Islam, new edition*, Leiden (Brill) 1960-.
Goldstein, Bernard R.
1967 *Ibn al-Muthannā's Commentary on the Astronomical Tables of al-Khwārizmī*, New Haven (Yale University Press).
al-Hāshimī, ʿAlī ibn Sulaymān
*The Book of the Reasons behind Astronomical Tables (Kitāb ʿilal al-zījāt)* (facsimile, translation by Fuad I. Haddad and Edward S. Kennedy, commentary by David Pingree and Edward S. Kennedy), New York (Delmar) 1981.
Hogendijk, Jan P.
1988 Three Islamic Lunar Crescent Visibility Tables, *Journal for the History of Astronomy* 19, pp. 29-44.
1989 The Mathematical Structure of Two Islamic Astrological Tables for "Casting the Rays", *Centaurus* 32, pp. 171-202.
1991 Al-Khwārizmī's Table of the "Sine of the Hours" and the Underlying Sine Table, *Historia Scientiarum* 42, pp. 1-12.
Kennedy, Edward S.
1956a A Survey of Islamic Astronomical Tables, *Transactions of the American Philosophical Society*, New Series vol. 46-2, pp. 123-177. Second edition: Philadelphia (American Philosophical Society) 1989 (page numbering from 1 to 55).
1956b Parallax Theory in Islamic Astronomy, *Isis* 47, pp. 33-53. Reprinted in *SIES*, pp. 164-184.
1964 Al-Khwārizmī on the Jewish Calendar, *Scripta Mathematica* 27, pp. 55-59. Reprinted in *SIES*, pp. 661-665.
1988 Two Medieval Approaches to the Equation of Time, *Centaurus*

31, pp. 1-8.

Kennedy, Edward S. & Janjanian, Mardiros
1965 The Crescent Visibility Table in al-Khwārizmī's *Zīj*, *Centaurus* 11, pp. 73-78. Reprinted in *SIES*, pp. 151-156.

Kennedy, Edward S. & Krikorian-Preisler, Haiganoush
1972 The Astrological Doctrine of Projecting the Rays, *al-Abhath* 25, pp. 3-15. Reprinted in *SIES*, pp. 372-384.

Kennedy, Edward S. & Muruwwa, Ahmad
1958 Bīrūnī on the Solar Equation, *Journal of Near Eastern Studies* 17, pp. 112-121. Reprinted in *SIES*, pp. 603-612.

Kennedy, Edward S. & Ukashah, Walid
1969 Al-Khwārizmī's Planetary Latitude Tables, *Centaurus* 14, pp. 86-96. Reprinted in *SIES*, pp. 125-135.

Kennedy, Edward S. & Waerden, Bartel L. van der
1963 The World Year of the Persians, *Journal of the American Oriental Society* 83, pp. 315-327. Reprinted in *SIES*, pp. 338-350.

King, David A.
1983 *al-Khwārizmī and New Trends in Mathematical Astronomy in the Ninth Century*, New York (New York University, Hagop Kevorkian Center for Near Eastern Studies).

1986 *A Survey of the Scientific Manuscripts in the Egyptian National Library*, Winona Lake IN (American Research Center in Egypt).

1987 Some Early Islamic Tables for Determining Lunar Crescent Visibility, in: *From Deferant to Equant: A Volume of Studies in the History of Science in the Ancient and Medieval Near East in Honor of E.S. Kennedy* (David A. King & George A. Saliba, eds.), New York (New York Academy of Sciences), pp. 185-225. Reprinted in David A. King, *Astronomy in the Service of Islam*, Aldershot GB (Variorum) 1993, chapter II.

Lesley, Mark
1957 Bīrūnī on Rising Times and Daylight Lengths, *Centaurus* 5, pp. 121-141. Reprinted in *SIES*, pp. 253-273.

Mercier, Raymond P.
1987 Astronomical Tables in the Twelfth Century, in: *Adelard of Bath. An English Scientist and Arabist of the Early Twelfth Century* (Charles Burnett, ed.), London (Warburg Institute), pp.

87-118.

Millás Vallicrosa, José Maria

1943 - 1950 *Estudios sobre Azarquiel*, Madrid / Barcelona (Instituto "Miguel Asin", Escuelas de Estudios Árabes de Madrid y Granada).

1947 *"El libro de los fundamentos de las tablas astronómicas" de R. Abraham ibn ʿEzra*, Madrid / Barcelona.

1963 La autenticidad del comentario a las tablas astronómicas de al-Jwārizmī por Aḥmad ibn al-Muṯannā', *Isis* 54, pp. 114-119.

Millás Vendrell, Eduardo

1963 *El comentario de Ibn al-Muṯannā' a las tablas astronómicas de al-Jwārizmī*, Madrid / Barcelona (Asociación para la Historia de la Ciencia Española).

Nallino, Carlo Alfonso

1899-1907 *al-Battānī sive Albatenii opus astronomicum*, 3 vols., Milan. Reprint of vols. 1 and 2: Frankfurt am Main (Minerva) 1969; reprint of vol. 3: Hildesheim (Olms) 1977.

1944 Al-Khuwārizmī e il suo rifacimento della Geografia di Tolomeo, *Raccolta di scritti editi e inediti*, vol. 5 (Roma) 1943, pp. 458-532.

Neugebauer, Otto E.

1956 Transmission of Planetary Theories in Ancient and Medieval Astronomy, *Scripta Mathematica* 22, pp. 165-192.

1962 *The Astronomical Tables of al-Khwārizmī. Translation with Commentaries of the Latin Version edited by H. Suter supplemented by Corpus Christi College MS 283*, Copenhagen (Det Kongelige Danske Videnskabernes Selskab).

1975 *A History of Ancient Mathematical Astronomy*, 3 vols., Berlin (Springer).

Neugebauer, Otto E. & Schmidt, Olaf

1952 Hindu Astronomy at Newminster in 1428, *Annals of Science* 8, pp. 221-228.

Pedersen, Fritz Saaby

1987 Canones Azarchelis: Some Versions, and a Text, *Cahiers de l'institut du moyen-âge grec et latin* 54, pp. 129-218.

1992 Alkhwarizmi's astronomical Rules: Yet Another Latin Version?, *Cahiers de l'institut du moyen-âge grec et latin* 62, pp. 31-75.

Pedersen, Olaf
1974 *A Survey of the Almagest*, Odense (Odense University Press).
Pingree, David
1965 The Persian "Observation" of the Solar Apogee in ca. A.D. 450, *Journal of Near Eastern Studies* 24, pp. 334-336.
1968a *The Thousands of Abū Maᶜshar*, London (Warburg Institute).
1968b The Fragments of the Works of Yaᶜqūb ibn Ṭāriq, *Journal of Near Eastern Studies* 27, pp. 97-125.
1970 The Fragments of the Works of al-Fazārī, *Journal of Near Eastern Studies* 29, pp. 103-123.
Salam, Hala & Kennedy, Edward S.
1967 Solar and Lunar Tables in Early Islamic Astronomy, *Journal of the American Oriental Society* 87, pp. 492-497. Reprinted in *SIES*, pp. 108-113.
Samsó, Julio
1992 *Las Ciencias de los Antiguos en al-Andalus*, Madrid (Editorial MAPFRE).
Sengupta, Prabodh Chandra
1934 *The Khaṇḍakhādyaka. An Astronomical Treatise of Brahmagupta*, Calcutta (University of Calcutta).
Sezgin, Fuat
1971-1984 *Geschichte des arabischen Schrifttums*, 9 vols, Leiden (Brill).
SIES E.S. Kennedy, Colleagues and Former Students, *Studies in the Islamic Exact Sciences*, Beirut (American University of Beirut) 1983.
Suter, Heinrich
1914 *Die astronomischen Tafeln des Muḥammed ibn Mūsā al-Khwārizmī in der Bearbeitung des Maslama ibn Aḥmed al-Madjrīṭī und der lateinischen Übersetzung des Adelard von Bath*, Copenhagen (Kongelige Danske Videnskabernes Selskab).
Toomer, Gerald J.
1964 Review of: O. Neugebauer, "The Astronomical Tables of al-Khwārizmī", Copenhagen 1962, *Centaurus* 10, pp. 202-212.
1968 A Survey of the Toledan Tables, *Osiris* 15, pp. 5-174.
1973 "al-Khwārizmī", in: *DSB*, vol. 7, pp. 358-365.

Vernet, Juan
1974 "al-Majrīṭī", in: *DSB*, vol. 9, pp. 39-40.
1976 "al-Zarqālī", in: *DSB*, vol. 14, pp. 592-595.
1978 "al-Khwārazmī", in: *EI²*, vol. 4, pp. 1101-1103.
1985 "al-Madjrīṭī", in: *EI²*, vol. 5, p. 1105.

Vernet, Juan & Catalá, M.A.
1965 Las obras matemáticas de Maslama de Madrid, *al-Andalus* 30, pp. 15-45. Reprinted in Juan Vernet, *Estudios sobre historia de la ciencia medieval*, Barcelona - Bellaterra 1979, pp. 241-271.

Waerden, Bartel L. van der
1960 - 1962 Ausgleichspunkt, "Methode der Perser" und indische Planetenrechnung, *Archive for History of Exact Sciences* 1, pp. 107-121.

Zinner, Ernst
1935 Die Tafeln von Toledo, *Osiris* 1, pp. 747-774.

V

# ORIGIN OF THE MEAN MOTION TABLES OF JAI SINGH

In two articles in this journal the relationship has been studied between the Persian *Zīj-i Muḥammad Shāhī*, completed in Jaipur around 1735 under Sawā'ī Jai Singh, and the astronomical tables of Philippe de La Hire, printed in Paris in 1727. Mercier concluded that Jai Singh's planetary tables were all taken directly from La Hire and do not depend on observations made in India. On the other hand, Sharma maintained that the planetary tables in the *Zīj-i Muḥammad Shāhī* are independent from those of La Hire.

In this article, the methods by which Jai Singh's tables for planetary mean motion were computed will be investigated in detail. First it will be shown that the initial mean positions in the *Zīj-i Muḥammad Shāhī* were calculated from La Hire's mean motion tables by adding the entries for the appropriate numbers of years, months and days and applying a correction for a difference in geographical longitude between Paris and Delhi of 73°30'.

Next, the mean motion parameters underlying Jai Singh's tables will be determined by means of recently introduced mathematical techniques and computer programs. It will be shown that the daily mean motions were calculated from particular values found in La Hire's tables. Our recomputations of Jai Singh's tables show only incidental differences of at most a couple of sexagesimal fourths of a degree. We conclude that the mean motion tables in the *Zīj-i-Muḥammad Shāhī* were in fact derived from the tables of La Hire, thus confirming Mercier's result.

Keywords: Least Number of Errors Criterion (LNEC), mean motion tables, Philippe de la Hire, Sawā'ī Jai Singh, *Tabulae astronomicae Ludovici Magni*, *Zīj-ī Muḥammad Shāhī*, *Zīj-ī- Sulṭānī*.

## Introduction

Sawā'ī Jai Singh (or *Savāī* Jayasiṃha, 1688-1743) was Maharajah of Amber from 1700 to his death.[1] He founded the city of Jaipur in 1727 and built large astronomical observatories with masonry instruments in Delhi, Jaipur, Ujjain, Mathura, and Benares.

One of the most well-known works produced by Jai Singh's astronomers is a Persian astronomical handbook with tables called the *Zīj-i Jadīd-i Muḥammad Shāhī* (c. 1735), which was named after the Moghul emperor who came to the throne in 1719.[2] In an article in this journal, Mercier (1984, pp. 157-158; see also Pingree 1987, pp.323-324) has shown that much of the trigonometrical and spherical-astronomical material in the *Zīj-i Muḥammad Shāhī* was taken over from the *Zīj-i Sulṭānī* of Ulugh Beg (c. 1440). Only the tables dependent on the obliquity of the ecliptic or on geographical latitude were recomputed, for which Jai Singh's newly observed obliquity value 23;28 and the latitudes of Delhi and Jaipur were used. On the other hand, Mercier (pp. 145-147) asserts that the planetary tables in the *Zīj-i Muḥammad Shāhī* are "identical" with those in the *Tabulae astronomicae Ludovici Magni* of Philippe de la Hire (second edition 1702, reprinted in 1727), apart from the use of the Islamic calendar instead of the Julian and a change of meridian from Paris to Delhi. He therefore concludes that the planetary mean motions in the *Zīj* are not the result of observations made in India. This last aspect is disputed by Sharma (1990 and 1995, pp. 243-250), who unsuccesfully tries to recompute Jai Singh's mean motion tables from those of La Hire and, on the basis of a comparison of the underlying mean motions in an Arabic year, states that the parameters in the two works are "totally different".

The purpose of this article is to establish the precise method by which the mean motion tables in the *Zīj-i Muḥammad Shāhī* were computed. First we will analyse the origin of Jai Singh's epoch values, the initial mean positions for February 20, 1719 (Gregorian). Then we will determine the values for the daily mean motion underlying the tables in the *Zīj* as accurately as possible and will investigate the way in which these parameters could have been obtained. We have made use of the manuscript of the *Zīj-i Muḥammad Shāhī* from the Arabic and Persian Research Institute in Tonk (Rajasthan). Because of lack of space, we cannot include the complete tables that are analysed in this article. Sharma reproduced Jai Singh's table for the mean motion of Venus in facsimile (1990, p. 38; 1995, p. 237) and transcribed part of his table for the solar mean motion (1995, pp. 239-240). He also included La Hire's table for Saturn (1990, pp. 35-36).[3]

## Epoch Positions

The epoch or starting point of the mean motion tables in the *Zīj-i Muḥammad Shāhī* is noon at Delhi on February 20, 1719 (Gregorian). This date corresponds to 1 Rabī' II 1131 Hijra and is the beginning of the Islamic month during which Muḥammad Shāh's predecessor Farrukhsiyar was deposed.[4] The mean positions at epoch are given to an accuracy of seconds, whereas most other tabular values are given to sexagesimal fourths (the few exceptions to this rule are listed in note 12). In general, all values found in mean motion tables are the result of computations based on a mean position at a certain time (not usually the epoch of the tables) and a value for the daily mean motion reduced from a set of observations. The fact that in the *Zīj-Muḥammad Shāhī* the accuracy of the epoch values is different from that of the other tabular values, points to two essentially different methods of computation. In particular, it seems possible that Jai Singh's epoch values were derived directly from another set of mean motion tables with values to seconds, whereas his other tabular values were computed on the basis of values of the underlying daily mean motions with at least four sexagesimal fractional digits.

The astronomical tables of La Hire display mean positions for noon at Paris on New Year's Day of the Julian years 0, 100, 200, ..., 1600 and the Gregorian years 1600 and 1700. These are accompanied by values of the mean motion in 1, 2, 3, ..., 19, 20, 40, 60, 80, 100, 200, 300, 400, 500 and 1000 years, in the months January to December, in 1 to 30 days, in 1 to 30 hours, and, for some of the mean motions, in minutes and seconds of an hour. In order to compute from La Hire's tables the mean position of a particular planet on any given Julian or Gregorian date, the mean motions in the periods of time elapsed since the beginning of the current century should be added to the mean position given for the beginning of the century. For example, to determine a mean position at the epoch of the *Zīj-i Muḥammad Shāhī* mentioned above, one would take the mean position for the Gregorian year 1700 and add to it the mean motion in 18 years, in the month of January, and in 19 days. Since the values in La Hire's mean motion tables are all displayed to seconds, it is possible that they were the source from which the epoch values in the *Zīj-i Muḥammad Shāhī* were calculated. This possibility will be investigated in greater detail in the following paragraphs.

It can be noted that many Arabic and Persian sets of mean motion tables (including that found in the *Zīj-i Muḥammad Shāhī)* display values for so-called "incomplete" years, months and days rather than for the "completed" periods used by La Hire. In such sets one looks up the mean motions with the *current* year, month and day instead of the numbers of years, months and days that have elapsed. Whereas many Islamic mean motion tables give two separate sets of values for the months in an ordinary Julian year and those in a leap year, La Hire simply gives instructions to add an extra day for the months March to December in a leap year.

The addition indicated above for determining a mean position at the epoch of the *Zīj-i Muḥammad Shāhī* from the tables of La Hire yields mean positions at noon in Paris. To obtain the mean positions at noon in Delhi (or any other locality), a correction has to be carried out for the difference in geographical longitude (or, equivalently, for the difference in time between noon at the desired locality and noon at Paris, where noon is to be understood as the time of culmination of the mean sun). For this purpose many Islamic mean motion tables include a column for the "differences between the two longitudes" (*mā bayna al-ṭūlayn),* which displays the corrections to the mean motions as a function of degrees of longitude difference. Since La Hire does not provide such a column, a longitude correction based on his tables must be carried out by converting the longitude difference into time (a difference of 15° in eastward direction corresponds to the culmination of the mean sun occurring on hour earlier) and then looking up the mean motion during that time in the tables for hours, minutes and seconds.

To illustrate the above, we will now compute the solar mean longitude at the epoch of the *Zīj-i Muḥammad Shāhī,* noon of February 20, 1719 (Gregorian), from the tables of La Hire. For this purpose we need to add the following values from the table for the solar mean motion in the *Tabulae astronomicae* (the superscript 's' indicates zodiacal signs);

| | |
|---|---|
| Mean position at noon of January 1, 1700 Gregorian | $9^s\ 10^\circ\ 52'27''$ |
| Mean motion in 18 (Julian or Gregorian) years | $11^s\ 29^\circ\ 38'42''$ |
| Mean motion in the month January | $1^s\ 0^\circ\ 33'18''$ |
| Mean motion in 19 days | $18^\circ\ 43'38''$ + |
| Mean position at noon of February 20, 1719 Gregorian | $10^s\ 29^\circ\ 48'5''$ |

The sum $10^s29^\circ48'5''$ is only 12'4" larger than the epoch value of Jai Singh's solar mean motion table. As a working hypothesis, we will now assume that the epoch positions in the *Zīj-i Muḥammad Shāhī* were computed from the tables of La Hire. Under that assumption, the difference of 12'4" found above must be due to a correction for the longitude difference between Paris and Delhi. We will thus proceed by determining the underlying longitude difference, which Mercier (1984, p. 146) found to be 73°30'.[5] Since the amount of the correction is proportional to the daily (or hourly) mean motion concerned, the largest corrections will be found for the mean motions in longitude of the Moon, Mercury, Venus, and the Sun, in that order. Since all epoch values in the *Zīj-i Muḥammad Shāhī* are given to seconds, this also implies that the most accurate determination of the longitude difference is possible from the differences between Jai Singh's epoch values for the above four mean motions and the corresponding mean positions at his epoch computed from the *Tabulae astronomicae*.

As an example, we will now compute the longitude difference on the basis of the lunar mean longitude. La Hire's lunar mean position at noon in Paris on February 20, 1719 Gregorian, computed similarly to the solar mean position above, is $11^s13^\circ58'31''$, which is $2^\circ41'25''$ larger than the value found in the *Zīj-i Muḥammad Shāhī*. This difference can be seen to be the mean lunar motion in longitude in precisely 4 hours and 54 minutes, since La Hire's tables give a motion of $2^\circ11'46''$ for 4 hours and a motion of 29'39" for 54 minutes. The same turns out to be true for the differences between Jai Singh's epoch values for Mercury, Venus, and the Sun and the corresponding values computed from the tables of La Hire:[6]

| Planet | Computed difference La Hire-Jai Singh | La Hire's motion in 4 hours | La Hire's motion in 54 minutes | Confidence level (time) |
|---|---|---|---|---|
| Moon | 2;41,25 | 2;11,46 | 0;29,39 | 0.9 seconds |
| Mercury | 0;50, 8 | 0;40,56 | 0; 9,12,27 | 2.9 seconds |
| Venus | 0;19,38 | 0;16, 2 | 0; 3,36,21 | 7.5 seconds |
| Sun | 0;12, 4 | 0; 9,51 | 0; 2,13 | 12.2 seconds |

If we add the motions in 4 hours and in 54 minutes and round the result to seconds, we obtain in each case precisely the difference between Jai Singh's epoch value and the position computed from La Hire's mean motion tables (i.e., the assumed longitude correction) given in the second column of the above table. This shows that Jai Singh's epoch values were in fact computed from the tables of La Hire with a correction for a longitude difference of $4^h54^m$ or 73°30'.

One may ask how accurate the above determination of the longitude difference is. In fact, differences sufficiently close to $4^h54^m$ will lead to the same corrections when these are rounded to seconds. The fifth column in the above table shows how many time seconds the longitude difference may be different from $4^h54^m$ before the rounded correction changes by one second. We note that even if the longitude difference differs by only one second from $4^h54^m$, the correction for the moon will not any more be equal to what we have found it to be. We therefore conclude that the longitude correction applied by Jai Singh was for *precisely* $4^h54^m$ or 73°30'. As was already noted by Mercier (1984, pp. 146-150), these values are in full agreement with the longitude difference between Paris and Delhi as found at various other places in the *Zīj-i Muḥammad Shāhī*, although they correspond better to the modern longitude difference between Paris and Jaipur.[7]

We can now complete our recomputation of the solar mean position at epoch in the *Zīj-i Muḥammad Shāhī* by subtracting from the value found above the longitude correction for 73°30', i.e., the mean motion in 4 hours and 54 minutes:

| | |
|---|---|
| Mean position at noon of February 20, 1719 Gregorian for the meridian of Paris: | $10^{s}29^{\circ}48'5''$ |
| Correction for difference in longitude between Paris and Delhi: | |
| Mean motion in 4 hours: | $9'51''$ |
| Mean motion in 54 minutes: | $2'13''$ |
| Mean position at noon of February 20, 1719 Gregorian for the meridian of Delhi: | $10^{s}29^{\circ}36'1''$ |

In agreement with our analysis above, the result is identical with Jai Singh's solar mean position at epoch.

Following the same computational scheme we recomputed the epoch values for all twenty mean motion tables in the *Zīj-i Muḥammad Shāhī*. For fourteen tables the recomputed epoch value equalled that in the *Zīj;* in one case (the motion of the aphelium of Venus) it was two seconds larger than the value given by Jai Singh and in five cases (aphelium and node of Saturn and Mars, node of Mercury) it was one second smaller. Since all six non-matches occur for aphelia and nodes of planets, the mean motions concerned are significantly smaller than one second per day. This implies that incorrect or omitted longitude corrections could not account for the differences. We suspect that the cause of the differences is possibly a scribal error for the aphelium of Venus and a misreading in the other five cases.[8]

## The Underlying Daily Mean Motions

We now proceed with an analysis of the remaining mean motion values given in the *Zīj-i Muḥammad Shāhī*. Jai Singh tabulates mean positions for the thirty Arabic years following his epoch, i.e., for the 1st Rabī‘ II of the so-called *extended* years 1132, 1133, 1134, ..., 1161 Hijra. In his 30-year cycle, the epoch year and the 3rd, 5th, 8th, 11th, 14th, 16th, 19th, 22nd, 24th, and 27th year after the epoch year are leap years of 355 days (instead of the ordinary 354). Thus Jai Singh's cycle does not run parallel with one of the cycles commonly used by medieval Muslim astronomers. In addition to the mean positions for a complete cycle, the *Zīj-i Muḥammad Shāhī* tabulates the mean motion in 30, 60, 90, ..., 300, 600, 900, and 1200 so-called *collected* years;[9] for current months * from Rabī' II of the following year,[10] and for current days from 1 to 61.[11] In the remainder of this article we will occasionally refer to the tables for extended and collected years, months, and days as the "sub-tables" of a mean motion table.

Because the tabular values in most of the mean motion tables in the *Zīj-i Muḥammad Shāhī* are displayed to sexagesimal fourths,[12] it is impossible that they were derived from the *Tabulae astronomicae* in the same way as the epoch values (as was noted above, La Hire's values are given to seconds only). We will therefore determine the underlying parameters of Jai Singh's mean motion tables directly from the tabular values without presupposing a relationship with any other tables. For this purpose we will make use of a recently introduced mathematical technique called the Least Number of Errors Criterion (LNEC). Descriptions of two slightly different forms of the LNEC can be found in Van Dalen 1993, Chapter 2.5, pp. 60-63 and in Mielgo 1996. The computer programs for DOS PC with which the analyses below were carried out can be obtained through the author's webpage (http://www.rz. uni-frankfurt.de/~dalen).

In brief, the LNEC determines the parameter underlying an astronomical table in such a way that the number of errors in the table (i.e., the number of differences between the table and a recomputation based on the parameter value found) is minimized. This can be done by calculating for any given tabular value the interval of parameter values for which that tabular value is correctly recomputed. By taking the intersection of all intervals thus obtained we find the range of parameter values for which all tabular values are correctly recomputed. If this intersection is empty, we instead use the range of parameter values for which the number of errors is as small as possible. As Mielgo indicated, this range can for instance be obtained by leaving out any intervals which do not intersect with the majority of intervals obtained. In each case we will assume that the historical parameter used for the computation of the table lies within the range of values for which the number of errors in the table is minimized.

The LNEC works particularly well for tables that have generally very few errors. Since calculations of mean motions can largely be performed by repeated additions of the mean motion per day or in certain other periods (cf. below), the probability of a computational error in mean motion tables tends to be smaller than in trigonometrically computed tables. On the other hand, an error in an intermediate addition can easily propagate through a whole mean motion table.

To illustrate the principle of the LNEC, we will determine the range of values of the daily mean motion of Saturn for which Jai Singh's mean motion of this planet in 30 Islamic years (10631 days) is correctly recomputed. The motion concerned is given in the *Zīj-i Muḥammad Shāhī* as $11^s26^\circ3'37''2'''54^{iv}$, and we may assume that this number was obtained by rounding to sexagesimal fourths the result of a multiplication of Jai Singh's daily mean motion of Saturn by 10631.[13] In case "modern" rounding were correctly applied, this result lay between $11^s26^\circ3'37''2'''53\frac{1}{2}^{iv}$ and $11^s26^\circ3'37''2'''54\frac{1}{2}^{iv}$; in the case of truncation, it lay between $11^s26^\circ3'37''2'''54^{iv}$ and $11^s26^\circ3'37''2'''55^{iv}$. A range within which the underlying daily mean motion of Saturn must lie can now be found by dividing the lower and upper bounds above by the number of days involved, 10631. For instance, under the assumption that modern rounding was used, we find that the daily mean motion must lie between $0;2,0,34,24,39,26,52^\circ$ and $0;2,0,34,24,39,27,12^\circ$. By intersecting this interval with intervals obtained in a similar way from other tabular values, the range within which Jai Singh's value for the daily mean motion of Saturn must lie can be further narrowed down.

Note that for the faster mean motions and longer periods we have to take into account the number of rotations (multiples of $12^s$ or $360^\circ$) that the planet has completed. These are not indicated in the *Zīj-i Muḥammad Shāhī* (as in most Islamic astronomical handbooks) and therefore need to be added separately to the tabular values.

When we analysed the mean motion tables of Jai Singh, we found that not all sub-tables were computed by multiplying an accurate value of the daily mean motion by the relevant numbers of days. For various tables it could be shown that the computations were performed by adding rounded intermediate results, instead of calculating an accurate multiple of the daily mean motion in each case.

**Table 1. Example of constant(left column) and non-constant (right column) tabular differences in mean motion tables of the types found in the *Zīj-i-Muḥammad Shāhī***

**Extended years**

| year | position | difference |
|---|---|---|
| 1 | 10$^s$29;36, 1 | |
| 2 | 10$^s$19;30,18 | 11$^s$19;54,17 |
| 3 | 10$^s$ 8;25,26 | 11$^s$18;55, 8 |
| 4 | 9$^s$27;20,34 | 11$^s$18;55, 8 |
| 5 | 9$^s$17;14,51 | 11$^s$19;54,17 |
| 6 | 9$^s$ 6; 9,59 | 11$^s$18;55, 8 |
| 7 | 8$^s$26; 4,16 | 11$^s$19;54,17 |
| 8 | 8$^s$14;59,24 | 11$^s$18;55, 8 |
| 9 | 8$^s$ 3;54,32 | 11$^s$18;55, 8 |
| 10 | 7$^s$23;48,49 | 11$^s$19;54,17 |
| 11 | 7$^s$12;43,57 | 11$^s$18;55, 8 |
| 12 | 7$^s$ 1;39, 5 | 11$^s$18;55, 8 |
| 13 | 6$^s$21;33,22 | 11$^s$19;54,17 |
| 14 | 6$^s$10;28,30 | 11$^s$18;55, 8 |
| 15 | 5$^s$29;23,38 | 11$^s$18;55, 8 |
| 16 | 5$^s$19;17,55 | 11$^s$19;54,17 |
| 17 | 5$^s$ 8;13, 3 | 11$^s$18;55, 8 |
| 18 | 4$^s$28; 7,20 | 11$^s$19;54,17 |
| 19 | 4$^s$17; 2,28 | 11$^s$18;55, 8 |
| 20 | 4$^s$ 5;57,36 | 11$^s$18;55, 8 |
| 21 | 3$^s$25;51,53 | 11$^s$19;54,17 |
| ⋮ | ⋮ | ⋮ |

**Extended years**

| year | position | difference |
|---|---|---|
| 1 | 10$^s$29;36, 1 | - |
| 2 | 10$^s$19;30,18 | 11$^s$19;54,17 |
| 3 | 10$^s$ 8;25,26 | 11$^s$18;55, 8 |
| 4 | 9$^s$27;20,35 | 11$^s$18;55, 9 |
| 5 | 9$^s$17;14,51 | 11$^s$19;54,16 |
| 6 | 9$^s$ 6;10, 0 | 11$^s$18;55, 9. |
| 7 | 8$^s$26; 4,17 | 11$^s$19;54,17 |
| 8 | 8$^s$14;59,25 | 11$^s$18;55, 8 |
| 9 | 8$^s$ 3;54,33 | 11$^s$18;55, 8 |
| 10 | 7$^s$23;48,50 | 11$^s$19;54,17 |
| 11 | 7$^s$12;43,59 | 11$^s$18;55, 9 |
| 12 | 7$^s$ 1;39, 7 | 11$^s$18;55, 8 |
| 13 | 6$^s$21;33,24 | 11$^s$19;54,17 |
| 14 | 6$^s$10;28,32 | 11$^s$18;55, 8 |
| 15 | 5$^s$29;23,41 | 11$^s$18;55, 9 |
| 16 | 5$^s$19;17,57 | 11$^s$19;54,16 |
| 17 | 5$^s$ 8;13, 6 | 11$^s$18;55, 9 |
| 18 | 4$^s$28; 7,23 | 11$^s$19;54,17 |
| 19 | 4$^s$17; 2,31 | 11$^s$18;55, 8 |
| 20 | 4$^s$ 5;57,39 | 11$^s$18;55, 8 |
| 21 | 3$^s$25;51,56 | 11$^s$19;54,17 |
| ⋮ | ⋮ | ⋮ |

**Collected years**

| years | motion | difference |
|---|---|---|
| 30 | 1$^s$ 8;24,53 | 1$^s$ 8;24,53 |
| 60 | 2$^s$16;49,46 | 1$^s$ 8;24,53 |
| 90 | 3$^s$25;14,38 | 1$^s$ 8;24,53 |
| 120 | 5$^s$ 3;39,31 | 1$^s$ 8;24,53 |
| 150 | 6$^s$12; 4,24 | 1$^s$ 8;24,53 |
| 180 | 7$^s$20;29,17 | 1$^s$ 8;24,53 |
| 210 | 8$^s$28;54,10 | 1$^s$ 8;24,53 |
| 240 | 10$^s$ 7;19, 3 | 1$^s$ 8;24,53 |
| 270 | 11$^s$15;43,55 | 1$^s$ 8;24,53 |
| 300 | 0$^s$24; 8,48 | 1$^s$ 8;24,53 |
| 600 | 1$^s$18;17,37 | 0$^s$24; 8,48 |
| 900 | 2$^s$12;26,25 | 0$^s$24; 8,48 |
| 1200 | 3$^s$ 6;35,13 | 0$^s$24; 8,48 |

**Collected years**

| years | motion | difference |
|---|---|---|
| 30 | 1$^s$ 8;24,53 | 1$^s$ 8;24,53 |
| 60 | 2$^s$16;49,46 | 1$^s$ 8;24,53 |
| 90 | 3$^s$25;14,39 | 1$^s$ 8;24,53 |
| 120 | 5$^s$ 3;39,31 | 1$^s$ 8;24,52 |
| 150 | 6$^s$12; 4,24 | 1$^s$ 8;24,53 |
| 180 | 7$^s$20;29,17 | 1$^s$ 8;24,53 |
| 210 | 8$^s$28;54,10 | 1$^s$ 8;24,53 |
| 240 | 10$^s$ 7;19, 3 | 1$^s$ 8;24,53 |
| 270 | 11$^s$15;43,56 | 1$^s$ 8;24,53 |
| 300 | 0$^s$24; 8,48 | 1$^s$ 8;24,52 |
| 600 | 1$^s$18;17,37 | 0$^s$24; 8,49 |
| 900 | 2$^s$12;26,25 | 0$^s$24; 8,48 |
| 1200 | 3$^s$ 6;35,13 | 0$^s$24; 8,48 |

For instance, the tables for months may be computed on the basis of rounded values for the motions in 29 and in 30 days; the tables for extended years were often computed on the basis of rounded values for 354 and for 355 days; and about half of the tables for collected years consist of plain multiples of the motion in 30 Arabic years. The use of rounded values for intermediate results can be easily recognized from the tabular differences of a mean motion table. If rounding to the number of digits of the table was performed at an intermediate stage of the calculation, the resulting tabular differences will be constant (i.e., they are all the same or, in the case of the tables for months and extended years, they assume only two different values for the two underlying periods). If no intermediate rounding was performed, the tabular differences will usually have two values differing by only one in the final digit (examples of constant and non-constant tabular differences for two types of sub-tables can be found in Table 1).

The probability that a mean motion table which was not computed on the basis of rounded intermediate results does have constant tabular differences is in fact very small.[14] Since more than half of the mean motion tables in the *Zīj-i Muḥammad Shāhī* have constant tabular differences in at least one or two of the sub-tables, we may assume that Jai Singh often rounded his intermediate results to the number of sexagesimal digits of his tables. Note that in all such cases we have determined the underlying daily mean motion from the rounded intermediate result only, rather than finding the parameter values for which all tabular values are correctly recomputed (i.e., using the LNEC). For instance, if the collected years have constant tabular differences, we have calculated the daily mean motion from the motion in 30 Arabic years. Below, an overview will be given of the various methods of computation that we have been able to recognize in Jai Singh's mean motion tables.

The third column of Table 2 presents our estimates of the daily mean motions underlying the tables of Jai Singh with an indication of their accuracy. An entry of the form $\mu \pm \nu$ indicates that the value of the daily mean motion used for the computation of the table concerned can be assumed to lie between $\mu-\nu$ and $\mu+\nu$. Note that the leading zeros of $\nu$ have been omitted, but that the final sexagesimal digit of $\nu$ always corresponds to the final digit of the estimate $\mu$. For instance, an entry 0;59,8,20,0±1,30 indicates that we expect the parameter

Table 2. Estimates of the mean motion parameters used by Jai Singh

| planet | motion | daily mean motion (in degrees) used by Jai Singh | daily mean motion multiplied by 365 | corresponding value of La Hire |
|---|---|---|---|---|
| Sun | longitude | 0;59, 8,19,46,50,57,27,27 ±10 | 359;45,40,19,59,59,32 | 359;17, 1 / 3 years |
| | apogee | 0; 0, 0,10, 6,34,34,34 ± 4 | 0; 1, 1,30, 0,20 | 0; 2, 3 / 2 years |
| Moon | longitude | 13;10,35, 1,22,11,28,38,30 ±17 | 129;23, 3,19,59,49 | 28; 9,10 / 3 years |
| | apogee | 0; 6,41, 4,29,35,20,29, 4 ±10 | 40;39,52,19,59,59,37 | 121;59,37 / 3 years |
| | node | −0; 3,10,38,18, 4,54,55, 4 ±10 | −19;19,42,59,59,54 | −19;19,43 / 1 year |
| Saturn | longitude | 0; 2, 0,34,24,39,27,12, 2 ±30 | 12;13,29,20, 0, 0 28 | 36;40,28 / 3 years |
| | aphelium | 0; 0, 0,13,28,46, 1,38, 9 ±17 | 0; 1,21,59,59,59,57 | 0; 1,22 / 1 year |
| | node | 0; 0, 0,11,45,12,19,44, 2 ±30 | 0; 1,11,30, 0, 0, 3 | 0; 2,23 / 2 years |
| Jupiter | longitude | 0; 4,59,16 | 30;20,32,20 | 91; 1,37 / 3 years |
| | aphelium | 0; 0, 0,15,30,25,23,51 ±10 | 0; 1,34,20, 4 | 0; 4,43 / 3 years |
| | node | 0; 0, 0, 2,19, 4,13,58 ±10 | 0; 0,14, 6, 0,45 | 0; 2,21 / 10 years |
| Mars | longitude | 0;31,26,39,13,58,18,25 ±10 | 191;17, 8,39,59,42 | 213;51,26 / 3 years |
| | aphelium | 0; 0, 0,10,57,32, 3,26 ±10 | 0; 1, 6,40, 0, 0,53 | 0; 3,20 / 3 years |
| | node | 0; 0, 0, 6, 4,55,53, 8 ±10 | 0; 0,36,59,59,58 | 0; 0,37 / 1 year |
| Venus | longitude | 1;36, 7,49,32, 3,17,25 ±10 | 224;47,36,20, 0, 0,57 | 314;22,49 / 3 years |
| | aphelium | 0; 0, 0,14,10,11,30,25,29 ±15 | 0; 1,26,12, 0, 0, 5 | 0; 7,11 / 5 years |
| | node | 0; 0, 0, 7,34,11,30,28 ±10 | 0; 0,46, 3, 0, 0,20 | — |
| Mercury | longitude | 4; 5,32,35,20,32,52,52 ±10 | 53;43,15, 0, 0, 2 | 53;43,15 / 1 year |
| | aphelium | 0; 0, 0,16,14,27,56,40,51 ±23 | 0; 1,38,47,59,59,48 | 0; 8,14 / 5 years |
| | node | 0; 0, 0,14, 0,39,27, 7,45 ±17 | 0; 1,25,14, 0, 0, 2 | 1;25,14 / 60 years |

concerned to lie between 0;59,8,18,30 and 0;59,8,21,30. The estimate of the daily mean motion of Saturn found above from the single tabular value $T\,(30^y)$ for 30 Arabic years would be expressed as $T\,(30^y)/10631 \pm$ 0;0,0,0,0,30/10631.[15]

As becomes clear from Table 2, we can in each case determine the underlying daily mean motion with an uncertainty of less than 10 in the seventh sexagesimal fractional digit. Whereas this accuracy has little meaning in an astronomical sense, we will see below that it helps us in investigating the relationship between Jai Singh's parameter values and those of La Hire and even in determining the precise way in which Jai Singh obtained his parameters. Using the values of the daily mean motions given in Table 2 and taking into consideration the particular methods of computation discussed below, we find that the errors in Jai Singh's mean motion tables generally amount to at most a sexagesimal fourth. Some very large errors that occur in most of the tables can be seen to be due to scribal errors or copying mistakes in the manuscript that we have used. In the least accurate table, that for the mean motion in longitude of Mercury, the 24 errors in the sub-table for extended years (which we have not been able to explain) range from – 4 to +1 sexagesimal fourths (see Table 5 below).[16]

Summarizing, we found that the following computational techniques have been used in the twenty mean motion tables in the *Zīj-i Muḥammad Shāhī*:

1. Accurate computation on the basis of *un-rounded* values for the daily mean motions (or, equivalently, for the mean motions in the periods underlying the respective sub-tables, namely, 29 and 30 days for the table for months, 354 and 355 days for the table for extended years, and 10631 days (30 Arabic years) for the table for collected years).
2. Computation on the basis of *rounded* values for the mean motions in the periods underlying the various sub-tables. Thus we found that in at least eight of the twenty mean motion tables the table for extended years was computed on the basis of rounded mean motion values for 354 and for 355 days. As a result of the rounding, in five of these tables the difference between the mean positions at the beginning of the years 31 and 1 is not equal to the mean motion in 30 years displayed in the table for collected years. In one more case in which the rounding was used, the value for the year 31 was modified in order to avoid such a disagreement (but resulting in an anomaly

within the table for extended years.) Furthermore, in ten tables the sub-table for collected years was based on a rounded value for 30 Arabic years. In all these cases the rounding of the underlying parameters could be recognized from the constant tabular differences of the sub-tables (cf. above).

3. In a number of cases the tables for months and days were computed with a heavily rounded value of the daily mean motion which is not compatible with the values used for the tables for extended and collected years.
4. In general it can be seen that the results of the mean motion computations were rounded to the number of digits of the tabular values in the "modern" way, i.e., sexagesimal digits 30 and higher were rounded upwards, digits 29 and lower were discarded. However, for the values in a couple of tables "truncation" appears to have been used, i.e., the sexagesimal fifths and below were simply discarded.
5. Some of the tables for extended years and for months have highly irregular error patterns, which we have not been able to explain. However, also in these tables the errors are never larger than four sexagesimal fourths (cf. Table 5).

A typical example for the variety of computational techniques used by Jai Singh is his table for the motion of the Ascending Node of Mars. A complete reconstruction of this table (including the derivation of the underlying value of the daily mean motion) will be given at the end of this article.

## Origin of Jai Singh's Values For The Daily Mean Motions

In order to investigate the relationship between the values for the daily mean motions in the *Zīj-i Muḥammad Shāhī* and those in the *Tabulae astronomicae*, it would be possible to analyse the tables of La Hire in the same way as those of Jai Singh and then compare the results. However, our estimates of Jai Singh's parameters are so accurate that we can verify a more general hypothesis concerning their origin, namely that they were derived from some other set of tables by taking for each type of mean motion one particular tabular value and dividing that by the corresponding (integer) number of days. According to this hypothesis, we expect that certain multiples of the parameters found from Jai Singh's tables are close to round numbers, i.e., numbers with only a limited number of non-zero sexagesimal fractional digits such as we find them in tables.

Table 3 shows a selection of multiples of the value of Saturn's daily mean motion in longitude that we derived from the *Zīj-i Muḥammad Shāhī*. Besides multiples from 1 to 30, those for the lengths of ordinary and leap years as well as average year lengths in some common calendars have been included. As can be seen from the table, the daily mean motion of Saturn multiplied by 365 is close to 12;13,29,20, i.e., an integer number of seconds divided by 3. Consequently, the daily mean motion of Saturn multiplied by 1095 is close to an integer number of seconds, namely 36;40,28. No other multiples are even approximately similarly close to a round number.

It turns out that this situation is typical for all mean motion parameters used by Jai Singh. The fourth column of Table 2 contains the 365-folds of the derived daily mean motions in the third column (multiples of 360° have been omitted). Except for the aphelium and node of Jupiter, all these 365-folds are close to a round number. We conclude that Jai Singh derived his parameters by dividing mean motion values for certain integer numbers of ordinary (i.e. non-leap) Julian years or Persian years by the corresponding multiples of 365.

Since most of the 365-folds are close to an integer number of seconds or to an integer number of seconds divided by 2 or 3, it is very probable that the mean motion values on which Jai Singh based his parameters were given to seconds. Under this assumption, the fifth column of Table 2 shows the mean motion values from which the parameters were derived. These turn out to be identical with values found in the *Tabulae astronomicae* of La Hire, except for the node of Venus, for which La Hire gives a motion of 0;15,22° in 20 Julian years, whereas Jai Singh used 0;15,21°. We therefore conclude that Jai Singh determined his values for the daily mean motions of the sun, moon, and planets by dividing mean motion values for certain numbers of Julian years from the *Tabulae astronomicae* by the appropriate multiples of 365.

**Table 3. Multiples of Jai Singh's value for Saturn's daily mean motion**

| multiplier | multiple |
|---|---|
| 1 | 0; 2, 0,34,24,39,27,12, 2 |
| 2 | 0; 4, 1, 8,49,18,54,24, 4 |
| 3 | 0; 6, 1,43,13,58,21,36, 6 |
| 4 | 0; 8, 2,17,38,37,48,48, 8 |
| 5 | 0;10, 2,52, 3,17,16, 0,10 |
| 6 | 0;12, 3,26,27,56,43,12,12 |
| 7 | 0;14, 4, 0,52,36,10,24,14 |
| 8 | 0;16, 4,35,17,15,37,36,16 |
| 9 | 0;18, 5, 9,41,55, 4,48,18 |
| 10 | 0;20, 5,44, 6,34,32, 0,20 |
| 11 | 0;22, 6,18,31,13,59,12,22 |
| 12 | 0;24, 6,52,55,53,26,24,24 |
| 13 | 0;26, 7,27,20,32,53,36,26 |
| 14 | 0;28, 8, 1,45,12,20,48,28 |
| 15 | 0;30, 8,36, 9,51,48, 0,30 |
| 16 | 0;32, 9,10,34,31,15,12,32 |
| 17 | 0;34, 9,44,59,10,42,24,34 |
| 18 | 0;36,10,19,23,50, 9,36,36 |
| 19 | 0;38,10,53,48,29,36,48,38 |
| 20 | 0;40,11,28,13, 9, 4, 0,40 |
| 21 | 0;42,12, 2,37,48,31,12,42 |
| 22 | 0;44,12,37, 2,27,58,24,44 |
| 23 | 0;46,13,11,27, 7,25,36,46 |
| 24 | 0;48,13,45,51,46,52,48,48 |
| 25 | 0;50,14,20,16,26,20, 0,50 |
| 26 | 0;52,14,54,41, 5,47,12,52 |
| 27 | 0;54,15,29, 5,45,14,24,54 |
| 28 | 0;56,16, 3,30,24,41,36,56 |
| 29 | 0;58,16,37,55, 4, 8,48,58 |
| 30 | 1; 0,17,12,19,43,36, 1, 0 |
| 354 | 11;51,23, 1,28,46,28,59,48 |
| 354;22 | 11;52, 7,14, 5,48,56,58,13 |
| 355 | 11;53,23,35,53,25,56,11,50 |
| 360 | 12; 3,26,27,56,43,12,12, 0 |
| 365 | 12;13,29,20, 0, 0,28,12,10 |
| 365;15 | 12;13,59,28,36,10,20, 0,11 |
| 366 | 12;15,29,54,24,39,55,24,12 |
| 1095 | 36;40,28, 0, 0, 1,24,36,30 |
| 1461 | 48;55,57,54,24,41,20, 0,42 |
| 10631 | 356; 3,37, 2,54,28,29, 6,22 |

It can be noted that Jai Singh was not completely systematic in the way in which he made use of mean motion values from the *Tabulae astronomicae.* It seems that he used La Hire's values for 2 years for the solar apogee and the node of Saturn, his values for 10, 20 or 60 years for the node of Jupiter, the aphelium of Venus, and the aphelium and node of Mercury, and his values for 3 years in the remaining 13 cases (in 4 of these cases he might also have used La Hire's value for 1 year, because the value for 3 years is exactly three times that for 1 year). Where Jai Singh made use of values for more than 3 years, he seems to have ignored the leap days in the Julian calendar and still performed the necessary divisions by a multiple of 365. Since this only occurs for the slow motions of the aphelia and nodes of the five planets, the resulting errors were relatively small. We do not know why Jai Singh used La Hire's mean motion values for 2 or 3, and occasionally for 10, 20 or 60 years to derive his own mean motion parameters. From a modern point of view, a more appropriate method would have been, for instance, to take the values for 1000 years and divide them by 365,250.

## Reconstruction of Jai Singh's Table for the Ascending Node of Mars

In conclusion of our analysis, we will now present a complete reconstruction of the table in the *Zīj-i Muḥamad Shāhī* for the motion of the ascending node of Mars. This table is a typical example for the variety of computational techniques used by Jai Singh and nicely illustrates the way in which he derived his mean motion parameters from the *Tabulae astronomicae* of Philippe de La Hire. Table 4 shows the tables for extended years, collected years and months with the errors according to two methods of computation which will be explained below:

I. an accurate computation on the basis of the underlying daily motion 0;0,0,6,4,55,53

II. a computation on the basis of the particular computational techniques that can be recognized from the table.

**Table 4. Recomputation of Jai Singh's table for the motion of the Ascending Node of Mars. Error I: differences between the table and an accurate computation based on a daily motion of 0;0,0,6,4,55,53°. Error II: differences between the table and a recomputation involving rounded values for ordinary and leap years and for 30 years**

Extended years

| year | position | error I | error II |
|---|---|---|---|
| 1 | 1$^{s}$17;36,29 | | |
| 2 | 1$^{s}$17;37, 4,59,11 | | |
| 3 | 1$^{s}$17;37,40,52,17 | +1 | |
| 4 | 1$^{s}$17;38,16,45,23 | +1 | |
| 5 | 1$^{s}$17;38,52,44,34 | +1 | |
| 6 | 1$^{s}$17;39,28,37,40 | +2 | |
| 7 | 1$^{s}$17;40, 4,36,51 | +2 | |
| 8 | 1$^{s}$17;40,40,29,57 | +2 | |
| 9 | 1$^{s}$17;41,16,23, 3 | +3 | |
| 10 | 1$^{s}$17;41,52,22,14 | +3 | |
| 11 | 1$^{s}$17;42,28,15,20 | +3 | |
| 12 | 1$^{s}$17;43, 4, 8,26 | +3 | |
| 13 | 1$^{s}$17;43,40, 7,36 | +3 | −1 |
| 14 | 1$^{s}$17;44,16, 0,43 | +4 | |
| 15 | 1$^{s}$17;44,51,53,49 | +4 | |
| 16 | 1$^{s}$17;45,27,53, 0 | +5 | |
| 17 | 1$^{s}$17;46, 3,46, 6 | +5 | |
| 18 | 1$^{s}$17;46,39,45,17 | +5 | |
| 19 | 1$^{s}$17;47,15,38,23 | +6 | |
| 20 | 1$^{s}$17;47,51,31,29 | +6 | |
| 21 | 1$^{s}$17;48,27,30,40 | +6 | |
| 22 | 1$^{s}$17;49, 3,23,46 | +7 | |
| 23 | 1$^{s}$17;49,39,16,52 | +7 | |
| 24 | 1$^{s}$17;50,15,16, 3 | +7 | |
| 25 | 1$^{s}$17;50,51, 9, 9 | +8 | |
| 26 | 1$^{s}$17;51,27, 8,20 | +8 | |
| 27 | 1$^{s}$17;52, 3, 1,26 | +8 | |
| 28 | 1$^{s}$17;52,38,54,32 | +8 | |
| 29 | 1$^{s}$17;53,14,53,43 | +9 | |
| 30 | 1$^{s}$17;53,50,46,49 | +9 | |
| 31 | 1$^{s}$17;54,26,39,55 | +9 | |

Collected years

| years | motion | error I | error II |
|---|---|---|---|
| 30 | 0;17,57,39,46 | | |
| 60 | 0;35,55,19,32 | +1 | |
| 90 | 0;53,52,59,18 | +1 | |
| 120 | 1;11,50,39, 4 | +2 | |
| 150 | 1;29,48,18,50 | +2 | |
| 180 | 1;47,45,58,36 | +2 | |
| 210 | 2; 5,43,38,22 | +3 | |
| 240 | 2;23,41,18, 8 | +3 | |
| 270 | 2;41,38,57,54 | +4 | |
| 300 | 2;59,36,37,40 | +4 | |
| 600 | 5;59,13,15,20 | +8 | |
| 900 | 8;58,49,53, 0 | +12 | |
| 1200 | 11;58,26,30,40 | +16 | |

Months

| month | motion | error I/II |
|---|---|---|
| Rabīʿ II | 0; 0, 0, 0, 0 | |
| Jumādā I | 0; 0, 2,56,23 | |
| Jumādā II | 0; 0, 5,58,51 | |
| Rajab | 0; 0, 8,55,14 | |
| Shaʿbān | 0; 0,11,57,42 | |
| Ramaḍān | 0; 0,14,54, 5 | |
| Shawwāl | 0; 0,17,56,33 | |
| Dhu'l-qaʿda | 0; 0,20,52,56 | |
| Dhu'l-ḥijja | 0; 0,23,55,24 | |
| Muḥarram | 0; 0,26,51,47 | |
| Ṣafar | 0; 0,29,54,14 | −1 |
| Rabīʿ I | 0; 0,32,50,38 | |
| ordinary year | 0; 0,35,53, 5 | −1 |
| leap year | 0; 0,35,59,10 | −1 |

Jai Singh started by computing his epoch value, the position of the ascending node of Mars on February 20, 1719 (Gregorian) from the tables of La Hire. For this purpose he added the appropriate tabular values for the beginning of the century and for the completed years, months, and days as shown below (cf. the earlier example for the solar mean longitude). Note that in this case it is not necessary to actually carry out the correction for the difference in geographical longitude between Paris and Delhi, because the daily motion (approximately 6"), and hence the correction, is significantly smaller than the accuracy of La Hire's tabular values (1"). For the motion in 19 days we have taken the value found for 20 days (2'), rather than that for 10 days (1', cf. note 8).

| | |
|---|---|
| Position at noon of January 1, 1700 Gregorian: | $1^s\ 17^\circ 25'20''$ |
| Motion in 18 (Julian or Gregorian) years: | $0^\circ 11'\ 4''$ |
| Motion in the month January: | $0^\circ\ 0'\ 3''$ |
| Motion in 19 days: | $0^\circ\ 0'\ 2''$ |
| Position at noon of February 20, 1719 Gregorian | ——— + |
| for the meridian of Paris or Delhi: | $1^s\ 17^\circ 36'29''$ |

Next Jai Singh took from La Hire's tables the motion of the ascending node of Mars in 3 Julian years, 0°1'51". He divided this value by 1095 and rounded the result to sexagesimal sixths to obtain as his value for the daily motion 0;0,0,6,4,55,53.[17]

In order to compute the table for *extended years,* Jai Singh multiplied the daily motion by 354 for an ordinary Hijra year and by 355 for a leap year and rounded the results to four sexagesimal fractional places, the accuracy of his table. Thus he obtained 0;0,35,53,6 (exactly: 0;0,35,53,5,42,42) for the motion in an ordinary year and 0;0,35,59,11 (exactly: 0;0,35,59,10,38,35) for the motion in a leap year. Following his intercalation scheme, he successively added these numbers to the epoch value and obtained a table with "constant" tabular differences (i.e., in this case, two different values for the tabular differences, one for every ordinary year and one for every leap year). In this way, the difference between the values for the years 31 and 1 became 19 x 0;0,35,53,6 + 11 x 0;0,35,59,11 = 0;17,57,39,55. By comparing this with the result of a multiplica-

**Table5. Recomputation of the sub-table for extended years of Jai Singh's table for the mean motion in longitude of Mercury (assumed daily mean motion: 4;5,32,35,20,32,53°)**

| year | mean position | error | year | mean position | error |
|---|---|---|---|---|---|
| 1 | 203;22,21, 0, 0 | | | | |
| 2 | 216;10,10, 6,34 | -1 | 17 | 7;12, 0,51,46 | -1 |
| 3 | 224;52,26,37,45 | -4 | 18 | 19;59,49,58,19 | -3 |
| 4 | 233;34,43, 9, 2 | -1 | 19 | 28;42, 6,29,35 | -1 |
| 5 | 246;22,32,15,35 | -2 | 20 | 37;24,23, 0,49 | -1 |
| 6 | 255; 4,48,46,51 | | 21 | 50;12,12, 7,24 | -1 |
| 7 | 267;52,37,53,24 | -2 | 22 | 58;54,28,38,35 | -4 |
| 8 | 276;34,54,24,39 | -1 | 23 | 67;36,45, 9,51 | -2 |
| 9 | 285;17,10,55,53 | -1 | 24 | 80;24,34,16,24 | -3 |
| 10 | 298; 5, 0, 2,26 | -2 | 25 | 89; 6,50,47,40 | -1 |
| 11 | 306;47,16,33,42 | | 26 | 101;54,39,54,13 | -3 |
| 12 | 315;29,33, 4,55 | -1 | 27 | 110;36,56,25,29 | -1 |
| 13 | 328;17,22,11,28 | -3 | 28 | 119;19,12,56,44 | |
| 14 | 336;59,38,42,44 | -1 | 29 | 132; 7, 2, 3,17 | -1 |
| 15 | 345;41,55,13,58 | -1 | 30 | 140;49,18,34,32 | |
| 16 | 358;29,44,20,31 | -2 | 31 | 149;31,35, 5,46 | |

tion of the accurate value of the daily motion by the number of days in 30 Arabic years (10631 x 0;0,0,6,4,55,53 = 0;17,57,39,45,35,43), Jai Singh probably realised that his value for the year 31 was too large by 9 sexagesimal fourths, but he did not adjust it. Note that the single error for this reconstruction ($36^{iv}$ instead of $37^{iv}$ for the year 13) can be explained as a scribal error.

Jai Singh computed the table for *collected years* by simply taking multiples of 0;17,57,39,46, i.e. the accurate value for the motion in 30 Arabic years rounded to the number of digits of the table. Since therefore the basic parameter of the table for collected years is the motion in 30 years, we have used the tabular value for 30 years rather than that for 300 or 1200 years to obtain the estimate presented in Table 2.

The table for *months* is generally accurate, except for three of the last four values, which are one fourth too small. These include the values for an ordinary

Hijra year and a leap year, which, therefore, do not agree with the values underlying the table for extended years. Note that the error in the value for Ṣafar (14 fourths instead of 15) could be explained as a scribal error.

Also the table for *days,* not displayed here, is generally accurate. Only the values for 11 days (0;0,1,0), 21 days (0;0,2,1), and 33 days (0;0,3,14) are one third too small. A larger error in the value for 61 days (0;0,6,0 should have been 0;0,6,5) can be explained as a scribal mistake.

As an example of a table which can be less exactly reconstructed, Table 5 displays part of Jai Singh's table for the mean motion in longitude of Mercury, along with the differences from an accurate computation on the basis of the daily mean motion listed in Table 2.

## Conclusion

We have shown that the astronomers of Jai Singh depended completely on the *Tabulae astronomicae* of Philippe de la Hire for computing the planetary mean motion tables in the *Zīj-i Muḥammad Shāhī.* Thus we have confirmed the conclusions of Mercier in his article "The Astronomical Tables of Rajah Jai Singh Sawā'ī," which appeared in this journal in 1984. We have been able to determine the precise method by which the mean motion tables in the *Zīj-i Muḥammad Shāhī* were computed. The epoch values were found by adding the appropriate values for centuries, years, months and days found in La Hire's tables. The daily mean motions were obtained by taking one particular mean motion value from La Hire's tables, mostly that for 3 Julian years, but incidentally the value for 2, 5, 10 or 60 years, and dividing it by the corresponding number of days. Finally, we have uncovered various details of the methods by which the astronomers of Jai Singh computed their mean motion tables, in particular the use of rounded values for the motion in the periods underlying the tables.

## Acknowledgements

The research laid down in this article has been conducted with financial support of the Alexander von Humboldt Foundation (Bonn, Germany), the Japan

Society for the Promotion of Science (JSPS), and the Dibner Institute (Cambridge, Massachusetts). I am grateful to Professor S. M. Razaullah Ansari for suggesting to me to carry out a mathematical analysis of Jai Singh's mean motion tables and for providing me with copies of manuscripts of the *Zīj-i Muḥammad Shāhī*. I benefited very much from discussions with Professors David Pingree and Raymond Mercier and Dr. Kim Plofker.

## NOTES AND REFERENCES

1 For the sources of the biographical information here presented and for additional information concerning life and works of Jai Singh, the reader is referred to the bibliography below.

2 According to Pingree (1976, p. 63), the *Zīj-i Muḥammad Shāhī* was most probably written by Abū al-Khayr Khayr Allāh Khān. However, for the sake of convenience we will refer to Jai Singh as the author of the work.

3 In particular for the solar apogee, the values given by Sharma differ from those we found in the manuscript of the *Zīj-i Muḥammad Shāhī* kept in Tonk.

4 Farrukhsiyar was deposed on March 1, 1719 (Gregorian), while Muḥammad Shāh came to power only on September 28 of that same year. In the seven intervening months, two puppet emperors sat on the throne. As Elphinstone (reference provided in Mercier 1984, note 8) indicates, these two emperors were left out from the official list of kings of the Moghul empire, so that the reign of Muḥammad Shāh was considered to have begun with the deposition of Farrukhsiyar. This explains how the epoch date of *Zīj-i Muḥammad Shāhī* can be connected to the accession of Muḥammad Shāh in spite of the time difference. Note that in traditional Persian calendars the beginning of the year was usually determined by the time of the vernal equinox. Also Akbar, Moghul emperor from 1556 to 1602, is known to have used a calendar based on the vernal equinox (David Pingree, personal communication). However, the vernal equinox of the year 1719 fell on 1 Jumāda I 1131 Hijra, so that it cannot explain the epoch date of the *Zīj-i Muḥammad Shāhī*.

5 Mercier's Table I contains a mistake in the line for the longitude of Venus: the epoch position at $t_L$ is 41;52,0 instead of 41;52,10. As a result, the calculated position at $t_J$ becomes 346;43,12.2 and the derived change in meridian 73;32. The misprint in the motion of the node of Venus in 30 Hijra years (correct is 0;22,21,15,13) was already noted by Mercier himself in his review of Sharma 1995 in *Isis* 88 (1997), p. 151.

6 We will use the standard notation for sexagesimal numbers, e.g., 2;41,25 stands for $2+\frac{41}{60}+\frac{25}{3600}$ or 2°41'25". Note that the values for the mean motion of Mercury and Venus in 54 minutes are given to sexagesimal thirds because they were obtained by dividing values from the respective tables for hours by 60.

7 In the Sanskrit translation of La Hire's *Tabulae astronomicae* by Kevalarāma, a correction is applied for a difference in geographical longitude between Paris and Jaipur of $4^h49^m$, i.e. 72°15' (Pingree, to appear). Note that this value is consistent with the difference 73°30' between Paris and Delhi found above and the longitude 1°15' west of Delhi assigned to Jaipur by Jai Singh (cf. Mercier 1984, p.150).

8 Because the motions of the aphelia and nodes of the planets amount to much less than a second per day, the same tabular value will be found for series of days if the motions are expressed in seconds. In such cases La Hire entered every value only once and left the remaining entries blank. As far as we know, he did not make it completely clear which value should be taken if none is found opposite the requested number of days. For our recomputation we have always taken the tabular value *above* the line for 19 days in cases where no value is given on that line. If we had always taken the tabular value *nearest* to the line for 19 days (so also possibly *below* it), the five differences of one minute mentioned above would have disappeared, but three new ones would have been introduced.

9 This range of years is precisely the same as that in the *Sulṭānī Zīj* of Ulugh Beg, but besides this there are no similarities between the two sets of tables: both the tabulated quantities and the underlying parameters are clearly different.

10 Since the tables are for "current" months, the first values (for Rabīʿ II) are always equal to zero, indicating that for a date in the first month of the year no mean motion from the table for months needs to be added to the mean position at the beginning of the year. The leap day is apparently inserted at the end of the last month of the year, Rabīʿ I, which thus receives 31 days in a leap year. However, in the manuscript of the Arabic and Persian Research Institute in Tonk, interlinear values facilitate insertion of the leap day at the end of Dhu'l-Ḥijja as in the common Hijra calendar.

11 Similarly to the months, the mean motion for the first day is zero in all cases. By shifting the sexagesimal point, the values for days can also be used for the Indian sexagesimal subdivisions of a day called *gharī* and *pal.*

12 Only the sub-table for extended years of the table for the mean solar longitude and that for collected years of the table for the motion of the solar apogee display values to five sexagesimal fractional digits. In the table for the mean longitude of Jupiter the fourths' position contains zeros only, because the underlying daily mean motion happens to be an integer number of sexagesimal thirds. In the sub-tables for extended and collected years of the table for the motion of Jupiter's aphelium, the fourths' position contains either 0 or 30, apparently the result of intermediate rounding. For all mean motions, the values in the second half of the table for days (and in some cases the whole table) are given to sexamesimal thirds instead of to fourths.

13 Whether this multiplication was effectively carried out as a multiplication or as a repeated addition of mean motions in larger periods such as Arabic ordinary and leap years, does not make a difference for the operation of the LNEC.

14 In the case of the type of mean motion tables in the *Zīj-i Muḥammad Shāhī*, the probability that the table for extended years has constant tabular differences is around 4.1%, whereas that for the table for collected years is 2.5%, and that for the table for months approximately 13.2%. Statisticians who are interested in the concept of randomness in connection with (deterministically computed) astronomical parameters and tables are referred to Van Dalen 1993, Section 1.2.4, pp. 15-19. The above probabilities are based on the assumption that the underlying values for the daily mean motion have a uniform distribution on a certain relevant interval.

15 This is the estimate obtained under the assumption that modern rounding was used. It is not the same as the entry for the longitude of Saturn in Table 2, because the sub-table for collected years of Saturn's mean motion table does not have constant tabular differences and hence we applied the LNEC to determine a more accurate estimate.

16 The errors of up to $1\frac{1}{2}$ degrees that can be seen in Sharma's recomputation of Jai Singh's table for the mean longitude of Venus (1990, p.40; 1995, p. 248) are due to an erroneous method of computation (cf. E.S. Kennedy's review of Sharma 1990 in *Mathematical Reviews* 1991m:01013 and R.P. Mercier's review of Sharma 1995 in *Isis* 88 (1997), pp. 150-151). Sharma calculated the motion in 20 Arabic years from La Hire's value for 20 Julian years by multiplying by 354;22/365;15 (1990, p.37; 1995, p. 247), but then distributed the result over what should have been 20 Arabic years using the intercalation scheme of the Julian calendar. Thus the values $T(n)$ ($n = 1, 2, 3, ..., 19$) in the fourth column of his Tables 3 and 11-4 (except $T(19)$, which does not seem to belong to this column at all) were effectively computed by the formula

$$T(n) = T(0) + \text{Trunc}\,(n.365\tfrac{1}{4}).\,(T(20) + 31.\,360 - T(0))/7305,$$

where $T(0) = 346;23,34$ is Jai Singh's epoch value, Trunc ($x$) denotes truncation of the fractional part of x, $T(20) = 181;31,54°$ equals $T(0)$ plus the motion in 20 Arabic years computed from La Hire's value for 20 Julian years, 31 is the number of rotations completed by Venus in 20 Julian years, and 7305 is the number of days in that period. Naturally there is no way in which the result could ever be close to Jai Singh's table.

17 That the exact quotient 0;0,0,6,4,55,53,25,28,46,... was rounded to sexagesimal sixths is made plausible by the fact that our analysis of Jai Singh's mean motion parameters yield an interval 0;0,0,6,4,55,53,8 ±10 of possible values for the daily motion of the ascending node of Mars (cf. Table 2). However, it should be mentioned that not all of the parameters used by Jai Singh can be precisely reconstructed in this way. It seems that in many cases the division of a value from La Hire's tables by the appropriate multiple of 365 was broken off at some point or the final digits were guessed to yield an approximate value only.

## BIBLIOGRAPHY

Virendra Svarup Bhatnagar. *Life* and *Times of Jai Singh (1688-1743)*, Delhi (Impex India) 1974.

Benno van Dalen. *Ancient and Mediaeval Astronomical Tables: Mathematical Structure and Parameter Values* (doctoral thesis), *University of Utrecht*, 1993.

Philippe de La Hire. *Tabularum astronomicarum Ludovici magni* (2nd ed.), Paris 1702 (repr. 1727).

Raymond P. Mercier. "The Astronomical Tables of Rajah Jai Singh Sawā'ī". *Indian Journal of History of Science* 19 (1984), pp. 143-171.

Honorino Mielgo. "A Method of Analysis for Mean Motion Astronomical Tables". *From Baghdad to Barcelona* (eds. Josep Casulleras and Julio Samso), Barcelona (Instituto Millás Vallicrosa) 1996, vol. 1, pp. 159-179.

David Pingree. *Census of the Exact Sciences in Sanskrit*, Series A, Vol. 3, Memoirs of the American Philosophical Society 111, Philadelphia 1976 (repr. in 1992).

David Pingree. "Indian and Islamic Astronomy at Jayasiṃha's Court." From *Deferent to Equant* ( eds. David A. King and George A. Saliba), New York (The New York Academy of Sciences) 1987, pp. 313-328.

David Pingree. "Philippe de La Hire at the Court of Jayasiṃha." To appear in *History of Oriental Astronomy, Proceedings of the Joint Discussion 17, held at the XXIIIrd IAU General Assembly, Aug*, 25-26, 1997 (ed. S.M. Razaullah Ansari), Dordrecht (Kluwer Academic Publisher).

Virendra Nath Sharma. "Zīj-i Muḥammad Shāhī and the Tables of de La Hire". *Indian Journal of History of Science* 25 (1990), pp. 34-44.

Virendra Nath Sharma. *Sawai Jai Singh and his Astronomy*, Delhi (Motilal Banarsidass) 1995.

**Correction**

V 48, line 3: 'from Rabīʿ II' → 'from Rabīʿ II to Rabīʿ II [of the following year]'.

VI

# The *Zīj-i Nāṣirī* by Maḥmūd ibn ᶜUmar

## The Earliest Indian-Islamic Astronomical Handbook with Tables and its Relation to the *ᶜAlāʾī Zīj* *

### 1 Introduction

Until recently, virtually nothing was known about the *Zīj-i Nāṣirī* by Maḥmūd ibn ᶜUmar, the earliest known Islamic astronomical handbook with tables that was written in India. Storey was the first western scholar to mention this work in the astronomical section of his *Persian Literature* [Storey 1958, p. 52]. He quoted an entry in [Āghā Buzurg Ṭihrānī 1936–78, vol. 8, p. 215] stating that the *Nāṣirī Zīj* was dedicated to Naṣīr al-Dīn Maḥmūd ibn Shams al-Dīn Iltutmish, sultan of Delhi from 1246 to 1265, and that a copy of the work was located in the important manuscript collection of Ḥusayn Āghā Nakhjawānī in Tabriz. Storey also mentioned a reference [*Oriens* 5 (1952), p. 193] to a letter of Muḥammad Qazwīnī published in *Nashriyya-yi Dānishkada-yi Adabīyāt-i Tabrīz* (*Revue de la Faculté des Lettres de Tabriz*) 2 (1328 H.S./A.D. 1949–50), pp. 119–126, which confirms the dedication of the zīj. Recent attempts to locate the manuscript of Nakhjawānī have been in vain. Furthermore, two small fragments of the *Nāṣirī Zīj* listed in the catalogues of the Mulla Firuz Library in Bombay turned out to be of little interest.

Our knowledge of the *Nāṣirī Zīj* has drastically improved with the appearance in 1994 of volume 23 of the catalogue of the Marᶜashī Library in Qum [Ḥusaynī & Marᶜashī 1994, p. 293]. This volume provides a one-page description of the Persian manuscript 9176 (165 folios) which contains a complete copy of the *Nāṣirī Zīj*. It was due to the efforts of Mr. Mohammad Bagheri (Encyclopaedia Islamica Foundation, Tehran), to whom we would like to express our deepest gratitude, and Prof. S. M. Razaullah Ansari (Aligarh Muslim University) that a photocopy of the whole manuscript was obtained. In the summer of 2002, Prof. Ansari

*First published in Ch. Burnett et al., *Studies in the History of the Exact Sciences in Honour of David Pingree*, Leiden (Brill) 2004, pp. 825–862. The original page numbers are shown in square brackets within the text.

**[p. 826]** and the present author had a chance to study various aspects of the *Nāṣirī Zīj* in Frankfurt. In the future, Prof. Ansari intends to publish a detailed account of the whole work. Below follow some preliminary results, concentrating on the tables for calculating planetary longitudes. It will be shown that nearly all of these tables derive directly or indirectly from the *ʿAlāʾī Zīj*, the latest of the six zījes written by the Caucasian astronomer al-Fahhād (ca. 1180), which is lost in its original form but influenced various later astronomical works. It partially survives in a * reworking for the Yemen by al-Fārisī (ca. 1270) and in a Byzantine recension by Gregory Chioniades (Constantinople, ca. 1305), which was in turn based on a Persian translation by Shams al-Dīn al-Bukhārī (Tabriz, 1295/96). In the course of our investigation it will be made plausible that Chioniades' version of the *ʿAlāʾī Zīj* contains the original planetary tables of al-Fahhād.

The *Nāṣirī Zīj* consists of two divisions (*rukn*), the first on "details" (*juzʾiyāt)* in 66 chapters (*bāb*, 121 folios), the second on "general principles" (*kulliyāt*) in 60 chapters (44 folios). The introduction (folios 1v–2v) states that Maḥmūd ibn ʿUmar studied the zījes of earlier "astronomers who made observations" (*aṣḥāb-i arṣād*) during thirty years and calculated from these works planetary conjunctions and solar and lunar eclipses for the purpose of comparison. The following astronomers are mentioned explicitly: Hipparchus, Ptolemy, Yaḥyā ibn Abī Manṣūr, Khālid al-Marwarrūdhī, Muḥammad ibn ʿAlī al-Makkī, Muḥammad ibn Mūsā ibn Shākir, al-Battānī, Sulaymān ibn ʿIṣma al-Samarqandī, al-Ṣūfī, Abu 'l-Wafāʾ, al-Bīrūnī, Ibn al-Aʿlam, Ḥabash al-Ḥāsib (these two are the only deviations from the chronological order), al-Khāzinī, and finally ʿAbd al-Karīm al-Shirwānī, i.e., al-Fahhād, the author of the *ʿAlāʾī Zīj*.

From the table of contents on folios 3r–v it becomes clear that the first division of the zīj deals with practical calculations, such as the conversion of dates in various calendars, spherical astronomical functions, planetary longitudes and latitudes, solar and lunar eclipses, and astrological quantities. The range of topics treated is quite extensive and includes, for example, the Jewish calendar, the latitude of the visible climate, tables of mean transfers (*wasaṭ al-taḥwīl*), equalization of the houses, projection of the rays according to the equator and the horizon, and *tasyīr* (*aphesis* or *directio*). The second division of the *Nāṣirī Zīj*, at which **[p. 827]** we have not yet looked in detail, presents explanations and proofs of the methods applied in the first division.

In this article we will limit ourselves to an investigation of the tables for calculating planetary longitudes. In order to make an attempt to place the *Nāṣirī Zīj* within the history of Islamic astronomy, we have used some preliminary results of the present author's project for the compilation of an updated survey of Islamic zījes. In 1956, Prof. E.S. Kennedy published *A Survey of Islamic Astronomical Tables*, in which he listed approximately 125 zījes, abstracted twelve of the most important of these works, and summarized the available historical information. During the last forty-five years, nearly one hundred additional works have come to light [cf. King & Samsó 2001]. The purpose of the author's project is first to compile a short list of all zījes now known with basic information on author, title, date, geographical origin, and available manuscripts, as well as references to the most important bio-bibliographical works and specific literature. Secondly, extensive studies on the treatment in zījes of topics such as the calculation of planetary positions, the prediction of solar and lunar eclipses, and mathematical astrology are envisaged on the basis of between 25 and 50 extant, mostly unpublished works. These studies will make it possible to describe the historical development of Islamic astronomy in more detail, whereas indexes of the mathematical characteristics of tables will facilitate the determination of the origin of unattributed materials. A preliminary database of the planetary parameters in approximately 25 zījes, relying heavily on a handwritten file by Prof. Kennedy, and an overview of the characteristics of planetary tables in around fifteen extant eastern-Islamic works predating the *Nāṣirī Zīj* have already been prepared and have been used for the present article.

Reference will be made to the following astronomers and zījes:

- The *Mumtaḥan Zīj* by Yaḥyā ibn Abī Manṣūr (Baghdad, 830), based on the observations carried out under the caliph al-Maʾmūn. Extant in a thirteenth-century revision in Escorial árabe 927 and published in facsimile in [Yaḥyā ibn Abī Manṣūr]. *
- The *Zīj* by Ḥabash al-Ḥāsib (Samarra, ca. 870), strongly related to the *Mumtaḥan Zīj* and extant in the manuscripts Istanbul Yeni Cami 784/1 (close to the original) and Berlin Ahlwardt 5750 (later recension). The introduction was edited **[p. 828]** and translated in [Sayılı 1955], and

a detailed summary presented in [Debarnot 1987].

- The *Ṣābiʾ Zīj* by al-Battānī (Raqqa, ca. 920). Extant as Escorial árabe 908 and published in [Nallino 1899–1907].
- The *ʿAḍūdī Zīj* by Ibn al-Aʿlam (Baghdad, d. 983). The author was well-known for his planetary observations, but his zīj is lost. Its parameters were reconstructed in [Kennedy 1977] and [Mercier 1985].
- The *Jāmiʿ Zīj* by Kūshyār ibn Labbān (Iran, ca. 1025). Highly popular and extant in more than twenty manuscripts of which Istanbul Fatih 3418 is the oldest complete one.
- Two zījes are attributed to Abu 'l-Wafāʾ (Baghdad, ca. 980): the *Wāḍiḥ Zīj* is lost, whereas part of *al-Majisṭī* is extant in Paris BNF arabe 2494 (with the tables left out).
- The *Ḥākimī Zīj* by Ibn Yūnus (Cairo, ca. 1000). Extant in Leiden Or. 143 (chapters 1–20) and Oxford Bodleian Hunt. 331 (chapters 21–44). The introduction and observation accounts were translated into French in [Caussin de Perceval 1804].
- *al-Qānūn al-Masʿūdī* by al-Bīrūnī (Ghazna, ca. 1030), a very extensive and learned work with complete explanations of the Ptolemaic planetary models and the determination of parameters. Extant in numerous copies and edited in [al-Bīrūnī].
- The *Sanjarī Zīj* by al-Khāzinī (Marw, ca. 1120). Extant in the manuscripts Vatican arabo 761 and London BL Or. 6669, both defective. The same author's *Wajīz* ("Summary") contains similar planetary tables. Editions and translations of various versions are being prepared by Prof. Pingree and his students. See already [Pingree 1999].
- The *ʿAlāʾī Zīj* by al-Fahhād (Shirwan, ca. 1180), the last of his six zījes. Highly influential but lost in its original version. All following works in this list are helpful in reconstructing al-Fahhād's tables and parameters.
- The *Shāmil Zīj* (anonymous, ca. 1240), said to be based on the planetary parameters of Abu 'l-Wafāʾ but also related to the *ʿAlāʾī Zīj*. Extant in more than ten manuscripts, of which Florence Laurenziana Or. 95 appears to be the oldest. The *Athīrī Zīj* by al-Abharī (Mardin, ca. 1240) can be considered to be a variant of the *Shāmil*, whereas
* the so-called *Utrecht Zīj* and the *Durr al-muntakhib* by the priest Cyriacus (Mardin, ca. 1480) were based on it. **[p. 829]**

- The Persian *Nāṣirī Zīj* by Maḥmūd ibn ʿUmar (Delhi, 1250), the earliest zīj written in India. Extant in the manuscript Qum Marʿashī 9176. It is studied for the first time in this article.
- The *Muẓaffarī Zīj* by al-Fārisī (Yemen, ca. 1270), a reworking of the *ʿAlāʾī Zīj* with adjustments for the author's geographical location and time. Extant in Cambridge Gg. 3.27/2.
- The *Zīj* by Jamāl al-Dīn Abu l-Qāsim ibn Maḥfūẓ al-Munajjim al-Baghdādī (1285), with materials going back to Ḥabash al-Ḥāsib and Abu 'l-Wafāʾ. Extant in Paris BNF arabe 2486.
- The Byzantine recension of the *ʿAlāʾī Zīj* by Gregory Chioniades (Constantinople, ca. 1305), based on a Persian version by Shams al-Dīn al-Bukhārī (Tabriz, 1295/96). Extant in Greek manuscripts in Florence and the Vatican, edited and translated in [Pingree 1985–86].
- The *Zīj* of al-Sanjufīnī (Tibet, 1366). Based on observations made by Muslim astronomers in Yuan China [see, for instance, van Dalen 2002], happens to contain a complete list of *ʿAlāʾī* parameters.

More information on these zījes can be found in [Kennedy 1956], [King & Samsó 2001], and, for sources related to the *ʿAlāʾī Zīj*, in [Pingree 1985–86, vol. 1, pp. 7–9 and 16–18].

The reader is assumed to be familiar with the general characteristics of Ptolemaic planetary theory and the setup and use of the tables for planetary mean motions and equations as found in the *Handy Tables* and Islamic zījes. Full explanations of theory and tables can be found in, for instance, [Neugebauer 1975, vol. 1, pp. 53–190] and [Pedersen 1974, Chapters 5, 6, 9, and 10], whereas Appendix 1 in [Neugebauer 1957] provides a brief but clear overview of the theories of eccentres and epicycles. The use of the *Handy Tables* is expounded in [Van der Waerden 1958]. An example of a full analysis of the mean motion tables in a particular zīj is [van Dalen 2000]. Information on the calendars involved can be found in the article TAʾRĪKH in the *Encyclopaedia of Islam, new edition*. Standard notation is used for sexagesimal numbers, e.g., 33;1,2 stands for $33+\frac{1}{60}+\frac{2}{60^2}$. A superscript "s" denotes a zodiacal sign, or also a general unit of 30° on the ecliptic, deferent, or epicycle. For example, $3^s$12;25 may denote 12;25° Cancer, but also an anomalistic planetary position of 102;25° measured from the apogee of the epicycle. By "first sign" I will refer to arguments 0 to 30° of a table for a planetary equation, by **[p. 830]** "second sign" to arguments 30 to 60°, etc. Superscripts "iv", "v" and "vi" stand for sexagesimal fourths, fifths and sixths.

## 2 The Daily Solar, Lunar, and Planetary Mean Motions

Folios 53r–58r of the *Nāṣirī Zīj* contain a set of tables for planetary mean motions. These include, for each of the periods of time listed below, the solar anomaly (*wasaṭ*, lit. "centrum"), the lunar longitude and anomaly, the elongation, the lunar node, and the centrum (*wasaṭ*) and anomaly of each of the five planets. All mean motions are tabulated to seconds for the following periods of time: 1, 2, ..., 10, 20, ..., 100, 200, ..., 1000 completed Persian years of 365 days (folios 53v–54r); current[1] Persian months Farwardīn, Urdībihisht, ..., Isfandār[mudh] plus a value for *tamām al-sana*, a "complete year" (folio 55r); current days 1, 2, 3, ..., 30 (folios 55v–56r); and 1 to 24 hours (folios 56v–57r). A table of the "difference [in mean motion] between the two longitudes" (*faḍl mā bayn al-ṭūlayn*) allows the adjustment of the mean positions found from the tables to geographical longitudes differing by 1, 2, ..., 10, 20, ..., 100 degrees from the base longitude of the zīj (folios 57v–58r). A separate table is provided for the motion of the solar and planetary apogees (folio 53r), which also lists the apogee positions at the epoch of the zīj (see below and Table 2). On folio 54v we find the motion of the pseudo-comet Kaid in the above-mentioned periods of years (which may have been inadvertently omitted from the general table of years) as well as the epoch positions at Delhi for all mean motions (also reproduced in Table 2).

Note that, unlike most Islamic zījes, the *Nāṣirī Zīj* does not include tables with actual mean *positions* for the beginnings of a certain collection of years; all tabulated values are mean *motions* in the given periods. For the actual positions we have to rely completely on the epoch positions listed on folios 53r (apogees) **[p. 831]** and 54v (all others). The only

[1] Sub-tables for months may be for "current" (*nāqiṣa*) or for "completed" (*tāmma*) months. In the former case, the tabulated motion for a given month is the motion that has taken place between New Year and the beginning of that month, which implies that the value for the first month is equal to zero. In the latter case, the tabulated motion is that between New Year and the end of the month, so that, for instance, to calculate a mean position during the month Ṣafar, the motion listed for Muḥarram must be used. Also the sub-tables for days may be for current or for completed days.

ways to correct scribal errors in these values are a check by means of the elementary relations between mean longitudes, centrums, and anomalies as described in the Appendix to this article and a comparison with the positions for Delhi at the Yazdigird, Hijra and Alexander epochs listed on folio 50v. As will be shown below, an attempt to reconstruct the epoch positions from those in other zījes will not prove succesful. In general, none of the zījes listed in the introduction to this article contain mean motion tables similar to the *Nāṣirī Zīj*, neither with respect to their setup nor their tabular values.

The parameters underlying the mean motion tables in the *Nāṣirī Zīj* were estimated by means of the Least Number of Errors Criterion (LNEC, introduced in [van Dalen 1993, Section 2.5]) which determines the range of parameter values for which the largest number of values in a given sub-table is correctly recomputed. A more extensive discussion of the method is presented in [Van Dalen 2000], whereas a slightly different approach is explained in [Mielgo 1996]. The application of the LNEC showed that the majority of the mean motion tables in the *Nāṣirī Zīj* were computed with a very high accuracy. As a matter of fact, most of the sub-tables do not contain any errors at all, others at most one or, very incidentally, two, which can partially be explained as scribal mistakes. Note that, due to the use of the Persian calendar with its constant year-length of 365 days, every sub-table is completely linear, consisting simply of multiples of the first value.[2] The ranges of the daily mean motions found by means of the LNEC are given in the second column of Table 1, where the notation $\mu \pm \epsilon$ indicates that all daily mean motions in the range $[\mu - \epsilon, \mu + \epsilon]$ produce the smallest possible number of **[p. 832]** errors in the sub-tables. In this procedure, the tabular values for the range 100, 200, …, 1000 years

[2] In estimating the underlying daily mean motions, I have treated the values for 1, 2, …, 10 years, those for 10, 20, …, 100 years, and those for 100, 200, …, 1000 years as three separate sub-tables, since it is possible that the value calculated for 10 or 100 years was first rounded to a lesser precision before being used as a constant multiplier for the next higher range of years. Since the later variant of the Persian calendar is used, in which the five epagomenal days are placed at the end of the year as opposed to after the eighth month Ābān, the sub-table for months displays the mean motions in 0, 30, 60, …, 360 days, and is therefore also completely linear. Note that the last value, in spite of the indication *tamām al-sana*, is not in fact the motion during a complete year, i.e., 365 days, but that between New Year and the beginning of the epagomenal days.

| motion | LNEC estimates | *ʿAlāʾī Zīj* |
|---|---|---|
| Apogee Longitude | 0; 0, 0, 8,57,37,29 ± $30^{vi}$ | 0; 0, 0, 8,57,46 * |
| Solar Longitude | | 0;59, 8,20,35,25 |
| Solar Anomaly | 0;59, 8,11,37,46,56 ± $5^{vi}$ | 0;59, 8,11,37,39 * |
| Lunar Longitude | 13;10,35, 1,55,32, 1 ± $2\frac{1}{2}^{vi}$ | 13;10,35, 1,55,32 |
| Lunar Anomaly | 13; 3,53,56,17,52, 3 ± $4^{vi}$ | 13; 3,53,56,17,51,59 |
| Lunar Elongation | 12;11,26,41,20, 7, 2 ± $3^{vi}$ | 12;11,26,41,20, 7 * |
| Lunar Nodes | −0; 3,10,37,35,29,15 ± $4\frac{1}{2}^{vi}$ | −0; 3,10,37,35,29,19 |
| Saturn Longitude | | 0; 2, 0,36, 4,33,33 |
| Saturn Centrum | 0; 2, 0,27, 6,55,32 ± $11^{vi}$ | 0; 2, 0,27, 6,47,33 * |
| Saturn Anomaly | 0;57, 7,44,30,51,26 ± $9^{vi}$ | 0;57, 7,44,30,51,27 * |
| Jupiter Longitude | | 0; 4,59,15,39,41 |
| Jupiter Centrum | 0; 4,59, 6,42, 2,57 ± $7^{vi}$ | 0; 4,59, 6,41,55 * |
| Jupiter Anomaly | 0;54, 9, 4,55,44, 4 ± $11^{vi}$ | 0;54, 9, 4,55,44 * |
| Mars Longitude | | 0;31,26,39,51,21 |
| Mars Centrum | 0;31,26,30,53,43, 2 ± $3^{vi}$ | 0;31,26,30,53,35 * |
| Mars Anomaly | 0;27,41,40,44, 3,57 ± $4\frac{1}{2}^{vi}$ | 0;27,41,40,44, 4 * |
| Venus Longitude | *same as solar longitude* | |
| Venus Centrum | *same as solar anomaly* | |
| Venus Anomaly | 0;36,59,28,43, 1,44 ± $5^{vi}$ | 0;36,59,28,43, 1,38 |
| Mercury Longitude | *same as solar longitude* | |
| Mercury Centrum | *same as solar anomaly* | |
| Mercury Anomaly | 3; 6,24,22, 7,59, 5 ± $4^{vi}$ | 3; 6,24,22, 7,59 |
| Kaid | −2°30′ / Persian year | |

**Table 1.** Second column: *LNEC-estimates of the daily mean motions underlying the tables in the Nāṣirī Zīj.* Third column: *The mean motion parameters of the ʿAlāʾī Zīj. Values with an asterisk are not listed by Chioniades but were recovered from his tables, the list in the Sanjufīnī Zīj, and other sources. Note that in the Nāṣirī Zīj the daily mean motions in centrum are all* $8^{v}$ *larger than those in the ʿAlāʾī Zīj due to the use of a different daily motion of the apogee.*

have been given the largest weight, since they are most significant and allow the most accurate determination of the underlying mean motion parameters.

From Table 1 we first note that, even though the parameter intervals typically have a width of approximately $10^{vi}$, nine out of fourteen (not counting the centrums of Venus and Mercury because they are identical to the solar anomaly and the motion of Kaid because it is based on a retrograde motion of precisely 2°30′ per year) include a round value to sexagesimal fifths. In the case of the mean anomaly of Mercury, such a round value is only narrowly missed. It is thus likely that most of the mean motion **[p. 833]** tables in the *Nāṣirī Zīj* were computed on the basis of daily mean motions to a precision of fifths.

A comparison of the estimated daily mean motions with values in Prof. Kennedy's parameter database of Islamic astronomy shows an obvious agreement with one particular set of daily mean motions, namely that listed in the Byzantine version of the *ʿAlāʾī Zīj* by Gregory Chioniades, in the Tibetan *Sanjufīnī Zīj*, and in a group of strongly related zījes including the anonymous *Shāmil Zīj*, the *Athīrī Zīj* by al-Abharī, the *Utrecht Zīj*, and the *Durr al-muntakhib* by the Priest Cyriacus. There is little doubt that this set of mean motions stems from the non-extant *ʿAlāʾī Zīj*, the latest of the six zījes by al-Fahhād (ca. 1180). In fact, the mean motion tables in Chioniades' reworking of this zīj, which was drawn upon a Persian version by Shams al-Dīn al-Bukhārī (Tabriz, 1295/96), are based on the daily mean motions concerned, as are those in the Yemeni *Muẓaffarī Zīj* by Muḥammad ibn Abū Bakr al-Fārisī (ca. 1270), who acknowledges the use of al-Fahhād's observations. The introduction to the *Shāmil Zīj* maintains that the author of the *ʿAlāʾī Zīj* had presented Abu 'l-Wafāʾ's mean motion parameters as the results of his own observations; the tables in the *Shāmil Zīj* are said to be based on Abu 'l-Wafāʾ's parameters, which is particularly interesting because this astronomer's planetary tables are completely lost. The mean motion tables in the *Shāmil Zīj* have a rather different structure from those in the Byzantine version of the *ʿAlāʾī Zīj* and the *Muẓaffarī Zīj* but the underlying daily mean motions never deviate by more than one sexagesimal fourth from the parameters in those works. That the list of *ʿAlāʾī* parameters also occurs in the *Sanjufīnī Zīj* appears to be a historical coincidence: the planetary mean motion tables in this work were not computed on the basis of the *ʿAlāʾī* parameters but from the tables in the *Huihuili*, a Persian zīj compiled in Mongol China in the 1270s and translated into Chinese in 1383.

A comparison of the estimates found above with the lists of *ʿAlāʾī* parameters allows us to determine exactly which daily mean motions were used to compute the tables in the *Nāṣirī Zīj* and which were the original parameters of the *ʿAlāʾī Zīj*. This comparison is complicated by the fact that some of the sources involved tabulate or list the solar and planetary *longitudes* and others the *centrums*, which differ by amounts of the daily apogee motion **[p. 834]** that are not the same for each source. Moreover, the lists in the works from the *Shāmil* group contain many scribal mistakes.

The clearest picture of the situation arises if we compare the daily mean motions from the Byzantine version of the *ʿAlāʾī Zīj*, which are listed without an asterisk in the third column of Table 1, with: 1) the actual mean motion tables in that same work; 2) the estimates derived from the *Nāṣirī Zīj*; and 3) the complete list of parameters in the *Sanjufīnī Zīj*. It turns out that each of the latter three sets of data can be derived from the basic set of *ʿAlāʾī* parameters listed in the Byzantine version. Both Chioniades (or possibly al-Fahhād himself) and Maḥmūd ibn ʿUmar needed to convert the daily mean motions in *longitude* from the basic set into daily mean motions in *centrum* in order to compute the tables that we find in their works. This conversion is carried out by subtracting the daily motion of the apogee from the daily mean motion in longitude. However, the daily motion of the apogee is not listed to the required precision of five or six sexagesimal fractional digits in Chioniades' version of the *ʿAlāʾī Zīj*, and hence possibly in the original work by al-Fahhād. This may have been the reason why Chioniades (or al-Fahhād) and Maḥmūd ibn ʿUmar used slightly different rates for the conversion. In fact, Chioniades consistently used the value 0;0,0,8,57,46 °/day underlying his table for the "equation of the solar apogee" [Pingree 1985–86, vol. 2, pp. 35–36], whereas Maḥmūd used a value within the range 0;0,0,8,57,37,40–38,0 °/day which we have not yet found in other sources but is quite close to a motion of 1° in 66 Julian years. Finally, the author of the list in the *Sanjufīnī Zīj* picked still another value for the apogee motion, namely, the very common 0;0,0,8,57,58 °/day, close to 1° in 66 Persian years, to calculate his daily mean motions in centrum. That the value used by Chioniades is the original parameter of the *ʿAlāʾī Zīj* is made plausible by the fact that the table for apogee motion in the *Muẓaffarī Zīj* by al-Fārisī is basically identical to that presented by Chioniades, the only difference being a change of the epoch for which the apogee positions are given from the year 541 Yazdigird to the year 631 (A.D. 1262 instead of 1172).

The parameter lists in Chioniades' work and in the *Sanjufīnī Zīj* contain a number of daily mean motions expressed to a precision of sexagesimal sixths (rather than fifths). These correspond very well to the intervals of estimates obtained from the *Nāṣirī* **[p. 835]** *Zīj* that do *not* contain a round value to fifths. For instance, the motion of the lunar node listed by Chioniades falls within the interval of estimates, as do the motions of Saturn in centrum and in anomaly that can be derived from the listed motion in longitude. The listed anomaly of Venus misses the interval of estimates by only a sixth, whereas the lunar anomaly, equal to Ptolemy's value, falls within the interval together with a round value to fifths. Thus we can conclude that, for the computation of his mean motion tables, Maḥmūd ibn ʿUmar based himself on exactly the values listed by Chioniades, but used a slightly different apogee motion to convert longitudes to centrums.

The works of the *Shāmil* group contain complete lists of daily mean motions in longitude, centrum and anomaly, but with values to fifths only. In general, the values are identical with, or extremely close to, those of the *ʿAlāʾī Zīj* in the third column of Table 1, except that the motions in centrum were determined from those in longitude by subtracting a daily apogee motion of 0;0,0,8,57,58°, as in the *Sanjufīnī Zīj*. In the *Shāmil* group, Saturn's daily mean motion in longitude is 0;2,0,36,4,35° (instead of $33^{v}33^{vi}$), in centrum 0;2,0,27,6,37 (instead of $35^{v}33^{vi}$), and in anomaly 0;57,7,44,30,21 (instead of $51^{v}27^{vi}$; since this is inconsistent with the longitudes of the Sun and Saturn, it could be a scribal error). If the introduction to the *Shāmil Zīj* is correct, we may thus conclude that both al-Fahhād and the author of the *Shāmil Zīj* used the mean motions of Abu 'l-Wafāʾ; otherwise, the *Shāmil* group must have copied the parameters of the *ʿAlāʾī Zīj* with a minor change in the motion of the apogee.

The only data we have to verify the statement in the *Shāmil Zīj* that the *ʿAlāʾī* parameters stem from Abu 'l-Wafāʾ are a number of incredibly precise values in the margins of the mean motion tables of Ḥabash al-Ḥāsib in the manuscript Berlin Ahlwardt 5750. It is not impossible that these are Abu 'l-Wafāʾ's, since the present author has noticed that from this thirteenth-century manuscript some of Ḥabash's original tables were scraped away with a knife or similar object (in many cases producing small holes in the paper) and replaced by tables of Abu 'l-Wafāʾ, which have a precision of thirds instead of minutes or seconds.[3] The **[p. 836]**

[3]Unpublished result. It may be noted, though, that at first sight the hands in which the marginal notes and the substituted table were written are not the same.

values attributed to Abu 'l-Wafāʾ in the margins are as follows (these were partially published in [Kennedy 1956, p. 169]):

| | |
|---|---|
| Solar Longitude | 0;59, 8,20,43,17,38,41,42,20, 5 |
| Lunar Longitude | 13;10,35, 1,55,37,39, 6,16,45,43 |
| Lunar Anomaly | 13; 3,53,56,17,50,25, 7,59,17,31 |
| Lunar Elongation | 12;11,26,41,12,20, 0,54,24,25, ? |
| Lunar Node | 0; 3,10,37,35,10, 1,51,42,13,28 |
| Saturn Longitude | 0; 2, 0,36, 4,27,58,33,41,41,42 |
| Jupiter Longitude | 0; 4,59,16,58,50,44,30,49,53,17 |
| Mars Longitude | *not included* |
| Venus Anomaly | 0;36,59,29, 7,49, 1,36, 9,21,59 |
| Mercury Anomaly | 3; 6,24, 6,59,45,22, 0,37,26,24 |

It will be clear that, even taking into account the possibility of multiple scribal errors, there cannot be a simple relationship between the parameters of the *ʿAlāʾī Zīj* in Table 1 and those attributed to Abu 'l-Wafāʾ.

As far as the origin of the mean motion parameters in the *ʿAlāʾī Zīj* is concerned, it appears that not all of them were based on new observations. In the introduction of the *Muẓaffarī Zīj*, al-Fārisī presents some interesting statements concerning observations made by al-Fahhād [Lee 1822, pp. 257–259]. For instance, he writes that al-Fahhād found the mean motions of the Sun and the Moon to be in agreement with the observations of the astronomers working under al-Maʾmūn, namely Yahyā ibn Abī Mansūr, Khālid al-Marwarrūdhī, al-ʿAbbās al-Jawharī, and Ḥabash al-Ḥāsib. In fact, it can be checked that the solar and lunar mean longitudes and the lunar mean anomaly found from the *ʿAlāʾī Zīj* differ by less than a minute of arc from those of Yahyā and Ḥabash. Al-Fahhād also found that the mean positions of Mars and Venus were in agreement with Ibn al-Aʿlam, whose *ʿAḍūdī Zīj* is likewise lost. From data in later works it can be seen that the *ʿAlāʾī* mean positions of these planets are indeed quite close to Ibn al-Aʿlam's (cf. [Mercier 1985]), although the agreement is not quite as good as in the case of the solar and lunar mean motions. Two more interesting statements by al-Fārisī to the extent that, unlike most astronomers, al-Fahhād found the apogee of Venus to be unequal to that of the Sun, and that the most accurate true positions of Mercury were produced by Ptolemy's tables, deserve further investigation. **[p. 837]**

## 3 The Solar, Lunar, and Planetary Epoch Positions

As mentioned above, the mean motion tables in the *Nāṣirī Zīj* display only motions, not actual mean positions. The motions obtained from the tables need to be added to corresponding mean positions listed on folios 53r (apogees) and 54v (all others). These positions have been reproduced in the second column of Table 2. From the description of the use of the mean motion tables (as well as from the values themselves) it becomes clear that the given positions are in fact for New Year (1 Farwardīn) of the Persian year 1 *before* Yazdigird. In this way, a mean position for, for example, the year 1372 Yazdigird can be obtained by simply adding the mean motions found for arguments 1000, 300, 70, and 2 in the table for years to the epoch position concerned.

The epoch positions to be used with the mean motion tables can be compared with a table on folio 50v that lists mean positions at Delhi for the Yazdigird, Hijra, and Byzantine (i.e., Dhu 'l-qarnayn or Alexander, 1 October 312 B.C.) epochs. Even though this list includes mean longitudes rather than centrums (although both are called *wasaṭ*) and does not include the anomalies of the superior planets,[4] it is an easy matter to verify that the two sets of positions are fully compatible. In nearly all cases the epoch positions of the mean motion tables can be obtained by subtracting the motion in a single Persian year from the positions for the Yazdigird epoch found in the table on folio 50v or derived from it by means of the elementary rules explained in the Appendix. It turns out that the positions for the Hijra epoch (Maḥmūd ibn ᶜUmar used the civil variant, i.e., Friday, 16 July, A.D. 622) and for the Byzantine epoch were likewise accurately computed using the mean motion tables in the *Nāṣirī Zīj*.[5] Thus we can reliably reconstruct a complete set of mean positions for Delhi at the Yazdigird epoch, which is found in the third column of Table 2. Values that were not taken directly from the table on folio 50v are marked with an asterisk. Scribal errors in the manuscript have been indicated in notes to the table.

[4] For the Yazdigird epoch the lunar elongation, the anomaly of the superior planets, and the centrum of the inferior planets are written in the margin.

[5] As a matter of fact, due to the large time span between the Byzantine and the Yazdigird epochs, it is even possible to conclude that not the original rate of apogee motion from the *ᶜAlāʾī Zīj*, 0;0,0,8,57,46 °/day, was involved, but that from the *Nāṣirī Zīj*, 0;0,0,8,57,38 °/day.

[p. 838]

| motion | 1 before Yazdigird | 1 Yazdigird |
|---|---|---|
| Solar Longitude | | 2$^s$26;57,24 |
| Solar Anomaly | 0$^s$ 2;32,45 | 0$^s$ 7;17,36 * |
| Solar Apogee | 2$^s$19;38,53 | 2$^s$19;39,48 |
| Lunar Longitude | 7$^s$23,31,36 | 0$^s$ 2,54,43 |
| Lunar Anomaly | 7$^s$ 7; 0, 0 | 10$^s$ 5;43, 7 [1] |
| Elongation | 4$^s$26;19,58 | 9$^s$ 5;57,19 |
| Lunar Node | 10$^s$13;36,39 | 2$^s$ 5;43,20 [2] |
| Saturn Longitude | | 7$^s$29; 7,58 |
| Saturn Centrum | 11$^s$17; 4,25 | 11$^s$29;17,10 * |
| Saturn Anomaly | 7$^s$10;17,20 | 6$^s$27;49,26 |
| Saturn Apogee | 4$^s$29;49,53 | 7$^s$29;50,48 |
| Jupiter Longitude | | 9$^s$ 2;47,10 |
| Jupiter Centrum | 2$^s$12;54,46 | 3$^s$13;14,22 * |
| Jupiter Anomaly | 7$^s$24;44,59 | 5$^s$24;10,14 |
| Jupiter Apogee | 5$^s$19;31,53 | 5$^s$19;32,48 |
| Mars Longitude | | 10$^s$10;35,56 [3] |
| Mars Centrum | 11$^s$18;45,50 | 6$^s$ 0; 2, 8 * |
| Mars Anomaly | 10$^s$27;52,55 | 4$^s$16;20,48 |
| Mars Apogee | 3$^s$10;32,53 | 4$^s$10;33,48 |
| Venus Longitude | | 2$^s$26;57,24 [4] |
| Venus Centrum | 0$^s$19;28,45 | 0$^s$19;13,36 * |
| Venus Anomaly | 8$^s$13;57, 6 | 3$^s$28;58,56 |
| Venus Apogee | 2$^s$ 2;42,53 | 2$^s$ 7;43,48 [5] |
| Mercury Longitude | | 2$^s$26;57,24 [6] |
| Mercury Centrum | 8$^s$ 6; 3,45 | 8$^s$ 6;48,28 * |
| Mercury Anomaly | 3$^s$ 9; 9,51 | 6$^s$ 3; 8, 6 [7] |
| Mercury Apogee | 6$^s$21; 7,53 | 6$^s$21; 8,48 |
| Kaid | 7$^s$17;30, 0 | 4$^s$15; 0, 0 [8] |

**Table 2.** *Solar, lunar, and planetary epoch positions for Delhi as found in the Nāṣirī Zīj.* Second column: *Positions for New Year of the year 1 before Yazdigird (folios 53r and 54v), to be used with the mean motion tables.* Third column: *Positions for New Year of the year 1 Yazdigird as found on folio 50v. Values indicated by an asterisk were reconstructed, values indicated by a dagger are written in the margin.*

Notes to the table: [1] The positions for the Hijra and Alexander epochs are in agreement with a value of 10$^s$5;44,7. [2] This is 360° minus the actual position of the ascending node, possibly pointing to a dependence on mean motion tables that tabulate the supplement of the nodal position rather than its actual position. [3] In the manuscript this value is corrected to 10$^s$10;36,36, which is in fact consistent with the tabulated anomaly and the solar longitude. [4] The other epoch positions for Venus and Mercury are in agreement with a mean longitude of 2$^s$27;11,38, differing from the tabulated longitude by the motion in nearly a quarter of a day. [5] The manuscript makes the apogee of Venus equal to that of the Sun for all three epochs. [6] See note 5. [7] The positions for the Hijra and Alexander epochs are in agreement with a value of 6$^s$0;8,6. [8] This is 360° minus the actual position of Kaid (cf. note 2).

**[p. 839]** It turns out that the epoch positions of the *Nāṣirī Zīj* cannot be easily derived from the mean positions in Chioniades' version of the *ʿAlāʾī Zīj* or in al-Fārisī's *Muẓaffarī Zīj*. Chioniades systematically maintains al-Fahhād's original epoch 541 Yazdigird (A.D. 1172) and his geographical longitude 84° (measured from the Fortunate Isles) for the region Shirwan in Azarbaijan. Thus he lists the longitudes of the apogees for the year 541 (besides for the Yazdigird and Hijra epochs) [Pingree 1985–86, vol. 2, p. 33], and his table for the "equation of the solar apogee" (see above) assumes its zero for that year. Al-Fārisī, on the other hand, changes the epoch to his own time, namely, to 631 Yazdigird (A.D. 1262), and consequently shifts the zero in his "equation of the solar apogee". Furthermore, al-Fārisī adjusts the original tables of the *ʿAlāʾī Zīj* for use in the Yemen (longitude 63°30′ from the Fortunate Isles, rounded to 64°). As a result, the mean positions given by al-Fārisī all differ from those in Chioniades by exactly the motion in $1^{h}20^{m}$, corresponding to a longitude difference of 20°.

As was mentioned above, the base meridian of the *Nāṣirī Zīj* is that of Delhi, to which Maḥmūd ibn ʿUmar attaches a longitude of 103°35′ (measured from the Western Shore of Africa) in an example in Division 1, Chapter 25 (folios 50v–51r) on the adjustment of planetary mean positions to different locations.[6] This corresponds to 113°35′ from the Fortunate Isles and hence to a longitude difference from Shirwan equal to 29°35′. This implies that Maḥmūd ibn ʿUmar would have needed to subtract the mean motion in nearly two hours from the epoch positions of the *ʿAlāʾī Zīj* in order to obtain the corresponding epoch positions for Delhi. An attempt to reconstruct Maḥmūd's epoch positions along these lines clearly showed that he did *not* use the epoch positions from the *ʿAlāʾī Zīj*. Whereas his solar longitude is in exact **[p. 840]** agreement

6 The longitude value 103°35′ for Delhi occurs in some more chapters of the *Nāṣirī Zīj*. It has not been found in other zījes but on various astrolabes (cf. [Kennedy & Kennedy 1987, p. 105]). The *Nāṣirī Zīj* also includes a geographical table (folios 32v–34r), which presents longitudes (likewise measured from the Western Shore of Africa) and latitudes for 109 localities distributed over eight climates. This table gives the longitude of Delhi as 104°29′, a value not yet known from other sources. The listed longitudes of Shirwan (57°30′, possibly a mistake for 67°30′) and Tabriz (73°10′) stem from al-Bīrūnī and are not compatible with al-Fahhād's longitude of 84° (measured from the Fortunate Isles) for Shirwan.

with Chioniades if we assume a correction for a longitude difference of 30° 15′, most of the other longitudes, centrums and anomalies are not even close to values thus reconstructed. The longitudes of the solar and planetary apogees, however, correspond with the values to minutes given by Chioniades [Pingree 1985–86, vol. 2, p. 33] with the exception of that of Mars, for which Chioniades has $4^s6;34$ instead of Maḥmūd's $4^s10;33,48$. In Chioniades, as well as in the *Nāṣirī Zīj*, the longitude of the apogee of Venus is different from that for the Sun, again pointing to a relation to al-Fahhād.

I have further compared the planetary positions in the *Nāṣirī Zīj* with the *Sanjarī Zīj* of al-Khāzinī (Marw, ca. 1120), the next earlier astronomer who is mentioned in the introduction, but likewise with a negative result. The epoch positions listed in the *Shāmil Zīj* and its relatives, which are said to be for a geographical longitude of 84°, turn out to be in full agreement with the *ʿAlāʾī Zīj* and hence not with the *Nāṣirī Zīj*. We conclude that Maḥmūd ibn ʿUmar either used positions based on still another zīj, or adjusted the epoch positions on the basis of his own observations.

## 4 The Solar, Lunar, and Planetary Equations

We will now turn our attention towards the tables for the solar, lunar, and planetary equations in the *Nāṣirī Zīj* and compare them with Chioniades' Byzantine version of the *ʿAlāʾī Zīj*, al-Fārisī's *Muẓaffarī Zīj* and the group of zījes related to the *Shāmil Zīj*. As will become apparent, the equations in all these works are to a smaller or larger extent related to each other, and in many cases show a dependency on the equation tables in the zījes of al-Khāzinī. Before comparing actual tabular values, we will look at more general characteristics of the tables, in particular the "displacements". Displaced equations have been described in [Salam & Kennedy 1967, for the lunar tables of Ḥabash], [Saliba 1976, for Cyriacus], [Kennedy 1977, in connection with Ibn al-Aʿlam], [Saliba 1977, for ʿAbd al-Raḥīm al-Qazwīnī], [Saliba 1978, again for Cyriacus], [van Dalen 1996, for Kūshyār's solar equation], [Van Brummelen 1998, for Kūshyār's planetary tables], and others. Here we will only present a brief general description of displacements.

**[p. 841]** In Ptolemy's *Almagest* and many Islamic zījes, the planetary equations need to be either added to, or subtracted from, specific quantities depending on whether their arguments fall in certain ranges. Displaced equations eliminate the conditional addition or subtraction by making the equations additive (or, sometimes, subtractive) throughout. This is done by increasing every equation value by a positive constant $c$ equal to, or larger than, the maximum equation. Thus, instead of a positive function $q(x)$ that has a maximum value $q_{\max}$ and is sometimes additive and sometimes subtractive, we obtain a positive, always additive function $c \pm q(x)$ with values between $c - q_{\max} \geq 0$ and $c + q_{\max}$. Since the equation now exceeds its actual value by the amount $c$, a mean motion $\mu$ to which the equation is added must be decreased by $c$ if the addition is to yield the same result; thus the mean motion is tabulated as $\mu - c$ instead of $\mu$. If $\mu$ is itself the argument of an equation, the arguments of this equation must be shifted by $c$, so that the equation for the original argument $\mu$ appears next to $\mu - c$. It is in particular in cases where this shift is necessary that $c$ is chosen as an integer number. In general, we define the displacement of a given planetary equation by a constant $c > 0$ as the operation in which every additive value of the equation is increased by $c$ and every subtractive value is subtracted from $c$, after which multiples of $12^s$ (360°) are discarded.

The planetary equations in the *Nāṣirī Zīj* are basically of standard type, except that they are displaced by 12 zodiacal signs. In accordance with the above definition, this means that subtractive values $q$ of the equation are represented as $12^s - q$, and that all tabular values can be *added* to the mean position concerned in order to obtain a true position. In the case of (originally) subtractive values, the result of this addition will almost always be larger than $12^s$, requiring $12^s$ to be subtracted again in order to obtain a number between 0 and 360°. As far as I know, the *Nāṣirī Zīj* is the earliest extant zīj in which displacements of 12 zodiacal signs occur, although displacements by the maximum equation (or next larger integer) were already used by Ḥabash al-Ḥāsib (ca. 850, only for the moon) and Kūshyār ibn Labbān (ca. 1025).[7] **[p. 842]** Since some

[7] [Kennedy 1977] and [Mercier 1989] disagree about the question whether the non-extant *ʿAḍūdī Zīj* by Ibn al-Aʿlam (ca. 970) utilized displaced equations. I have re-inspected the main primary source for information on Ibn al-Aʿlam's planetary tables,

of the tables in Chioniades' version of the *ʿAlāʾī Zīj* and in al-Fārisī's *Muẓaffarī Zīj* are also displaced by 12 signs, the *ʿAlāʾī Zīj* by al-Fahhād may have been the original source for this type of displacement.

The tables for the planetary equations in the Byzantine version of the *ʿAlāʾī Zīj* and in the *Muẓaffarī Zīj* are partially very different from those in the *Nāṣirī Zīj* but very similar to each other. The tables of the solar equation and of the planetary equations of anomaly are displaced by the longitude of the apogees concerned, so that the equation must always be added to the tabulated mean centrum (rather than to the mean longitude) in order to obtain the true longitude. Furthermore, the tables for the lunar and planetary equations of anomaly are not only displaced but also of what I will call "mixed type", which means that the first halves of the tables display values of the equation for one position of the epicycle on the deferent, the second half for another. The mixed equations eliminate the conditional addition or **[p. 843]** subtraction that occurs in most cases

---

namely, the *Ashrafī Zīj* by Muḥammad ibn Abī ʿAbd Allāh Sanjar al-Kamālī of Yazd, known as Sayf-i Munajjim (1302). The only surviving manuscript of this work is Paris, Bibliothèque Nationale de France, supplement persane 1488. Al-Kamālī first presents his own planetary equations, which are essentially displaced copies of those of Ḥabash al-Ḥāsib, and then enables the reader to calculate planetary positions according to twelve well-known zījes by including all planetary equations from those works that are different from his own. For Ibn al-Aʿlam, al-Kamālī tabulates the solar equation (without displacement), the equations of centrum for Saturn and Jupiter (with displacements in concordance with al-Kamālī's own equations of anomaly for these planets), and both equations for Mercury (without displacements). By comparing all alternative equations presented in the *Ashrafī Zīj* with the zījes from which they originate (in so far as these are extant), I found that al-Kamālī in general correctly reproduces the actual equation values, but that in various cases displacements and shifts were omitted, modified, or introduced. Since every single alternative equation has a maximum value different from that of al-Kamālī himself, it is thus clear that the author's purpose was to accurately represent the magnitude of the equations but not necessarily their displacements. If we make the plausible assumption that Ibn al-Aʿlam's planetary equations were all either of the displaced type or of the standard type, it follows that al-Kamālī removed the displacements from the Mercury equations (and possibly the solar equation), or that he introduced the displacements of the equations of centrum for Saturn and Jupiter. In my opinion, the latter possibility is more likely, since it made it possible to use Ibn al-Aʿlam's equations of centrum for Saturn and Jupiter in combination with al-Kamālī's own displaced equations of anomaly for these planets. I will therefore for the time being assume that Ibn al-Aʿlam did not use displaced equations.

in the course of the Ptolemaic interpolation involved in the calculation of the equation of anomaly (see below). As a result, the only operations left in the calculation are one multiplication and two additions. However, this requires additional tables for the equation itself and for the interpolation function. An extensive explanation of the "mixed equation of anomaly" will be found in a forthcoming publication by the present author on the characteristics of tables for calculating planetary longitudes in Islamic zījes. This publication will also discuss characteristics not treated here such as terminology, exact layout of the tables, etc. As far as I know, Chioniades' version of the *ʿAlāʾī Zīj* and the *Muẓaffarī Zīj* are the earliest extant works that utilize equations of anomaly of mixed type. Given that the equation tables in the two works share such a highly peculiar characteristic, whereas we have already seen that al-Fārisī's tables for planetary mean motions can be derived from those of Chioniades, we conclude with reasonable certainty that Chioniades' tables, which are set up for the longitude of al-Fahhād's location Shirwan and his epoch A.D. 1172, are the tables from the original *ʿAlāʾī Zīj*.

*Solar Equation*

The solar equation in the *Nāṣirī Zīj* is tabulated to seconds of a degree as a function of the mean anomaly and is displaced by 12 zodiacal signs (see above). It assumes a maximum value of 1°59′0″, which corresponds to a solar eccentricity of 2;4,35,30 units and originally goes back to the Mumtaḥan observations. The solar equation tables for this parameter by Yaḥyā ibn Abī Manṣūr and Ḥabash al-Ḥāsib were still highly inaccurate, whereas al-Battānī and Kūshyār used the minimally different eccentricity value 2;4,45. However, Abu 'l-Wafāʾ (ca. 980) provided a table, extant in the Berlin manuscript of Ḥabash's *Zīj* and in the *Zīj* by Jamāl al-Dīn ibn Maḥfūẓ al-Baghdādī (1286), with maximum 1°59′, values to sexagesimal thirds, and errors of at most 2 thirds. Other early Muslim astronomers, such as Ibn al-Aʿlam and al-Bīrūnī, observed maximum equations slightly different from the Mumtaḥan value, whereas al-Khāzinī used the clearly larger 2°12′23″.

Instead of the plain solar equation, Chioniades' version of the **[p. 844]** *ʿAlāʾī Zīj* and the *Muẓaffarī Zīj* tabulate the true solar longitude to seconds for every degree of the mean anomaly. Thus the tabulated function is $\lambda_A + \bar{a} \pm q(\bar{a})$, where $\lambda_A$ is the longitude of the solar apogee, $\bar{a}$ the mean anomaly, and $q(\bar{a})$ the solar equation. Here Chioniades uses a longitude of the solar apogee for the year 541 Yazdigird, namely $2^s27°50'43''$, which is presumably al-Fahhād's original value. Al-Fārisī adjusts this longitude to his own epoch 631 Yazdigird to obtain $2^s29°12'30''$. In fact, Chioniades' and al-Fārisī's tables differ throughout by exactly the apogee motion in 90 Persian years, $1°21'47''$.[8] In both sources, a correction for the motion of the apogee is necessary for years other than the epoch year. This correction is carried out by means of the "equation of the apogee" (see above), which needs to be added to, or subtracted from, the true solar longitude depending on whether the desired year follows or precedes the respective epochs.

It is a simple matter to reconstruct the actual solar equation values from the tables of Chioniades and al-Fārisī, so that they can be compared with other sources. (The same holds for lunar and planetary equations that are displaced by the apogee longitude or involve a mixed equation.) In the remainder of this article, whenever I compare tables with different displacements or formats, I will tacitly assume that they have been reduced to a standard form. It turns out that Chioniades' solar equation contains only eight errors (in 180 tabular values) of at most $1''$. Five of these errors match with errors in al-Fārisī and Maḥmūd ibn ʿUmar, whose tables have a total of nine and eleven errors respectively. It is therefore probable that the three tables come from a common source, for which al-Fahhād is the most likely candidate. Chioniades would then have copied the original form of the table directly from the *ʿAlāʾī Zīj* without adjusting the apogee longitude, whereas al-Fārisī did carry out such an adjustment. Furthermore, Maḥmūd ibn ʿUmar would have extracted the standard form of the solar equation from al-Fahhād's table of the true solar longitude. Because of the small number of errors, it is not possible to decide whether al-Fahhād used the highly accurate table of Abu 'l-Wafāʾ or performed an independent **[p. 845]** computation. It is quite certain, though, that the undisplaced solar equation table in the

[8] There are only eight deviations (out of 360 tabular values) from this constant difference, five of which can easily be explained as scribal errors.

*Shāmil Zīj* [see van Dalen 1989, pp. 106–113], which is tabulated for every 12′ of the mean anomaly and has 17 errors for integer arguments but only one in common with the *ʿAlāʾī Zīj*, constitutes an independent computation.

*Lunar Equation of Centrum*

In the *Nāṣirī Zīj*, the lunar equation of centrum is tabulated as a function of the elongation, which is somewhat less common than the double elongation. The table has values to minutes for every degree of the argument and is displaced by 12 zodiacal signs. The maximum equation is Ptolemy's standard value 13°8′, corresponding to his lunar eccentricity of 10;19 units.

In Chioniades' version of the *ʿAlāʾī Zīj* and in the *Muẓaffarī Zīj*, the lunar equation of centrum is displaced by the maximum equation, 13°8′, and shifted upwards by 5°, in agreement with the displacement of the equation of anomaly (see below). A comparison of tabular values does not yield much information in this case, since all three tables differ in at most five or six places from Ptolemy's table in the *Handy Tables* (which also has the elongation as its argument). The lunar equation of centrum in the *Shāmil Zīj*, on the other hand, has the double elongation as its argument and a clearly larger number of deviations from Ptolemy.

*Lunar Equation of Anomaly*

In Ptolemy's *Almagest* and *Handy Tables*, as in practically all Islamic zījes, the lunar and planetary equations of anomaly are calculated by performing so-called "Ptolemaic interpolation" between values of the equation for two or three particular positions of the epicycle on the deferent. This interpolation is carried out by means of a non-trivial function of the position of the epicycle, the "interpolation function". In the case of the moon, the tables provided for the calculation of the equation of anomaly are usually the equation at the apogee of the deferent (often called the "second equation"), the differences in the equation between apogee and perigee, and the interpolation function.

In the *Nāṣirī Zīj*, the three tables to be used in the calculation of the lunar equation of anomaly display values to minutes for **[p. 846]** every degree of the respective arguments. They are of standard type, except that the second equation is displaced by 12 zodiacal signs. The equation at apogee reaches a Ptolemaic maximum of 5°0′, the *ikhtilāf* ("difference [in the equation]") of 2°39′.

The structure and use of the tables for the lunar equation of anomaly in the *ʿAlāʾī Zīj* and the *Muẓaffarī Zīj* are completely different from the *Nāṣirī*, since they are not only displaced but also of "mixed type". Thus the table of the second lunar equation contains (originally) subtractive values for the equation of anomaly at the perigee of the deferent in its first half (arguments 0 to 180°), and (originally) additive values for the equation of anomaly at the apogee in its second half (arguments 180 to 360°). All values are displaced by 5°, and, since the subtractive values in the first half of the table assume a maximum of 7°39′, another $12^s$ is added to those subtractive values which have an absolute value larger than 5° (we can thus say that the table is displaced by $12^s5°$). It seems highly probable that the insufficient displacement of 5° (instead of 8°, the next larger integer of the maximum equation) provides a historical clue as to the origin of this table. Earlier tables for the lunar equation of anomaly with a displacement of 5° include those of Ḥabash al-Ḥāsib and al-Bīrūnī, whereas Kūshyār used a displacement of 8°. As will be shown below, the underlying values of the equation in Chioniades' version of the *ʿAlāʾī Zīj* and in the *Muẓaffarī Zīj* point to a dependence on Ḥabash, since al-Bīrūnī used a different parameter and tabulated the equation to seconds instead of to minutes.

Of course, the Ptolemaic interpolation on values in the first half of the table of the second lunar equation now needs to be carried out differently from the second half. In fact, the table of differences in the equation of anomaly is accompanied by two interpolation functions that are supplementary, i.e., whose values sum up to one. The standard interpolation function as found in most zījes is used with the additive values of the equation, i.e., in the second half of the table of the second equation, whereas the supplementary function is used when the equation is taken from the first half of the table. The result is that also the Ptolemaic interpolation is *always* carried out by means of an addition and never requires a subtraction. That this procedure in fact yields the correct equation of anomaly, will be shown in my forthcoming publication on the characteristics of tables for calculating planetary **[p. 847]** longitudes in Islamic zījes.

Even though the structure of the tables for the lunar equation of anomaly in Chioniades and al-Fārisī is so different, it turns out that the underlying values are basically the same as those in the *Nāṣirī Zīj*. A comparison with the tables in earlier extant zījes shows that this is much more significant than in the case of the equation of centrum, since many Muslim astronomers appear to have calculated the lunar equation of anomaly anew. The only tables to which those in the *Nāṣirī Zīj* and the *ʿAlāʾī Zīj* are really close are those of Yaḥyā ibn Abī Manṣūr and Ḥabash al-Ḥāsib, which, in turn, seem to have been independent of Ptolemy's. Since we have noted above that also the insufficient displacement of 5° may point to a dependence on Ḥabash (whereas we had already seen in the section on the daily mean motions that al-Fahhād found the lunar parameters of the Mumtaḥan astronomers to be the most correct), we may conclude that the table for the second lunar equation in the *ʿAlāʾī Zīj* most probably derives from that of Ḥabash.

The situation is more complicated with the differences in the lunar equation of anomaly between the apogee and perigee of the deferent. The tables for this function in the *Nāṣirī*, *ʿAlāʾī*, and *Muẓaffarī* Zījes are all three basically identical with the table in the *Handy Tables* or in al-Battānī. However, the differences that can be reconstructed from the two halves of the mixed table of the equation of anomaly in Chioniades and al-Fārisī show more than 40 deviations from the explicitly given differences and exhibit a shift in the tabular values between arguments 60 and 90° which also occurs in the *Mumtaḥan Zīj* and with Ḥabash al-Ḥāsib. Thus again a dependence on Ḥabash is plausible.

A comparison of the tables for the lunar interpolation function in various sources is more cumbersome because these usually consist of numerous repetitive stretches of the same numbers between 0 and 60 minutes, and are therefore extremely sensitive to mistakes in copying. Furthermore, their argument may be the elongation or the double elongation and will be shifted in the case of displaced lunar equation tables. We therefore only note that again the interpolation functions in the *Nāṣirī Zīj*, Chioniades' version of the *ʿAlāʾī Zīj*, and the *Muẓaffarī Zīj* are basically identical, and that they are close to that of Ḥabash al-Ḥāsib. **[p. 848]**

| planet | maximum equation | eccentricity |
|---|---|---|
| Saturn | 6;31 | 3;25 |
| Jupiter | 5;15 | 2;45 |
| Mars | 11;25 | 6; 0 |
| Venus | 1;59 | 1; 2 |
| Mercury | 3; 2 | 3; 0 |

**Table 3.** *Parameters of the planetary equations of centrum in the Nāṣirī Zīj, the Byzantine version of the ʿAlāʾī Zīj, and the Muẓaffarī Zīj.*

*Planetary Equation of Centrum*

In the *Nāṣirī Zīj*, the planetary equations of centrum are tabulated to minutes for each degree of the mean centrum and are displaced by 12 zodiacal signs. Except for Venus, the maximum equations, and hence the underlying eccentricities, are equal to Ptolemy's values in the *Almagest* and the *Handy Tables* (cf. Table 3). The new Islamic maximum equation for Venus, 1°59′, like that of the Sun, stems from the observations made in Baghdad under the caliph al-Maʾmūn, and occurs in the zījes of Yaḥyā ibn Abī Manṣūr, Ḥabash al-Ḥāsib, al-Battānī, and others. The only other zījes that include the displacement of 12 signs are Chioniades' version of the *ʿAlāʾī Zīj* and the *Muẓaffarī Zīj*; Kūshyār uses displacements equal to the maximum equation or next higher integer degree.

As far as the tabular values for the equations of centrum are concerned, it appears that most Muslim astronomers up to the thirteenth century simply copied those from the *Handy Tables* (except, of course, for Venus). Exceptions to this rule are found, in particular, with Ibn al-Aʿlam, who observed new maximum equations for Saturn, Jupiter, and Mercury; Kūshyār, who adjusted the maximum equation for Mars to 11°30′; Ibn Yūnus, who modified the maximum equation for Mercury to 4°2′; and al-Bīrūnī, who introduced an error in the computation for Mercury [cf. Yano 2002] and in general carried out a different type of interpolation between values from the *Almagest* rather than using the *Handy Tables* directly. For Venus, both al-Bīrūnī and al-Khāzinī reverted to a Ptolemaic maximum equation of 2°23′, whereas most other astronomers stayed with the Mumtaḥan value, 1°59′.

The values for the equations of centrum in the *Nāṣirī Zīj* are generally close to those in the *Handy Tables*, and show hardly any differences from the tables in Chioniades' version of the *ʿAlāʾī Zīj* **[p. 849]** and the *Muẓaffarī Zīj*. In a few cases, certain deviations from the *Handy Tables* allow us to draw more detailed conclusions about the origin of the tables in the *ʿAlāʾī*, *Muẓaffarī*, and *Nāṣirī* Zījes. For instance, the tables for Saturn in these three works contain a peculiar interpolation pattern between arguments 0 and 18°, which is further only found with al-Khāzinī. Also the tables for Mars in the *ʿAlāʾī Zīj* and the *Muẓaffarī Zīj* are clearly closer to the table of al-Khāzinī than to the *Handy Tables*; however, in this case the *Nāṣirī Zīj* contains a shift of the tabular values in the fourth sign (arguments 90–120°) which is furthermore only found in the *Shāmil Zīj*.

*Planetary Equation of Anomaly*

As was explained above, the lunar and planetary equations of anomaly are usually calculated by means of Ptolemaic interpolation between two or three fixed positions of the epicycle on the deferent. In the case of the planets, these positions are the apogee, the perigee, and a "central position" at which the distance of the epicycle centre from the earth is precisely 60 units, and hence the equation is independent of the planetary eccentricity.[9] Tables are usually provided for: the central equation of anomaly (the "second equation"); the differences in the equation between the central position and the apogee, which I will call "decrements" since they need to be subtracted from the central equation; the differences in the equation between the central position and the perigee, the "increments"; and an interpolation function. We will now discuss the tables in the *Nāṣirī Zīj* of each of these types and their relations to tables in other works.

The tables in the *Nāṣirī Zīj* for calculating the planetary equation of anomaly are of standard type, except that the central equation is displaced by 12 zodiacal signs. All functions are tabulated to minutes for every degree of the respective arguments. Table 4 displays for each planet the maximum values of the decrements, the central equation of anomaly, and

[9] Even though the Mercury model is somewhat different from that for the other planets, the setup and use of the tables for its equations are basically the same. A setup of the tables for the planetary equation of anomaly in which only two reference equations (namely, for the apogee and the perigee) are used, is found in some later zījes, for example, al-Kāshī and Ulugh Beg.

| planet | maximum values decrements | central eq. | increments | epicycle radius | apogee longitude |
|---|---|---|---|---|---|
| Saturn | 0;21 | 6;13 | 0;25 | 6;30 | $8^s$ 8; 2 |
| Jupiter | 0;30 | 11; 3 | 0;34 | 11;30 | $5^s$27;44 |
| Mars | 5;38 | 41; 9 | 8; 3 | 39;30 | $4^s$14;45 |
| Venus | 1;42 | 45;59 | 1;52 | 43;10 | $2^s$15;55 |
| Mercury | 3;12 | 22; 2 | 2; 1 | 22;30 | $6^s$29;20 |

**Table 4.** *Parameters of the planetary equation of anomaly in the Nāṣirī Zīj, the Byzantine version of the ʿAlāʾī Zīj, and the Muẓaffarī Zīj (the apogee longitudes are as found in the ʿAlāʾī Zīj for al-Fahhād's epoch 541 Yazdigird). Note that the decrements and increments for Mercury in the ʿAlāʾī Zīj and the Muẓaffarī Zīj (maximum values 2;50 and 2;13 respectively) are not in agreement with the parameters that underlie the equation of centrum and the central equation of anomaly.*

the increments, and **[p. 850]** the underlying radius of the epicycle (the eccentricities are already listed in Table 3; the last column of Table 4 shows the apogee longitudes for 541 Yazdigird that will be seen below to be involved in the equation of anomaly tables in Chioniades' version of the *ʿAlāʾī Zīj*). All maximum values, and hence the underlying eccentricities and epicycle radii, are Ptolemaic, which implies that the Maʾmūnic value of the eccentrity of Venus (cf. above) was *not* taken into account in the decrements and increments.

The planetary equation of anomaly in the Byzantine version of the *ʿAlāʾī Zīj* and in the *Muẓaffarī Zīj* is again implemented by means of a "mixed equation" (cf. the description of the lunar tables above). In this case, two mixed tables are needed, one displaying the equation of anomaly at the apogee of the deferent in its first half (arguments 0 to 180°) and the central equation in its second half (arguments 180 to 360°), and one displaying the central equation in its first half and the equation at perigee in its second half. Both tables are accompanied by a table with respectively decrements and increments of the central equation of anomaly and two interpolation functions whose values add up to one. The mixed tables are displaced by the longitude of the planetary apogee, so that the equation found from them can simply be added to the true centrum of the planet in order to obtain its true longitude. Again, Chioniades maintains the original apogee longitudes of al-Fahhād, whereas al-Fārisī adjusts them to his own time. The maximum values of the decrements, central equation, and increments in the *ʿAlāʾī Zīj* and in the *Muẓaffarī Zīj* are **[p. 851]** the same as those in the *Nāṣirī*, except for the decrements and increments of Mercury (respectively 2;50 instead of 3;12 and 2;13 instead of 2;1; cf. below).

In general, also the tables for the central equation of anomaly in Islamic zījes up to the thirteenth century were simply copied from the *Handy Tables*. Ibn Yūnus reproduced the original tables particularly accurately; the typically 20 deviations found in many other zījes can mostly be explained as scribal errors or from some small adjustments of the interpolation pattern, in particular around the maximum equation. Also for the equation of anomaly, al-Bīrūnī performed entirely new interpolations between the values from the *Almagest*. Clear evidence for the fact that the tables in the *Nāṣirī Zīj*, the *ʿAlāʾī Zīj*, and the *Muẓaffarī Zīj* are strongly related to each other, and all depend on al-Khāzinī, is provided by the tables for Jupiter and Mars. Where these differ in all four sources in around 20 (Jupiter) or even 50 values (Mars, due to large systematic differences in the fifth and sixth signs) from the *Handy Tables*, they do not differ in more than five values among each other.

The tables of decrements and increments of the central equation of anomaly show much less computational variation. This is partly because they were calculated by means of "distributed linear interpolation" [cf. Van Brummelen 1998, p. 278] and are thus monotonically increasing before their maximum and monotonically decreasing thereafter.[10] Furthermore, since the decrements and increments are tabulated to only minutes, whereas, except for Mars, their maximum values are at most a little more than one degree, the tables concerned consist for the most part of repetitions of the same values.

Similar to the equation of centrum and the central equation of anomaly, most of the tables of planetary decrements and increments in Islamic zījes up to the thirteenth century are based on the *Handy Tables*. The variations that we find are mostly of two types: 1) adjustments of the interpolation pattern, for instance in order to evenly spread out tabular differences in a linear part of the table, or to smoothen the section surrounding a maximum; 2) shifts of tabular values by one (sometimes two)

[10]It can be verified that if the decrements and increments were computed as the actual differences of accurately computed equation of anomaly values at two different epicycle positions they would not generally be monotone.

row(s) upwards **[p. 852]** or downwards. These shifts are often scribal mistakes, in which case they may well extend to the bottom of the column concerned, where the error would finally be noticed by the copyist. Shifts are particularly common in sections of the tables where values are repeated at least four times in a row. It seems that many of the deviations from the *Handy Tables* in tables of decrements and increments of the central equation of anomaly arise from this kind of scribal errors, the number of deviations gradually increasing in the course of the centuries, sometimes to even more than half of the tabular values.

It is precisely the shifts in large parts of columns that allow us to draw conclusions about the relations between tables of decrements and increments of the planetary equation of anomaly in Islamic zījes. For instance, the decrements and increments for Saturn, the increments for Jupiter, and the decrements and increments for Venus in the *ʿAlāʾī*, *Muẓaffarī*, and *Nāṣirī* Zījes have long shifts in common of up to a total of 63 deviations from the *Handy Tables*, which are further only found with al-Khāzinī. Also most of the other decrement and increment tables in the *Nāṣirī Zīj* are very close to the *ʿAlāʾī* and *Muẓaffarī* Zījes, but here the evidence for a dependence on al-Khāzinī is less conclusive. The increments for Mars in the *Nāṣirī Zīj* display an upward shift of one row through arguments 158 to 177° (due to the omission of the value for 158°), but besides there are only two differences from the *Muẓaffarī Zīj* and seven from the *ʿAlāʾī*. A special case is Mercury, for which al-Khāzinī follows the *Handy Tables*, whereas the *ʿAlāʾī Zīj* and the *Muẓaffarī Zīj* tabulate a completely different function which is in agreement with the equations of anomaly at the apogee, central position and perigee as found in the mixed tables (as a matter of fact, as opposed to the lunar equation of anomaly, the decrements and increments of the planetary equations of anomaly are in each case in full agreement with the equations given in the mixed tables). I do not currently have an explanation for these deviating decrements and increments. In any case, the author of the *Nāṣirī Zīj* apparently restored the original functions from the *Handy Tables* or one of its direct descendants (the number of deviations from al-Khāzinī's tables is here clearly larger than from the *Handy Tables*). In this process he introduced a shift of the complete first sign of the decrements by inserting an extra zero for argument 1°.

**[p. 853]** It is remarkable that the new value for the eccentricity of Venus that occurred for the first time in the *Mumtaḥan Zīj* and was used in the tables for the central equation of anomaly in various later zījes, was not incorporated into the equation decrements and increments in those zījes except in the *Mumtaḥan Zīj* itself and, independently, in the *Ḥākimī Zīj*. All other zījes that used the new value simply copied the decrements and increments from the *Handy Tables*, thus introducing errors of up to 20′ in the final equation of anomaly.[11] Since Ibn Yūnus observed a new epicycle radius for Venus besides a new eccentricity, and also adjusted the parameters of the Mercury model, he had to compute the equations for both planets completely anew. Kūshyār experimented with a variant of Ptolemaic interpolation in which the roles of the strong and weak variables were interchanged [cf. Van Brummelen 1998]; his tables for decrements, increments, and the interpolation function therefore cannot be compared with other zījes.

The tables of the interpolation function for the planetary equation of anomaly, whose values lie exclusively between 0′ and 60′, share the property with the tables for decrements and increments that they consist largely of repetitions of the same values and are therefore extremely susceptible to accidental shifts of parts of columns.[12] Most interpolation tables appear to be variants of those in the *Handy Tables* with a smaller or larger number of shifts, presumably introduced by careless copyists. Some of these shifts are so peculiar that they allow us to establish dependences between tables of the interpolation function. As a matter of fact, a comparison of the deviations from the *Handy Tables* in the interpolation tables in the two zījes of al-Khāzinī and in the *ʿAlāʾī Zīj*, the *Muẓaffarī Zīj*, and the *Nāṣirī Zīj* shows very **[p. 854]** clear relations between all these

[11] In his *Ḥākimī Zīj*, Ibn Yūnus notes the inconsistency in various zījes between the new Mumtaḥan value of the eccentricity of Venus and the Ptolemaic decrements and increments of the equation of anomaly, and states that this first occured with Yaḥya ibn Abī Manṣūr [Caussin de Perceval 1804, p. 74 (58 in separatum)]; the eccentricity value 2;3,35 that Ibn Yūnus associates with Yaḥyā is undoubtedly a scribal mistake for 2;4,35. Since the decrements and increments for Venus in the extant recension of the *Mumtaḥan Zīj* were correctly computed for the new eccentricity value, they may not stem from the original zīj.

[12] The only exceptions to this rule are tables with values to seconds as are found in Ptolemy's *Almagest* and in al-Bīrūnī's *Masʿūdic Canon*.

works, especially in the cases of Jupiter and Venus. That the exact dependences may be complicated is shown by the Jupiter table. The *ʿAlāʾī*, *Muẓaffarī*, and *Nāṣirī* Zījes copy four longer shifts from the *Sanjarī Zīj*, whereas the *Wajīz* omits one of these and adds still another. The *ʿAlāʾī* and *Muẓaffarī* add three more shifts of which one, however, is not found in the *Nāṣirī Zīj*. In the case of Venus, Chioniades' version of the *ʿAlāʾī Zīj* provides an additional shift of the complete fourth sign (arguments 90–120°) and some more smaller ones, none of which is found in the *Muẓaffarī* and *Nāṣirī* Zījes.

Al-Fahhād presumably introduced one peculiar type of shift in the interpolation functions that was not yet found with al-Khāzinī. For all planets except Mercury, the central position of the epicycle center corresponds to an angular distance of approximately 88° from the apogee of the deferent. This means that the interpolation coefficients for values of the true centrum from 1 to 87° are to be used with the equation of anomaly decrements, and those for values from 88 to 180° with the increments. Al-Khāzinī included the two ranges of coefficients in separate tables. He thus needed four columns for the coefficients that are used with the increments, of which the first contains only values for arguments 88 and 89°. Apparently al-Fahhād wanted to avoid this waste of space and squeezed the values for 88 and 89° into the column for the fourth sign (arguments 90–120°). At the same time he spread out the values in the last section of the table that was used with the decrements in order to fill up the free space for arguments 88 and 89°. The result is that al-Fahhād's interpolation functions for all planets except Mercury have a clear distortion with respect to the *Handy Tables* in the neighborhood of 90 degrees. This distortion can be clearly recognized in the Byzantine version of the *ʿAlāʾī Zīj*, the *Muẓaffarī Zīj*, and the *Nāṣirī Zīj*, as well as in the *Shāmil Zīj*.

## 5 Summary and Conclusions

In our study of the *Nāṣirī Zīj* by Maḥmūd ibn ʿUmar we have clearly established the dependence of this work on the *ʿAlāʾī Zīj* by al-Fahhād. We have seen that Maḥmūd computed accurate mean motion tables on the basis of the daily mean motions in longitude and in anomaly listed in the Byzantine version of the **[p. 855]** *ʿAlāʾī Zīj* by Gregory Chioniades and in the further unrelated *Sanjufīnī Zīj*. Since these parameters also underlie the mean motion tables in Chioniades and in al-Fārisī's *Muẓaffarī Zīj*, which is said to be based on al-Fahhād's observations, we may conclude that they stem from the original *ʿAlāʾī Zīj*. For calculating the mean motions in centrum, Maḥmūd utilized a slightly different value of the daily apogee motion. There is no direct relation between the epoch positions in the *Nāṣirī Zīj*, which are for the longitude of Delhi ($103°35'$ from the Western Shore of Africa) and the first day of the year 1 before Yazdigird, and those in the *ʿAlāʾī Zīj*; thus it is still unclear whether Maḥmūd found his epoch positions from new observations or from still another zīj.

The solar, lunar and planetary equations in the *Nāṣirī Zīj* are all displaced by 12 zodiacal signs, which means that subtractive values $q$ are tabulated as $12^s - q$ and thus become additive. We have seen that this type of displacement stems from al-Fahhād; the more complicated type of displacement by the maximum equation or next higher integer is already found with Ḥabash al-Ḥāsib and Kūshyār ibn Labbān. The values of the equations in the *Nāṣirī Zīj* agree very well with those in Chioniades' version of the *ʿAlāʾī Zīj* and in al-Fārisī's *Muẓaffarī Zīj*. It is thus very probable that Maḥmūd reconstructed his solar equation from the table for the true solar longitude in the *ʿAlāʾī Zīj* and his plain equations of anomaly from the mixed and displaced tables in that work. In the equation of centrum and the equation of anomaly increments for Mars, Maḥmūd or a later scribe introduced shifts in a whole column by omitting one particular value at the beginning. For Mercury, Maḥmūd restored the decrements and increments of the central equation of anomaly from the *Handy Tables*, again accidentally shifting a whole column by one row; for reasons not yet known to us, the decrements and increments in the *ʿAlāʾī Zīj* are completely different functions (see also below).

In the course of our investigation, we have been able to establish to a large extent the original form of the planetary tables in the *ʿAlāʾī Zīj*. Because of the close similarity of the tables of Chioniades with those of al-Fārisī, and since Chioniades uses what is very probably al-Fahhād's epoch year A.D. 1172 and his base longitude of 84° for Shirwan, we may conclude that the Byzantine version of the *ʿAlāʾī Zīj* faithfully reproduces al-Fahhād's planetary **[p. 856]** tables. In the *Muẓaffarī Zīj*, al-Fārisī adjusted the planetary equations to his own epoch year A.D. 1262 and the mean positions to the longitude of Yemen (64° or 63°30′), but further left the structure of the tables unchanged. We have seen that the solar and lunar mean motions in the *ʿAlāʾī Zīj* were in agreement with those of the Mumtaḥan astronomers, as stated by al-Fārisī in the introduction to his zīj. Furthermore, we have verified the statement that the mean motions of Mars and Venus are quite close to those of Ibn al-Aʿlam.

None of the planetary equations in the original *ʿAlāʾī Zīj* were given in the standard form. Instead of the solar equation, al-Fahhād tabulated the true solar longitude, which, for years other than the epoch year, needed to be corrected for the motion of the apogee. The lunar equation of centrum was displaced by the maximum equation, 13°8′. The lunar equation of anomaly was displaced by the maximum equation at apogee, 5°0′, but was also of "mixed type", which means that one half of the table displayed the equation at apogee, the other at perigee. In this way, the conditional addition or subtraction that occurs in the calculation of the general equation of anomaly by means of Ptolemaic interpolation, is eliminated at the cost of some extra tables. The planetary equations of centrum were all displaced by 12 zodiacal signs and hence of the same form as found in the *Nāṣirī Zīj*. The planetary equations of anomaly, finally, were of mixed type and displaced by the longitude of the apogees concerned. Also here, a correction for the motion of the apogee was necessary.

As far as the origin of the tables for the planetary equations in the original *ʿAlāʾī Zīj* is concerned, we have seen that the solar equation, based on the Mumtaḥan maximum of 1°59′0″, was independently computed or possibly rounded from Abu 'l-Wafāʾ's table. The lunar equations have their displacements in common with the *Zīj* of Ḥabash al-Ḥāsib, with which they also show the highest coincidence of the tabular values. Most of the planetary equations ultimately derive from the *Handy Tables*, but numerous peculiar error patterns make clear that al-Fahhād's direct sources for the tables for Saturn, Jupiter, Mars and Venus must have been the *Sanjarī Zīj* and the *Wajīz* of al-Khāzinī. The one exception to this rule is the equation of centrum for Venus, for which the *ʿAlāʾī Zīj* seems to have used the table of Kūshyār with the Mumtaḥan maximum of 1°59′, whereas al-Khāzinī performed **[p. 857]** an independent computation for the Ptolemaic maximum 2°23′. Although the decrements and increments of the equation of anomaly also depend on the maximum equation of centrum, al-Fahhād here sticked to al-Khāzinī's tables for Ptolemy's parameters, thus introducing an inconsistency that leads to errors in the longitude of Venus of up to 20′. Contrary to common usage, al-Fahhād took the longitude of the apogee of Venus to be different from that of the Sun. In the interpolation functions of all planets except Mercury, al-Fahhād introduced a peculiar distortion in the neighborhood of the central position of the epicycle on the deferent. In agreement with al-Fārisī's statement that al-Fahhād found his observations of Mercury to be in best agreement with Ptolemy, the *ʿAlāʾī Zīj* includes the tables of the equation of centrum, the central equation of anomaly, and the interpolation function for this planet from the *Handy Tables*. However, the decrements and increments of the central equation of anomaly are represented by a completely different function that we have not been able to explain.

We have also presented some scattered information on the planetary tables in the *Shāmil Zīj*. The statement in its introduction that the *ʿAlāʾī Zīj* and the *Shāmil Zīj* use the mean motions of Abu 'l-Wafāʾ could not be confirmed. The daily mean motions listed in the *Shāmil Zīj* are very close to those in the the *ʿAlāʾī Zīj*, and the epoch positions are in full agreement. The solar equation also has a maximum value 1°59′0″, but was independently computed for every 12′ of the mean anomaly. The lunar equation of centrum is tabulated as a function of the double elongation and with clearly more deviations from the *Handy Tables* than most other zījes. The lunar equation of anomaly at apogee was undoubtedly copied from Kūshyār's *Jāmiʿ Zīj*. The tables for the planetary equations are all of standard type without displacements. The tabular values exhibit some similarities to other zījes, such as Kūshyār's *Jāmiʿ Zīj*, the *ʿAlāʾī Zīj*, and the *Nāṣirī Zīj*. In particular, the equations of Venus were taken from Kūshyār, whereas the interpolation functions include the peculiar distortion around the central position on the deferent that was introduced by al-Fahhād. **[p. 858]**

## Appendix: Elementary Relations between Mean Motions

Because of the way in which the Ptolemaic solar, lunar and planetary models are set up, certain elementary relations exist between the mean motions and positions. These have been used in this article to verify the consistency of the epoch positions given in the *Nāṣirī Zīj* and to correct scribal errors in them. Note that the relations hold for actual mean positions as well as for mean motions in any given period. They are the following:

1. The difference of the solar mean longitude and the solar mean anomaly is the longitude of the solar apogee; the difference of the planetary mean longitude and the planetary mean centrum is the longitude of the apogee of the planet concerned.
2. The lunar elongation is the difference of the lunar and solar mean longitudes, the double elongation is twice that difference.
3. The sum of the mean longitude and mean anomaly of the superior planets equals the solar mean longitude; the sum of the mean centrum and mean anomaly of the superior planets equals the solar mean centrum.
4. The mean longitude of the inferior planets is equal to the solar mean longitude; the mean centrum of the inferior planets is equal to the solar mean centrum.

## Bibliography

Āghā Buzurg Ṭihrānī, Muḥammad Muḥsin 1936–78. *al-Dharīʿat ilā taṣānīf al-Shīʿa* (*Bibliography of Shiʿit'es Literary Works*, in Arabic), 25 vols., Najaf / Tehran.

al-Bīrūnī, Abū Rayḥān Muḥammad. *al-Qānūnu'l-Masʿūdī (Canon Masudicus). An Encyclopaedia of Astronomical Sciences*, 3 vols., Hyderabad (Osmania Oriental Publications Bureau), 1954–56.

Caussin de Perceval, Armand Pierre 1804 (an XII de la République). Le livre de la grande table hakémite, *Notices et extraits des manuscrits de la Bibliothèque Nationale* 7, Paris **[p. 859]** (Bibliothèque Nationale), pp. 16–240 (pp. 1–224 in separatum).

Dalen, Benno van 1989. A Statistical Method for Recovering Unknown Parameters from Medieval Astronomical Tables, *Centaurus* 32, pp. 85–145.

——. 1993. *Ancient and Mediaeval Astronomical tables: mathematical structure and parameter values* (doctoral dissertation), Utrecht (Utrecht University, Mathematical Institute).

——. 1996. al-Khwārizmī's Astronomical Tables Revisited: Analysis of the Equation of Time, in *From Baghdad to Barcelona. Studies in the Islamic Exact Sciences in Honour of Prof. Juan Vernet* (Josep Casulleras & Julio Samsó, eds.), Barcelona (Instituto Millás Vallicrosa de Historia de la Ciencia Arabe), vol. 1, pp. 195–252.

——. 2000. Origin of the Mean Motion Tables of Jai Singh, *Indian Journal of History of Science* 35, pp. 41–66.

——. 2002. Islamic Astronomical Tables in China: The Sources for the *Huihui li*, in *History of Oriental Astronomy. Proceedings of the Joint Discussion-17 at the 23rd General Assembly of the International Astronomical Union, organised by the Commission 41 (History of Astronomy), held in Kyoto, August 25–26, 1997* (S. M. Razaullah Ansari, ed.), Dordrecht (Kluwer), pp. 19–31.

Debarnot, Marie-Thérèse 1987. The Zīj of Ḥabash al-Ḥāsib: A Survey of MS Istanbul Yeni Cami 784/2, in *From Deferent to Equant. A Volume of Studies in the History of Science in the Ancient and Medieval Near East in Honor of E.S. Kennedy* (David A. King & George A. Saliba, eds.), New York (The New York Academy of Sciences), pp. 35–69.

Ḥusaynī, Aḥmad & Marʿashī, Maḥmūd 1994. *Fihrist-i nuskhahā-yi khaṭṭī-yi Kitābkhānah-yi ʿumūmī-yi Ḥaḍrat-i Āyatallāh al-ʿUẓmā Marʿashī Najafī* (Catalogue of the manuscripts of the Public Library Marʿashī, in Persian), vol. 23, Qum (Chāpkhānah-yi Mihr-i Ustuwār). **[p. 860]**

Kennedy, Edward S. 1956. A Survey of Islamic Astronomical Tables, *Transactions of the American Philosophical Society, New Series* 46-2, pp. 123–177. Reprinted in 1989 with page numbers 1–55.

——. 1977. The Astronomical Tables of Ibn al-Aʿlam, *Journal for the History of Arabic Science* 1, pp. 13–23.

Kennedy, Edward S. & Kennedy, Mary Helen 1987. *Geographical Coordinates of Localities from Islamic Sources*, Frankfurt am Main (Institute for the History of Arabic-Islamic Science).

King, David & Samsó, Julio 2001. Astronomical Handbooks and Tables from the Islamic World (750-1900): an Interim Report, *Suhayl* 2, pp. 9–105.

Lee, Samuel 1822. Notice of the Astronomical Tables of Mohammed Abibekr al-Farsi, *Transactions of the Cambridge Philosophical Society* 1, pp. 249–265. Reprinted in Fuat Sezgin (ed.), *Islamic Mathematics and Astronomy*, volume 77, Frankfurt (Institute for the History of Arabic-Islamic Science), 1998, pp. 315–331.

Mercier, Raymond P. 1989. The Parameters of the *Zīj* of Ibn al-Aʿlam, *Archives Internationales d'Histoire des Sciences* 39, pp. 22–50.

Mielgo, Honorino 1996. A Method of Analysis for Mean Motion Astronomical Tables, in *From Baghdad to Barcelona. Studies in the Islamic Exact Sciences in Honour of Prof. Juan Vernet* (Josep Casulleras & Julio Samsó, eds.), Barcelona (Instituto Millás Vallicrosa de Historia de la Ciencia Arabe), vol. 1, pp. 159–179.

Nallino, Carlo Alfonso 1899–1907. *al-Battani sive Albatenii opus astronomicum (al-Zīj al-Ṣābiʾ)*, 3 vols, Milan (Ulrich Hoepli).

Neugebauer, Otto 1957. *The Exact Sciences in Antiquity*, second edition, Providence (Brown University Press).

——. 1975. *A History of Ancient Mathematical Astronomy*, 3 vols., Berlin (Springer). **[p. 861]**

Pedersen, Olaf 1974. *A Survey of the Almagest*, Odense (Odense University Press).

Pingree, David 1985–86. *The Astronomical Works of Gregory Chioniades. Part I: The Zīj al-ʿAlāʾī*, 2 vols., Amsterdam (Gieben).

——. 1999. A Preliminary Assesment of the Problems of Editing the Zīj al-Sanjarī of al-Khazini, in *Editing Islamic Manu-scripts on Science. Proceedings of the Fourth Conference of Al-Furqān Islamic Heritage Foundation. 29th–30th November 1997* (Yusuf Ibish, ed.), London (Al-Furqān Islamic Heritage Foundation).

Salam, Hala & Kennedy, Edward S. 1967. Solar and Lunar Tables in Early Islamic Astronomy, *Journal of the American Oriental Society* 87, pp. 492–497.

Saliba, George A. 1976. The Double-Argument Lunar Tables of Cyriacus, *Journal for the History of Astronomy* 7, pp. 41–46.

——. 1977. Computational Techniques in a Set of Late Medieval Astronomical Tables, *Journal for the History of Arabic Science* 1, pp. 24–32.

——. 1978. The Planetary Tables of Cyriacus, *Journal for the History of Arabic Science* 2, pp. 53–65.

Sayılı, Aydin 1955. The Introductory Section of Ḥabash's Astronomical Tables Known as the "Damascene" Zīj (in Turkish, English and Arabic), *Ankara Üniversitesi Dil ve Tarih-Coğrafya Fakültesi dergisi* 13, pp. 132–151.

Storey, Charles Ambrose 1958. *Persian Literature: a Bio-Bibliographical Survey*, vol. 2, London (Luzac & Co.).

Van Brummelen, Glen 1998. Mathematical Methods in the Tables of Planetary Motion in Kūshyār ibn Labbān's *Jāmiʿ Zīj*, *Historia Mathematica* 25, pp. 265–280.

Waerden, Bartel L. van der 1958. Die Handlichen Tafeln des Ptolemaios, *Osiris* 13, pp. 54–78.

Yaḥyā ibn Abī Manṣūr. *The Verified Astronomical Tables for the Caliph al-Maʾmūn (al-zīj al-maʾmūnī al-mumtaḥan). Facsimile* **[p. 862]** *of Escorial Library MS árabe 927, with an introduction by E.S. Kennedy*, Frankfurt am Main (Institute for the History of Arabic-Islamic Science), 1986.

Yano, Michio 2002. The First Equation Table for Mercury in the *Huihui li*, in *History of Oriental Astronomy. Proceedings of the Joint Discussion-17 at the 23rd General Assembly of the International Astronomical Union, organised by the Commission 41 (History of Astronomy), held in Kyoto, August 25–26, 1997* (S. M. Razaullah Ansari, ed.), Dordrecht (Kluwer), pp. 33–43.

**Additions**

VI 2, line 9; 4, line –11; 5, line 11; 9, line 11; etc.: Thanks to Sonja Brentjes, I later learned that there is a partial copy of the original version of the *ʿAlāʾī Zīj* by al-Fahhād al-Shirwānī in the library of the Salar Jung Museum in Hyderabad (ms H 17). Since this original is in Persian, there is also no reason any more to assume an intermediate (oral or written) Persian version by Shams al-Dīn al-Bukhārī.

VI 3, line –6: Later I discovered a second recension of the *Mumtaḥan Zīj* in the University Library in Leipzig (ms Vollers 821); see article VII in this collection.

VI 4, line –2: Recent unpublished research by Mohamed Abu Zayed, Jan P. Hogendijk and the present author has shown that the so-called *Utrecht Zīj* is a copy of a similar second zīj by al-Abharī rather than a separate work by a different author.

Sayılı, Aydın 1955. The Introductory Section of Ḥabash's Astronomical Tables Known as the "Damascene" Zīj (in Turkish, English and Arabic), *Ankara Üniversitesi Dil ve Tarih-Coğrafya Fakültesi Dergisi* 13, pp. 132–151.

Storey, Charles Ambrose 1958. *Persian Literature: a Bio-Bibliographical Survey*, vol. 2, London (Luzac & Co.).

Van Brummelen, Glen 1998. Mathematical Methods in the Tables of Planetary Motion in Kūshyār ibn Labbān's *Jāmiʿ Zīj*, *Historia Mathematica* 25, pp. 265–280.

Waerden, Bartel L. van der 1958. Die Handlichen Tafeln des Ptolemaios, *Osiris* 13, pp. 54–78.

Yaḥyā ibn Abī Manṣūr, *The Verified Astronomical Tables for the Caliph al-Maʾmūn (al-Zīj al-mumtaḥan al-Maʾmūnī)*. Facsimile (ms. 8631 of El Escorial Library, ...), with an introduction by E.S. Kennedy, Frankfurt am Main (Institute for the History of Arabic-Islamic Science) 1986.

Yano, Michio 2002. The First Equation Table for Mercury in the Ḥāmid Zīj, in *History of Oriental Astronomy. Proceedings of the Joint Discussion-17 at the 23rd General Assembly of the International Astronomical Union, organised by the Commission 41 (History of Astronomy), held in Kyoto, August 25–26, 1997*, (S.M. Razaullah Ansari, ed.), Dordrecht (Kluwer), pp. 33–43.

## Additions

VI 2, line 9; 4, line –11; 6, line 11; 8, line 11, etc.: Thanks to Sonja Brentjes I later learned that there is a partial copy of the original version of the *Jāmiʿ Zīj* by al-Fahhād at Shirwānī in the library of the Salar Jung Museum in Hyderabad (ms H 17). Since this original is in Persian, there is also no reason any more to assume an intermediate (oral or written) Persian version by Shams al-Dīn al-Bukhārī.

VI 8, line –5: Later I discovered a second recension of the *Mumtaḥan Zīj* in the University Library in Leipzig (ms Vollers 821); see article VII in this collection.

VI 8, line –2: Recent unpublished research by Mohammed Abu Zayed, Jan P. Hogendijk and the present author has shown that the so-called [illegible] *Zīj* is a copy of a similar second *zīj* by al-Abharī rather than a separate work by a different author.

VII

# A Second Manuscript of the *Mumtaḥan Zīj*

## 1 Introduction

Until now the *Mumtaḥan Zīj* by Yaḥyā ibn Abī Manṣūr, one of the earliest extant Islamic astronomical handbooks with tables,[1] has been known only from a single manuscript of a late recension. Yaḥyā's zīj was based on the results of the observational program carried out in Baghdad on the order of the Abbasid caliph al-Ma'mūn during the years AD 828–829. It consists of a mixture of materials of Indian and Iranian origin and the first systematic attempts to calculate Ptolemaic tables for planetary motions and spherical astronomical quantities on the basis of updated values for the underlying parameters. What has been thought to be the only surviving copy of the *Mumtaḥan Zīj* is contained in MS Escorial árabe 927. It is not complete, and is supplemented by texts and tables of a later date, for instance by Abu 'l-Wafā' (Baghdad, second half of the tenth century) and Kūshyār ibn Labbān (Iran, ca. 1000). One of the most important objectives of previous research on the Escorial manuscript, in particular by Vernet, Kennedy and Viladrich, has been to distinguish between the original material stemming from Yaḥyā and later additions.

In June 2004, during a visit to the Universitätsbibliothek in Leipzig, I had the chance to look at the manuscript Vollers 821 (formerly DC 120) in some detail. Since both the first and the last folios of this volume deal with calendar conversion, it was catalogued in section 41 ("Chronologie und Kalender") of Vollers 1906 as a purely chronological work. However, it

[1] Somewhat over one hundred Islamic zījes were briefly described in the standard work Kennedy 1956a. Meanwhile more than two hundred works in Arabic and Persian are known, of which more than one hundred are extant. In the second issue of this journal (2001), King and Samsó published a broad overview of Islamic zījes and the categories of tables they contain. An extensive new survey of Islamic astronomical handbooks with detailed information on all extant works is currently being prepared by the present author.

turned out that the manuscript is in complete disorder, and that in between the chronological sections there is a whole range of materials of the kind that is usually found in zījes. On first inspection, I noticed that the manuscript contains the same planetary mean motions and equations as found in the Escorial copy of the *Mumtaḥan Zīj*, and that various of its tables and chapters are attributed to Yaḥyā ibn Abī Manṣūr, al-Maʾmūn, and the important tenth-century astronomer Ibn al-Aᶜlam, who was famous for the accuracy of his observations but whose zīj is lost.

Two months later, an extensive investigation of a microfilm of the Leipzig manuscript confirmed that it contains much of the material in the Escorial copy that has been recognized as belonging to the original *Mumtaḥan Zīj*. Besides the planetary mean motions and equations, it includes an elementary section on chronology, a long section on spherical astronomy with some tables, and the highly interesting material on solar and lunar eclipses that has been associated with Yaḥyā. Moreover, the Leipzig manuscript contains some chapters and tables attributed to al-Maʾmūn and Yaḥyā that are not present in the Escorial copy, such as tables for calculating lunar eclipses by *manzila*s and a table for year transfers. Although no title page, preface, and colophon are found anywhere in the Leipzig manuscript, a passage on fols. 47ʳ–47ᵛ confirms that it constitutes a recension of the *Mumtaḥan Zīj*. This passage concludes a long introductory chapter on the planets and their spheres with the statement that Ibn al-Aᶜlam's tables for Mars were included in "this modified Maʾmūnic zīj", because his observations were the most correct (see section 3L and Figures 3a and 3b below).

Besides the original Maʾmūnic material, the Leipzig manuscript turned out to contain chapters and extracts from other early Muslim astronomers, namely, Ḥabash al-Ḥāsib, who was present in Baghdad and Damascus at the time of the observations made for al-Maʾmūn but only finished his impressive zīj after AD 860 in Samarra, and the famous observer from Raqqa, al-Battānī, who wrote his *Ṣābiʾ Zīj* around the year 900. The inclusion of an oblique ascension table for Mayyāfāriqīn and some nearby localities suggests that the compiler of the particular recension extant in Leipzig lived in what is now southeastern Turkey. That the manuscript was at least to some extent used in this region is confirmed by one of the very few marginal notes in a different hand, which gives the latitude of Isᶜird (or: Siᶜird) as 37°18′ and its longitude as 76°28′36″ above a text on solar eclipses on fol. 40ᵛ.

All in all, the Leipzig manuscript includes more materials from the early period of Islamic astronomy than the Escorial copy and none that are essentially later than Ibn al-Aᶜlam. It thus constitutes a highly important source for our knowledge of Islamic astronomy in the ninth and tenth centuries and for the reconstruction of the original zījes of Yaḥyā ibn Abī Manṣūr and Ibn al-Aᶜlam. Since the Leipzig and Escorial manuscripts both replace Yaḥyā's mean motions and equations for Mars by those of Ibn al-Aᶜlam, they probably ultimately go back to the same recension of the *Mumtaḥan Zīj* (presumably compiled in the tenth century), but the Leipzig copy seems to have reached its present form at an earlier stage than the Escorial one.

The purpose of this article is to make the contents of the Leipzig manuscript known to the extent that any scholars of the history of Islamic astronomy can determine which parts of the work may be of interest to them. Since the manuscript is in great disorder, in Section 2 an attempt is made to reconstruct the order of the original manuscript. In Section 3, the complete contents of the manuscript, arranged by topic, is described in some detail. A summary at the end of this section gives an overview of the sources of the materials in the Leipzig manuscript and the most important differences from the Escorial manuscript. The Appendix provides a complete table of contents in the present order of the folios with indications of which folios originally belonged together and where interrupted texts and tables are continued.

Throughout the remainder of this article, I will use the siglum **L** for the Leipzig manuscript and **E** for the Escorial copy. Because of the confusing folio numbering in **E** (the numbers appear on the verso sides of the folios and number 30 has been left out), I refer for this manuscript to the page numbers in the facsimile edition Sezgin 1986. An overview of the contents of **E**, also indicating the folio numbers, can be found in Vernet 1956. Occasionally I will refer to the most original of the two manuscripts of the zīj of Ḥabash al-Ḥāsib, namely MS Istanbul Yeni Cami 784/2, fols. $69^v$–$229^r$, by the siglum **H**. This manuscript is abstracted in Debarnot 1987, with a comparison with the later recension MS Berlin We. 90 (Ahlwardt #5750) in an Appendix. Although Ḥabash's zīj was to a large extent based on the observations made under al-Maʾmūn, it has only very little material in common with the *Mumtaḥan Zīj*. Finally, the *Ṣābiʾ Zīj* of al-Battānī, extant in the unique copy MS Escorial árabe 908, is edited and commented upon in Nallino 1899–1907.

Only scattered references are included in this article to treatments of topics discussed in zījes. A general overview of the subject matter of these works can be found in Kennedy 1956a, pp. 139–145, with some additions in King & Samsó 2001. Numerous studies on technical subjects from zījes are collected in Kennedy *et al.* 1983 and King 1986. For further information on the astronomers mentioned, the reader is referred to Rosenfeld & İhsanoğlu 2003, Sezgin 1978, the *Dictionary of Scientific Biography*, and the *Biographical Encyclopedia of Astronomers* (to appear with Kluwer in 2004 or 2005).

## 2 Rearranging the manuscript

The Leipzig manuscript is written on fragile paper, now partially damaged. The size of the pages is 17 × 25 cm, of which text and tables cover approximately 12 × 18,5 cm. The written part of the pages is surrounded by a double frame, but the lines of script (between 17 and 19 on a page) are often not straight. Except for the last two folios, the whole manuscript is written in a somewhat sloppy naskh, presumably of northern Iraqi provenance, which can be dated to roughly the year AD 1200. The text has relatively few diacritical dots, and especially *alif* and *wāw* are mostly connected to the following letters. Non-final *kāf* s frequently occur with a miniature *kāf* above the vertical stroke, and numbers are almost always written out in words (in **E**, the same numbers are mostly written in *abjad* or Hindu numerals). Another typical feature of the manuscript is that the punctuation marks used at the end of chapters or "paragraphs", large Arabic letters *hāʾ* with a dot in the middle, are filled with red ink. Red is further used for nearly every chapter title and for certain columns in the tables; the red is very clear in the manuscript and can be read reasonably well from the microfilm that I have used.

As has been mentioned in the introduction, **L** is bound in complete disorder. For instance, the chronological chapters and tables at the beginning and end of the work, which led Vollers to believe that the whole contents was chronological, are in fact part of one larger section on date conversion. Below an attempt is made to reconstruct the original order of the manuscript, which is made particularly difficult since the chapters (and tables) are not numbered and no catchwords are provided on the verso pages. In many cases, texts and tables can be seen to continue in very different parts of the manuscript. In some cases, the beginning or end of a chapter (and in only one case of a table) appears to be missing completely.

In the suggested original order of the folios presented below, a somewhat thicker and longer line between ranges of folios indicates a *definite* break in the manuscript: the text or table preceding such a break does not end at the end of the last folio of the range but has no continuation anywhere in the manuscript, and similarly the beginning of the text or table on the first folio after such a break is not contained in the manuscript. A thinner and shorter line preceded by a question mark indicates a *possible* break in the manuscript: a text or table ends at the end of the last folio before such a break and a new text or table starts at the beginning of the first folio after it. I have used two criteria for combining ranges of folios starting or ending with possible breaks: consecutiveness of folio numbers and consistency of topics. However, since also originally **L** appears to have been a hodgepodge of different materials from different sources thrown together in a more or less random order, in particular the second criterion is a rather arbitrary one. Asterisks before or after the brief descriptions of the topics in the list below indicate that the section concerned is continued from, or continues on, a folio at a different place in the manuscript.

| | |
|---|---|
| missing: | • title page, preface, first part of section on chronology |
| 113–129 | • *chronology (with extract from al-Battānī) |
| | • general introduction on the heavenly sphere, divisions of the ecliptic, multiplication and division of parts of degrees (associated with al-Battānī) |
| | • on the setup of an almanac (*taqwīm*) |
| | • lunar latitude* |
| 131–151 | • *lunar latitude |
| | • planetary visibility |
| | • spherical astronomy |
| | • "applications" (*ittiṣālāt*) |
| | • treatise on a sundial for seasonal hours |
| | • chronology, with table for date conversions* |
| 1–2 | • *chronology: tables for date conversions* |
| 34 | • *chronology: table for date conversion |
| | • introduction to planetary motions* |
| 45–50 | • *introduction to planetary motions |
| | • mean motions, true longitudes, planetary stations* |
| 35 | • *planetary stations, lunar latitude, lunar node |
| | • planetary visibility* |
| ... | • (one folio missing?) |
| 51–56 | • *planetary visibility |
| | • trigonometry, declination and lunar latitude |
| ? ——— | |
| 57–63 | • lunar eclipses by a method attributed to Yahyā |
| | • table of planetary stations |
| | • planetary latitudes, with tables |
| ? ——— | |
| 64–65 | • solar declination and lunar latitude, with table |
| ? ——— | |
| 66–99 | • tables for planetary mean motions and equations |
| | • trigonometry: sine table |
| ? ——— | |
| 31 | • trigonometry: tangent table; fixed star table* |
| 153–155 | • *fixed star table, geographical table* |

152 • *geographical table
• lunar nodes, lunar eclipse circle*

---

5–8 • *last two words of preceding section: *wa-kawākibihi*
• year transfers
• spherical astronomy: azimuth
• world ascendant, year transfers
• table for year transfers based on Ibn al-Aʿlam
• time of a year transfer

?————

9–12 • chronology: Christian feasts and fasts, with tables*
29–30 • *chronology: table for Christian feasts and fasts
• lunar eclipses*
13–20 • *lunar eclipses, solar eclipses*

---

111 • parallax table, solar eclipse table

?————

32–33 • solar and lunar eclipse tables

?————

110 • hourly motion of Sun and Moon, lunar crescent visibility*
36–44 • *lunar crescent visibility
• lunar eclipses (Yaḥyā's method of *manzila*s), with tables
• solar eclipses, with tables
• year transfers, entry of Sun into signs, rising of Sirius*
112 • *rising of Sirius
• chronology: conversion from Byzantine to Persian years
• lunar crescent visibility: list of visibility limits
• correction of the ascendant by the method of Vettius Valens*
21–22 • *correction of the ascendant, ascendant of the *qubba*
• table of the apogee of Mars
• two tables for the motion of the planetary apogees

?————

23 • simplified method for oblique ascensions, with small tables
• table for the duration of planetary stations, retrogradations, etc.

?————

24–28 • year transfers according to al-Maʾmun, with table
• right ascension table, table of the sine of the declination
• conjunctions and oppositions
• universal oblique ascension table*

---

| | |
|---|---|
| 130 | • *time of day or night, oblique ascensions* |
| ... | • (one folio missing) |
| 100–109 | • spherical astronomy: various topics |
| | • right ascension table, oblique ascension tables* |
| 4 | • *oblique ascension table |
| | • shadow from altitude, time of night from fixed star* |
| 3 | • *time of night, reference to tangent table, planetary rays |
| | • Hijra month beginnings (8-year cycle), with table |
| 156–157 | • chronology, in different hand |

Note that there are five chapters in **L** for which I have not been able to find the beginning, four chapters for which I have not found a continuation, and one table whose second half is very probably missing (these can be recognized in the list above by asterisks immediately following or preceding definite breaks):

*Chapters whose beginning has not been found in the manuscript:*

| | |
|---|---|
| $3^{r}$ | timekeeping (time of night in seasonal and equal hours) |
| $5^{r}$ | last two words of an unspecified chapter: *wa-kawākibihi* |
| $51^{r}$ | planetary visibility (by means of calculation) |
| $113^{r}$ | chronology (weekday of month beginning in unspecified calendar) |
| $130^{r}$ | timekeeping (determination of the seasonal hours of daylight) |

*Chapters and one table whose end has not been found in the manuscript:*

| | |
|---|---|
| $4^{v}$ | timekeeping (time of the night measured by a fixed star) |
| $20^{v}$ | eclipses (determination of the ascendant for solar eclipse phases) |
| $28^{v}$ | universal oblique ascension table (only latitudes 1 to 33° present) |
| $35^{v}$ | planetary visibility (standard method by means of tables) |
| $152^{v}$ | eclipses (lunar eclipses by means of the "eclipse circle") |

## 3 Contents of the Leipzig manuscript arranged by subject

### *A. Chronology*

There are two longer sections on chronology in **L**, which are both part of the longest range of folios of which the original order can be reliably restored (see Section 2). These two sections were separated in the original manuscript by a number of completely different topics without any particular order or structure.

The first larger section on chronology (fols. 113–123$^{v}$, incomplete at the beginning) gives a full treatment of the Arabic, Persian, Byzantine and Coptic calendars. It contains various of the chronological chapters that are also found in the first part of **E** (pp. 20–21, 6–10, 34–35). These include the *mujarrad* tables for the month beginnings in the Arabic, Byzantine and Persian calendars with instructions for their use. In his *Chronology*, al-Bīrūnī (973–1048, active in Khwarazm and Ghazna) attributes the *mujarrad* table for the Hijra calendar to Ḥabash al-Ḥāsib, but the table in **H**, fol. 86$^{v}$, does not in fact carry that name (cf. Debarnot 1987, p. 41). **L** omits some of the most basic chapters in **E** giving the names and lengths of the months in the four calendars, but on the other hand includes topics missing from **E**, such as the method for calculating the Byzantine *aṣl* (number of days since epoch, cf. van Dalen 1998). **L** inserts an extensive extract (fols. 117$^{r}$–121$^{v}$) from chapter 32 of the *Ṣābi*ʾ *Zīj* of al-Battānī without the corresponding tables. This section doubles much of the presumably original Mumtaḥan material, but is generally more thorough in its definitions and formulation of algorithms. Because in **E** the chronological material starts immediately after the introduction, and in **L** there is no introduction or beginning of the above chronological section, it is very well possible that folio 113$^{r}$ is the first folio of the original manuscript which is still extant. It is in fact more damaged than most of the other folios, as one might expect for the first sheet in a volume.

The second larger section on chronology in **L** (fols. 147$^{r}$–151, 1–2, 34$^{r}$) consists of explanations of the use of two tables for the conversion of dates expressed in seven different calendars and with respect to nine different epochs: the Flood, Nabonassar, Philippus, the Two-Horned (Alexander), August, Diocletian, Hijra, Yazdigird, and the not very commonly used epoch of the Abbasid caliph al-Muʿtaḍid, 11 June 895. A triangular table on fols. 151$^{v}$, 1$^{r}$ displays for each of these epochs the day of the week at which they fell and the type of year (Persian, Byzantine, Arabic) with which they were used. Furthermore, for each pair of epochs the difference between them is expressed in Persian years and days and as total number of days (written in Hindu numerals). Since the latest epoch included in the table is that of al-Muʿtaḍid, and the epoch of Malikshāh (15 March 1079) is not yet mentioned, we may conclude that this table was very probably compiled in the tenth or eleventh century. In fact, it is very similar to that in al-Bīrūnī's *Chronology* (Sachau 1879, p. 133), which also ends with the epoch of al-Muʿtaḍid and additionally inserts the epoch of Antoninus. However, the eras of the Coptic calendar (August and Diocletian) are quite different between the two sources.[2]

[2] The table in **L** implies the following dates for epochs that were not unambiguously defined

The triangular table is followed by another table (fols. 1$^{v}$–2$^{v}$, 34$^{r}$) that can be used for calendar conversions. This table displays the dates with respect to the epochs of the Flood, Nabonassar, Philippus, the Two-Horned, August, Diocletian and the Hijra which correspond to the beginnings of the collected Yazdigird years 1, 41, 81, ..., 801 (extended up to 1001 by a different hand in the margin), as well as the numbers of years and days in each of the calendars concerned that are equal to 1, 2, 3, ..., 40 Persian years and to each of the Persian months. Only for the Hijra calendar do the given values come with minutes and seconds of a day. A very similar table is found in the thirteenth-century zīj of Jamāl al-Dīn Abu 'l-Qāsim ibn Maḥfūẓ al-Baghdādī (MS Paris BNF arabe 2486, fols. 21$^{v}$–23$^{v}$), who, for many parts of his work, depended strongly on Ḥabash al-Ḥāsib and, for instance, also included the *mujarrad* tables mentioned above. The explanation for the use of the two conversion tables in **L** is highly repetitive and only explains the operations in general terms without giving any specific examples that might allow dating.

One more somewhat larger chronological section is found on fols. 9–12, 29$^{r}$, and consists of instructions for the determination of the Christian feasts and fasts with two extensive tables. The same material is found in **E**, pp. 141–148, and the tables also in the zīj of al-Baghdādī (MS Paris BNF arabe 2486, fols. 17$^{r}$–18$^{v}$, 14$^{r}$–15$^{r}$; the explanatory text on fols. 28$^{v}$–30$^{r}$ is very different from that in **L** and **E**). The use of the tables has been explained in Saliba 1970, pp. 187–188, 193, 201–202. The method in **L** and **E** prescribes a substraction of 1204 from the Byzantine year, making it plausible that it was written down around the year AD 900.

Some more chronological material is scattered throughout **L**, and in one or two cases seems to have been used as filling material. An approximate method with table for finding the beginning of the Arabic months on the basis of an 8-year cycle is found on fol. 3$^{v}$ (the explanatory text in **E**, pp. 11/119, is different). A very peculiar method for the conversion of Hijra years to Byzantine years (fol. 43$^{v}$; **E**, p. 12) involves the years of the *fatra* (period between the prophets Jesus and Muhammad), 612, and the years of the Seven Sleepers (*aṣḥāb al-kahf*), 307. The method boils down to adding a constant, 919, to the Hijra years, which led to correct results in the period from AD 1038 to 1068. Finally, a section on a standard conversion of Byzantine dates into Persian ones is hidden between chapters on the heliacal rising of Sirius and on lunar crescent visibility on fol. 112$^{r}$.

in medieval sources: Flood: Thursday, 17 February 3102 BC; August: Thursday, 13 November 30 BC; Diocletian: Thursday, 11 November 284.

### *B. Trigonometry*

**L** contains relatively little material concerned with trigonometry. On fols. 52ᵛ–54ᵛ, there are five chapters giving definitions of the sine and the versed sine, which are illustrated by means of a diagram. Reference is made to a sine table with values for every degree up to ninety, and the use of this table for finding the sine and versed sines of a given arc and vice versa (by means of linear interpolation) is explained. These five sections, also contained in **H**, fols. 124ᵛ–125ʳ, are followed by the basic spherical astronomical material discussed in section C. Immediately after the tables for the planetary equations, on fols. 98ᵛ–99ᵛ, there is another, differently worded section on the determination of sines and versed sines from tables (**E**, pp. 171–172), accompanied by a sine table with values to seconds for every degree for a radius of the base circle equal to 60. This table does not appear to be related to various other early Islamic sine tables that I have checked, and it is not directly derived from Ptolemy's table of chords either. The table has 38 errors, which are nearly all positive and are only incidentally larger than 1″.

**L** includes a cotangent table with values to seconds for a gnomon length of 12 units (fol. 31ʳ), which immediately precedes the star table. This table appears to be related to the table in **E**, p. 120, but not to any other early Islamic tables that I have checked. Note that there is a reference to a tangent table for radius 60 on fol. 4ᵛ (cf. section C), whereas al-Battānī's chapter on shadows and altitudes on fol. 135ᵛ makes use of a cotangent table for gnomon length 12.

### *C. Spherical Astronomy and Timekeeping*

**L** contains two larger sections dealing with spherical astronomical topics. The first of these (fols. 130, ..., 100–109, 4), incomplete at the beginning as well as at the end and with a lacuna in the middle, contains explanatory text and tables that are also found in **E**, pp. 162–168, 90–96. The lacuna can be assumed to consist of one folio covering the text in **E** from p. 164, line 7 to p. 165, line 19 (line 2 from the bottom of the page), all part of a chapter on oblique ascensions. The remaining text deals with the following topics: seasonal hours of day and night, oblique ascensions, half arc of daylight and seasonal hours, length of daylight, hour length, conversion of equal hours into seasonal hours, "sine of daylight", time of day from a measurement of the Sun, solar midday altitude, conversion of equal and ascensional degrees, ortive amplitude, ascendant, and the twelve houses. On fol. 130ᵛ, it is said that the obliquity of the ecliptic, "according to what was found by Ibn al-Aᶜlam,

Yaḥyā ibn Abī Manṣūr al-Ḥāsib, and Abu 'l-Ḥasan al-Ṣūfī", was equal to 23°33′.

I have carried out brief mathematical investigations of the tables for right and oblique ascensions that follow the explanatory text.[3] The normed right ascension on fols. 103ʳ–104ʳ (**E**, pp. 93–94) is based on the Ptolemaic obliquity value 23°51′ and was calculated by applying so-called *distributed linear interpolation* (cf. Van Brummelen 1998, p. 278) between accurate values for every 10 degrees. The oblique ascension table for Baghdad (latitude 33°) on fols. 104ᵛ–106ʳ, accompanied by "parts of the hours", is *not* the same as that in **E**, pp. 95–97, which has tabular differences instead. In fact, the table in **L** turns out to be based on the Indian obliquity value 24°0′ and latitude 33°0′, whereas the table in **E** was computed for obliquity 23°35′ and the latitude value 33°25′ used by al-Nayrīzī and Abu 'l-Wafā' in the tenth century. An oblique ascension table for the latitude 36° of Raqqa (fols. 106ᵛ–108ʳ in **L**, pp. 90–92 in **E**) is basically identical with that in the *Ṣābi' Zīj* of al-Battānī. Finally, oblique ascension tables for the location of the compiler were added to both of the recensions extant as manuscripts **L** and **E**. In the case of **E**, this was Mosul, to which an unrealistically precise latitude 35°55′48″ was attached. The compiler of **L** added an oblique ascension table for "Mayyāfariqīn, Āmid, Arzan, Badlīs, Khilāṭ (or: Akhlāṭ) and each locality whose latitude is 37°30′", and can hence be assumed to have lived in the region of Diyār Bakr in modern southeastern Turkey.

After the ascension tables (fol. 4ᵛ), **L** continues with a section on the determination of the shadow from a given solar altitude by means of a tangent (not cotangent!) table for radius 60. This table is not included in **L** but could very well be that on p. 105 in **E**, which is surrounded by tables that very probably stem from the original *Mumtaḥan Zīj* or another early Islamic source. Folio 4ᵛ in **L** ends with an incomplete section on the time of night from a measurement of a fixed star, which is different from **E**, pp. 185–186, and is *not* continued by the very similar text on fol. 3r.[4]

One would expect that the surviving section on spherical astronomy described above was preceded by chapters on the solar declination, right as-

[3] Here and elsewhere, parameters that are not explicitly said to be mentioned in the manuscript and that are not included in the titles of tables as listed in the Appendix were determined by the present author by means of the methods described in his doctoral thesis van Dalen 1993 and with the aid of his computer programs TA and MM.

[4] Folio 3 begins with the continuation of a chapter on the time of night expressed in seasonal and in equal hours, and then calculates the ascendant for the time found. Next follow a brief chapter on planetary rays (see section K below) and the approximate method for Hijra month beginnings based on an 8-year cycle (see above).

cension and equation of daylight, and possibly by chapters on trigonometry. However, the material dealing with these topics on fols. 52$^v$–56$^v$ is of a very different nature, providing extensive definitions and methods of calculating the quantities as well as finding them from tables. The chapter on the solar declination on fols. 54$^v$–55$^r$ presents the results of Ibn al-Aᶜlam's observations of the solar altitude at the solstices, leading to an obliquity of 23°33′ and a latitude for Baghdad of 33°21′.[5] The following chapters spell out the conditions for ascendance (*suᶜūd*) and descendance (*hubūṭ*) of the Sun and the Moon in detail. A table for the solar declination and lunar latitude on fols. 64$^v$–65$^v$ (accompanied by instructions on fol. 64$^r$) is based on obliquity 23°35′ and maximum latitude 4°46′ and hence is not fully compatible with the above material, which involves three different early obliquity values but not the common 23°35′. The lunar latitude values are basically the same as those given together with the lunar equations in both **L** and **E**. We must thus conclude that the material originally preceding fol. 130 in **L** is lost.

The second larger section on spherical astronomy in **L** (fols. 133$^r$–141$^v$) is found in between longer sections on planetary visibility and on "applications" (*ittiṣālāt*). It deals with the following topics: solar declination (the obliquity is said to be 23°35′ "according to al-Battānī, Ḥabash al-Ḥāsib, Yaḥyā ibn Abī Manṣūr, and others"), right ascensions, altitude of the equatorial pole from the length of the longest day, increases in the length of daylight, calculation of altitude and shadow from each other, azimuth and shadow from altitude, local meridian, oblique ascensions, solar midday altitude. The chapters concerned run parallel to Chapters 4 to 15 of the *Ṣābiʾ Zīj* of al-Battānī and also share with that work the use of chords for the calculations. Only one longer section entitled *Taṣnīf al-ᶜamal bi-l-maṭāliᶜ* (fols. 139$^r$–141$^r$) appears to derive from a different source. This rather elementary chapter describes the use of general sets of tables for the normed right ascension and the oblique ascensions of the climates for determining ascensions and their inverse, arcs of daylight and of nighttime, equal hours, seasonal hours of day and night, etc.

Further material in **L** on spherical astronomical topics is found on fols. 23$^r$–28$^v$. This section starts with a simplified method for calculating oblique ascensions by means of two small tables (to three sexagesimal places and based on obliquity 23°35′) for the right ascension and the sine of the right ascension for every 10 degrees of arc.[6] After a table for the duration of

[5] Exactly the same data are mentioned on the flyleaf of an early copy of the *Īlkhānī Zīj* by the famous polymath Naṣīr al-Dīn Ṭūsī (Maragha, ca. 1270); see King 2000, pp. 225–228, esp. footnote 55.

[6] These tables are identical with those on fol. 178$^r$ of the Escorial manuscript of the zīj of

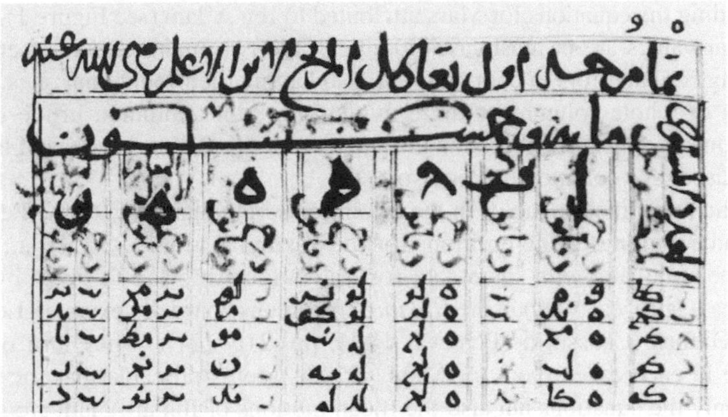

Figure 1: Table of equations for Mars, attributed to Ibn al-Aᶜlam (**L**, fol. 90ʳ)
(reproduced with kind permission of the Universitätsbibliothek in Leipzig)

the planetary phases and one for year transfers attributed to Yaḥyā ibn Abī Manṣūr, fols. 25ᵛ–26ᵛ present a table of right ascensions with values to seconds for every degree. This table is also found in **H**, fols. 132ᵛ–133ᵛ, and was computed for obliquity 23°35′ by means of linear interpolation between values for multiples of three degrees. However, the following highly accurate table for the tangent of the declination (*fuḍūl al-maṭaliᶜ*, fol. 27ʳ), a function that can be used for the easy computation of right ascensions and the equation of daylight, is based on obliquity 23°33′. This table is practically the same as that in MS Paris BNF arabe 2520, fol. 69ᵛ (cf. King 2004, p. 151, Section 7.1.9). These two tables are followed by a chapter on conjunctions and oppositions that ends with the calculation of the ascendant for the time of a syzygy by means of a universal oblique ascension table; the first part of this table, based on the Indian obliquity value 24°, survives on fol. 28ᵛ.

## *D. Planetary longitudes*

Fols. 66ʳ–98ᵛ of **L** contain the same set of tables for planetary mean motions and equations that is also found in **E**, pp. 25–31, 36–37, 32–33, 38–89,

al-Battānī. In Nallino 1899–1907, vol. 2, p. 58, various of the tabular values for the right ascension were corrected in ways that cannot be explained by ordinary scribal errors. I have not checked whether the explanatory text also occurs in the *Ṣābi*ᵓ *Zīj*. Similar data were already given by Ptolemy in the *Almagest*; cf. Neugebauer 1975, vol. 2, pp. 980–982. References to these two secondary sources are contained in King 2004, p. 152, Section 7.2.

including the equations for Mars attributed to Ibn Aᶜlam (see Figure 1). The tabular values are in most cases identical, the two manuscripts sometimes having scribal errors in common, but also having their individual ones. The shifts of whole columns of digits which are rather common in the mean motion tables in **E** are not found in **L**. We conclude that the two sets of tables are based on a common predecessor.

The mean motion tables in the two manuscripts, laid out for the Persian calendar and presenting positions for the Yazdigird years 1, 21, 41, ..., 601, can be compared with some data for Yaḥyā ibn Abī Manṣūr given by Ibn Yūnus (Cairo, ca. 1000) in the *Ḥākimī Zīj*, which is partially extant in Leiden and Oxford (Caussin de Perceval 1804, pp. 216–221). They turn out to agree in each case except for Mars. It thus seems probable that for Mars not only the equations but also the mean motions of Ibn al-Aᶜlam are given in **L** and **E**. Note that the mean motion tables for Mars in both manuscripts have values to seconds only, whereas the tables for all other planets have values to thirds. Furthermore, in **E** the equations for Mars come without the columns for the first and second station which are present for all the other planets; in **L** these stations were apparently added (see also section F below). Nevertheless, the equations for Mars in **L** and **E** are all basically the same as found in the *Handy Tables*, which would mean that Ibn al-Aᶜlam only made significant modifications to the mean motions for this planet.[7] The mean motion parameters for Mars attributed to Ibn al-Aᶜlam in **L** (fol. 61ʳ) as well as in **E** (p. 67, in an empty space of the mean motion table) are in full agreement with the tables in the two manuscripts.

The solar equation in **L** is basically identical with that in **E**, and hence does not contain the errors spotted by Ibn Yūnus in the solar equation in the

[7] The *Ashrafī Zīj*, compiled in the early fourteenth century by Sayf-i munajjim-i Yazdī and extant in the unique MS Paris BNF suppl. persane 1488/1, gives the elements of planetary motion for more than ten different zījes, including the one by Ibn al-Aᶜlam. From the fact that for Ibn al-Aᶜlam only the solar equation, the lunar equation of anomaly, and the equation of centrum for Saturn, Jupiter and Venus are presented, we may conclude that his equations for Mars were basically the same as those in the *Handy Tables*. The earliest documented change of one of the equations for Mars by a Muslim astronomer is that of the equation of centrum by Kūshyār ibn Labbān (ca. AD 1000; cf. Van Brummelen 1998, pp. 268). The mean motion in longitude that is associated with Ibn al-Aᶜlam in the *Ashrafī Zīj* is in almost perfect agreement with the tables for Mars in **L** and **E**; for the mean motion in anomaly there is a small deviation.
The planetary parameters from Ibn al-Aᶜlam's zīj were reconstructed from various sources in Kennedy 1977 and Mercier 1989. To the results it may be added that the mean motion tables for Mars in **L** and **E** are fully compatible with the table for the Hijra calendar in the zīj of al-Baghdādī (MS Paris BNF arabe 2486, fols. 85ᵛ–86ʳ). Byzantine materials related to Ibn al-Aᶜlam were studied in Tihon 1989.

*Mumtaḥan Zīj*, which are in fact present in the zīj of Ḥabash al-Ḥāsib (**H**, fols. 90ʳ–91ʳ). The longitudes of the solar and planetary apogees mentioned in the titles of the equation tables in **L** are the same as those in **E** and agree with the values mentioned by Ibn Yūnus, except for Mars, for which **L** gives an apogee longitude of 128°50′, whereas Ibn Yūnus attributes to Yaḥyā ibn Abī Manṣūr the value 124°33′ (Caussin de Perceval 1804, pp. 220–221). This very value is indeed found in the list of apogee longitudes on fols. 60ᵛ–61ʳ of **L**.[8] It is reasonable to assume that the table for the apogee of Mars on fol. 21ᵛ, immediately preceding the table for apogee motion based on Ibn al-Aᶜlam's parameter of 1° in 70 Persian years and also itself computed for this value, displays the apogee longitudes according to Ibn al-Aᶜlam. This table displays the longitude given with the equation table for the year 180 Yazdigird (AD 811), around 20 years too early for Yaḥyā. Its values are also in relatively good (but not perfect) agreement with those in the *Ashrafī Zīj*.[9] Note that **L**, fol. 22ᵛ, also includes a table for apogee motion based on the parameter of "al-Battānī and al-Maʾmūn", 1° in 66 Julian years.

The instructions for the use of the tables for mean motions and equations (**L**, fols. 47ᵛ–50ᵛ; **E**, pp. 157–159) are in concordance with the tables. From **E**, the instructions for finding the mean motions and the true solar longitude are missing. In neither manuscript are the instructions found in the direct vicinity of the planetary tables. On fols. 124ʳ–124ᵛ of **L** we find two inconsequential fragments on mean motions in anomaly (similar to **H**, fol. 102ᵛ) and a brief section on the equation of time (involving a constant also used by Ḥabash; cf. Debarnot 1987, p. 42) in between a chapter on retrogradation and a general introduction to astronomy attributed to al-Battānī.

### *E. Planetary Latitudes*

The only section on the latitudes of the five planets in **L** appears on fols. 60ᵛ–63ᵛ. It starts immediately after the tables for the planetary stations (see below) with a brief explanation of the determination of planetary latitudes according to "the method of Ptolemy, simplified". The procedure described

[8] It can be noted that the equation tables of Ḥabash mention the same apogee longitudes as the *Mumtaḥan Zīj* except for Saturn (242°50′ vs. 244°30′; see **H**, fols. 89ʳ, 103ʳ, 107ʳ, 111ʳ, 115ʳ, and 119ʳ).

[9] Cf. footnote 7. Interestingly enough, the epoch value of the table for the apogee of Mars in **L** is the same as that given on fols. 2ʳ–2ᵛ of the unique MS Paris BNF arabe 5968 of an eleventh-century Ismāᶜīlī zīj called *Dastūr al-munajjimīn*. On the other hand, the apogee longitude of Mars given for the Hijra epoch in the *Baghdādī Zīj* (fol. 42ᵛ) does not appear to be reconcilable with any of the other surviving values.

yields the planetary latitude directly from a single table as a function of the distance between the planet and one of the nodes. A similar method is found on **E**, p. 115, except that there for the inferior planets the node is subtracted from *al-ḥiṣṣa al-thāniya*, presumably the true anomaly. As has been shown in Viladrich 1988, the latitude tables in **E**, pp. 109–114, are sinusoidal functions, for which the underlying maximum values were taken from the *Handy Tables*. In **L**, the tables are even simpler because, different from **E**, the northern and southern latitudes for the superior planets are basically identical. The maximum latitudes of the five planets are: Saturn 5°0′, Jupiter 2°0′, Mars 3°45′, Venus 2°26′, Mercury 2°28′. For Saturn and Mars these maxima happen to be identical with those of the second latitude as given by al-Khwārizmī (cf. Suter 1914, pp. 138–167; the tables as such are *not* identical), but in general the origin of the latitude tables in **L** is completely unclear to me. The following longitudes of the planetary nodes, given together with the instructions (fol. 60$^v$, lines 6–11, repeated in lines 11–15), might very well be corrupt: Saturn 120°, Jupiter 80°, Mars 60°, Venus 40°, Mercury 20°.

The planetary latitude tables are immediately followed by a combined table for the solar declination for obliquity 23°35′ and the lunar latitude with maximum 4°46′, accompanied by instructions for its use (fols. 64$^r$–65$^v$). The lunar latitude is basically identical with that found in the last column of the table for the lunar equations, whereas the solar declination (calculated by means of interpolation on intervals of 6 degrees) is less accurate than all other declination tables surviving from the early Islamic period. It is remarkable, in particular, that the combined solar declination and lunar latitude table in the zīj of Ḥabash al-Ḥāsib (**H**, fols. 99$^r$–100$^r$), based on the same parameters, is clearly different from the table in **L**.

There are scattered other sections concerning lunar latitudes in **L**. On fols. 35$^r$–35$^v$, chapters on finding the lunar latitude and the true longitude of the node from tables are embedded between texts on planetary stations and planetary visibility. On fols. 55$^v$–56$^r$, the solar declination and lunar latitude are found by calculation and from a combined table involving a maximum lunar latitude of 5°0′. On fols. 129$^v$, 131$^r$, the lunar latitude is found from the seventh column of the table for the lunar equations; hence this chapter, surrounded by longer texts on general principles of almanacs and planetary visibility (see sections L and G below), may very well be from the original *Mumtaḥan Zīj*. Finally, on fol. 152$^v$, following the geographical table and preceding the chapter on the use of the eclipse circle (see section H), there is a trivial chapter on finding the positions of the lunar nodes.

### *F. Planetary stations*

A set of tables for the first stations of the five planets occurs on fols. 57$^{v}$–60$^{r}$ of **L** in between a simple method for lunar eclipses attributed to Yaḥyā ibn Abī Manṣūr and non-Ptolemaic material on planetary latitudes. For each planet, the values in these tables are basically identical with those from Ptolemy's *Handy Tables*, which were also adopted by al-Battānī in his *Ṣābiʾ Zīj*. (In contrast, Ḥabash al-Ḥāsib included the first stations as found in the *Almagest*, which were tabulated as a function of a different argument.) The tables for the stations that are included with the planetary equations in **L** (fols. 77$^{v}$–98$^{r}$) are basically the same as the separate tables, except for Mars, for which the tabulated stations are completely different from Ptolemy or any known early Islamic sources (note that **E** does not include the stations for Mars at all). It seems possible that the compiler of the recension in **L** added the stations from the zīj of Ibn al-Aᶜlam to the latter's equations for Mars.

There is scattered other material related to planetary stations in **L**. A table on fol. 23$^{v}$ displays for each of the five planets the duration (in days and minutes of a day) of their direct motion, first station, retrograde motion, second station, again direct motion, and the sum of these five quantities. Standard instructions for the use of tables for the stations are found on fols. 50$^{v}$, 35$^{r}$ (here explicit reference is made to columns for the first and for the second station, probably those in the tables for the planetary equations), and on fols. 123$^{v}$–124$^{r}$ (here the second station is found as 360° minus the first station).

### *G. Lunar and Planetary Visibility*

Two methods for determining the visibility of the lunar crescent after new moon are given on fols. 110$^{r}$–110$^{v}$, 36$^{r}$ in **L**. They appear to differ in their details from other known early methods, but are clearly based on the Indian criterion, which lets the visibility depend on the difference in setting time between the Sun and the Moon. The only other material on lunar crescent visibility in **L** is a list of visibility limits for the twelve zodiacal signs on fols. 112$^{r}$–112$^{v}$. This list also occurs in **E**, p. 12, and is discussed in King 1987, pp. 213–214.

**L** contains an extensive section on the visibility of the planets on fols. 131$^{r}$–133$^{r}$. This starts by giving definitions and some simple criteria for visibility. It then presents a "more correct" method that involves the following arcs of visibility: Saturn 14°0′, Jupiter 12°45′, Mars 14°30′, Venus 5°40′, Mercury 11°30′. For the calculation of the difference in setting time between the Sun and the planet, taking the latter's latitude into account, reference is made to a chapter on lunar crescent visibility. The text continues with a method for finding the number of days that have passed since, or will elapse until, a first or last visibility, and ends with instructions for determining the visibility of a planet from the table for the fourth climate in the *Almagest*.

A set of standard planetary visibility tables is present on fols. 51ᵛ–52ʳ in **L**. The instructions for their use start on fol. 35ᵛ, which ends with taking a tabular value for the zodiacal sign of the planet, and continue after a lacuna on fol. 51ʳ with the calculation of the time of first or last visibility using the difference between solar and planetary velocity. Further information on planetary visibility in Islamic sources can be found in Kennedy & Agha 1960.

## *H. Eclipses*

**L** contains various sections on solar and lunar eclipses, not all of which are also included in **E**. The method (fols. 40ᵛ–41ʳ) of calculating solar eclipses by means of tables for *al-samt* (fols. 41ᵛ–42ʳ) and *ʿarḍ al-shams* (fol. 42ᵛ) as well as a solar eclipse table as a function of *ʿarḍ al-qamar al-muḥkam* (fol. 43ʳ), is also found in **E** (pp. 15–17, 19/138, 22–24). In Kennedy & Faris 1970, p. 20, the association of this material with Yaḥyā ibn Abī Manṣūr was established. The explanatory text in **L** does not mention the geographical latitude 35°55′48″, which also occurs elsewhere in **E** in connection with the city of Mosul, but it appears to be more complete than the version in **E** and might hence help to solve the few remaining problems in the method.

In **L**, the method above is preceded by an unusual method for calculating lunar eclipses (fols. 36ʳ–40ᵛ), which is associated in the text with Yaḥyā ibn Abī Manṣūr and Ḥabash al-Ḥāsib and is not contained in **E**. Depending on the value of the true lunar anomaly, the magnitude and duration of the eclipse and, if applicable, the duration of totality are taken from one of six similar tables, called the first to six *manzila* (not to be confused with the lunar mansions). For higher accuracy, minutes given together with the *manzila*s as a function of the true anomaly allow the performance of linear interpolation between the six tables. A brief section on the colours of lunar eclipses, similar to that on pp. 14/171 of **E**, has been added to the explanatory text on fol. 36ᵛ.

Both **L** (fol. 57ʳ) and **E** (p. 14) include a simple method for the calculation of lunar eclipses that is associated with Yaḥyā ibn Abī Manṣūr and involves three small tables headed *al-bāb al-awwal* (magnitude as a function of the lunar distance from a node), *al-bāb al-thānī* (duration of the eclipse as a function of the magnitude), and *al-bāb al-thālith* (duration of totality as a function of the magnitude).

Fols. 29$^{v}$–30, 13–20 of **L** contain another extensive set of instructions for calculating lunar and solar eclipses, incomplete at the end. They have the peculiar characteristic that the ascendant and upper midheaven are explicitly calculated for each of the phases of the eclipse that are found. The following tables in **L** are associated with the method for lunar eclipses presented in this section: eclipse tables for nearest and farthest distance with argument columns for the distance of the moon from the node numbered "1" to "4" as well as columns for the magnitude and duration (fols. 32$^{v}$–33$^{r}$); and an eclipse equation as a function of the true anomaly (fol. 33$^{v}$). The following tables are associated with the method for solar eclipses presented in this section: lunar parallax (fol. 111$^{r}$); latitude component of lunar parallax (*ikhtilāf ʿarḍ al-qamar li-l-ruʾya*, fol. 32$^{r}$, **E** p. 137);[10] an eclipse table as a function of *ʿarḍ al-qamar al-muḥkam* (fol. 111$^{v}$), and the eclipse equation also used for lunar eclipses (fol. 33$^{v}$). The section on conjunctions and oppositions (fols. 27$^{v}$–28$^{r}$; **E**, pp. 159–161), which also ends with the determination of the ascendant and upper midheaven (using a small universal table of oblique ascensions, only partially extant for latitudes 1° to 33°) can be assumed to be from the same source as the eclipse material mentioned above.

Another interesting, incomplete section on lunar eclipses occurs on fol. 152$^{v}$. It gives instructions for drawing the configuration of an eclipse by means of an "eclipse circle" (*dāʾirat al-kusūf*). The complete text is available in **H**, fols. 205$^{v}$–206$^{r}$. I have not been able to piece together the remaining texts and tables in **L** related to eclipses: a small section on the hourly motion of the Sun and Moon (fol. 110$^{r}$; **E**, pp. 9/16), a simplified method for lunar eclipses referring to a table that has the lunar latitude as argument (fol. 43$^{v}$), and a solar eclipse table as a function of the distance between moon and node (fol. 44$^{r}$). The only eclipse material in **E** not contained in **L** is a theoretical treatise on pp. 176–179.

### *I. Fixed stars*

**L** contains on fols. 31$^{v}$, 153$^{r}$ the table with ecliptical and equatorial positions of 18 fixed stars for the year 380 Yazdigird (AD 1011) that is also found in **E**, pp. 189–190. As has been shown in Girke 1988, the ecliptical coordinates in this table are based on those in the table of 24 stars said to be derived from the Mumtaḥan observations at Baghdad in 214 Hijra (AD 829), likewise extant

[10] The same table is found in the extant Latin recension of the zīj of al-Khwārizmī; see Suter 1914, pp. 191–192 and Neugebauer 1962, pp. 121–123. For a general description of Yaḥyā ibn Abī Manṣūr's parallax theory, see Kennedy 1956b, pp. 44–46.

in **E**, p. 188, but in a later hand. Note that the *Īlkhānī Zīj* of Naṣīr al-Dīn al-Ṭūsī attributes coordinates for the same 18 stars, derived from the table in **L** and **E**, to Ibn al-Aʿlam (cf. Kunitzsch 1964, pp. 397–398).

Besides the star table, the only other material in **L** dealing specifically with fixed stars is a chapter on the rising of Sirius (fols. 44ᵛ, 112ʳ), which gives the date, the time of day or night and the ascendant for the moment of the heliacal rising as a function of the remainder of the division of the incomplete Byzantine years plus six by four. The date of rising is either 19 or 20 July.

### *J. Geography*

**L** contains a table of geographical coordinates for 180 localities said to be taken from *Kitāb Ṣūrat al-arḍ* (fols. 153ʳ–155ᵛ, 152ʳ). Given the early Islamic character of much of the material in the manuscript, one might expect that this title refers to the book by Muḥammad ibn Mūsā al-Khwārizmī (edited in von Mžik 1926), or to the work by the geographers of al-Maʾmūn in general (cf. Sezgin 2000, vol. 10, chapter I-D and p. 148). A first investigation of the table reveals that its direct source is most likely the geographical table in al-Battānī's *Ṣābiʾ Zīj* (Nallino 1899–1907, vol. 2, pp. 33–54 (edition) and vol. 3, pp. 234–241 (Arabic)), whose title includes both the "awsāṭ al-buldān" ("centres of the regions") and the *Kitāb Ṣūrat al-arḍ* that are also mentioned in the title of the table in **L**. In fact, basically all localities and regions listed in **L** are also found with al-Battānī and most of the differences in coordinates are due to copying mistakes. The data for Baghdad, Raqqa and Harran are indicated to have been "verified" (*mumtaḥan*), as in the *Ṣābiʾ Zīj*. However, the localities in **L** occur in a haphazard order and only incidentally include consecutive stretches in the same order as found in the zīj. In three or four cases columns of degrees or minutes of one of the coordinates were shifted, resulting in nonsensical data. For some localities coordinates in **L** different from al-Battānī's are identical with those in the work on the qibla by the 13th-century Egyptian scholar Zayn al-Dīn al-Dimyāṭī (QBL in Kennedy & Kennedy 1987, pp. 443–448), but this latter source does not include the coordinates for a large number of regions, which are typical for al-Battānī.

### *K. Astrology*

Fols. 24ᵛ–25ʳ of **L** display an eternal table for finding year transfers, which is attributed to al-Maʾmūn (see Figure 2). The table in fact gives the ascendant of the year transfer as a function of that for the previous year. It is preceded

Figure 2: Table of year transfers associated with al-Maʾmūn (**L**, fol. 24ᵛ)
(reproduced with kind permission of the Universitätsbibliothek in Leipzig)

by instructions, which mention the excess of revolution as 86°45′, and a brief section on determining the transfer of the world year.

On fol. 44ᵛ, following a set of eclipse tables, there is a somewhat isolated chapter on finding the ascension of the year transfer from a given ascendant. The excess of revolution here given in words, 86°43′39″, underlies a table for the same purpose on **E**, p. 142. This chapter is followed by a list of the solar mean positions corresponding to the entry of the true Sun in the zodiacal signs. Like most of these lists, the values are not particularly accurate (cf. Viladrich 1996), but they correspond reasonably well to the Maʾmūnic apogee longitude 82°39′ and an eccentricity (not maximum equation!) of 1;59ᵖ.

A third, larger section on year transfers is found on fols. 5ʳ–8ʳ. At first (fols. 5ʳ–5ᵛ), the transfer of the world year, i.e., the time of the true vernal equinox, is found by calculating the solar longitude at noon of the day on which the equinox occurs and then correcting for the difference between this longitude and 0° Aries, if desired by means of an iterative procedure. On fols. 5ᵛ–6ᵛ, the use of the table on fols. 7ᵛ–8ʳ for calculating the transfers of nativities and world years is explained. This table, said to be according to the observations of Ibn al-Aᶜlam, displays the difference in the ascension of the ascendant (*faḍl al-ṭāliᶜ*) and in the date and time of the transfer (*faḍl al-taʾrīkh*) for 1 to 90 years. The underlying excess of revolution, slightly less than 87°13′, can be read directly from the table. This value is highly accurate and, as far as I know, hitherto unattested.

The above material is interspersed with chapters of a more spherical astronomical character: on the calculation of the azimuth from the altitude and the ascendant (fols. 6$^{v}$–7$^{r}$), of the "ascendant of the world" or "centre of the earth" from the ascendant of the world year at Baghdad (fol. 7$^{r}$), and of the time of day or night (*dāʾir min al-falak*) of a year transfer at Baghdad and other localities (fol. 8$^{v}$). A small chapter (fol. 7$^{r}$) on finding the transfer of world years is identical with that on fol. 24$^{r}$.

A long chapter on *ittiṣālāt* ("applications", various types of relative positions of the planets in longitude and latitude) on fols. 141$^{v}$–144$^{r}$ of **L** deals with aspects and rays adjusted for latitudes and with nativity *tasyīr*s. It has a similar title and is concerned with similar topics as Chapter 54 of the *Ṣābiʾ Zīj*, although the two texts do not seem to be identical.

On fol. 3$^{r}$ of **L** there is a brief chapter on the projection of the rays according to an author whose name is written as "Drīnūsh". This is very probably Dorotheus, normally rendered in Arabic as Dhūruthiyūs, one of the most important Greek authorities for early Muslim astrologers (see, for instance, Sezgin 1979, pp. 32–38, and the supplement volume of the *Dictionary of Scientific Biography*). The method described is based on right ascensions and therewith of a type usually associated with pre-Islamic Iranian astrology. On fols. 112$^{v}$, 21$^{r}$–21$^{v}$ a section entitled "Chapter of *namūdhār*s" describes the correction of the time of a nativity by means of the method of Wālīs (Vettius Valens). This is followed on fol. 21$^{v}$ by a brief chapter on the "ascendant of the *qubba*" (cupola, central meridian of the world), which is calculated by adding twenty degrees to the oblique ascension of the ascendant of the world year at Baghdad and then taking the inverse right ascension.[11]

### *L. Various*

This section lists the few remaining sections and tables that do not fit so easily in one of the preceding categories. On fol. 123$^{v}$ of **L**, between the first larger section on chronology and a chapter on retrogradations, there is a small section on linear interpolation between two values from a table for an equation (*taʿdīl daqāʾiq al-ḥiṣṣa*, "the equation for minutes of the argument"). A treatise on a sundial for seasonal hours on fols. 144$^{r}$–147$^{r}$ is similar to Chapter 56 of the *Ṣābiʾ Zīj*.

A chapter without title but with a lacuna (fols. 124$^{v}$–125, ..., 126–128$^{r}$),

[11] Since the centre of the world is given a longitude of 90°, this method implies a longitude for Baghdad of 70°, as found with al-Khwārizmī, rather than 80° as in the zīj of al-Battānī and the geographical table in **L**.

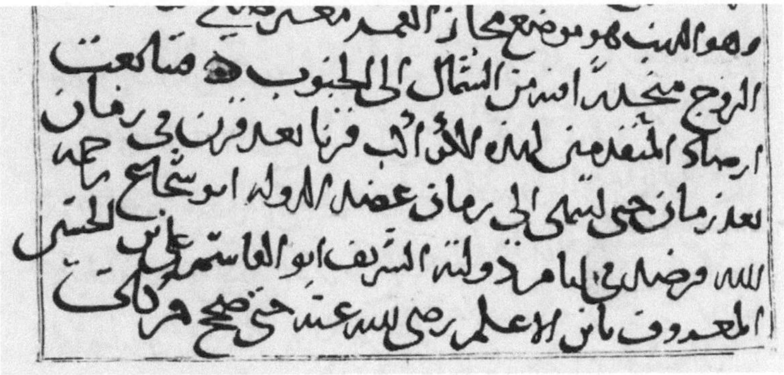

Figure 3a: First part of section mentioning the "Maᵓmūnic Zīj" (**L**, fol. 47ʳ, bottom)
(reproduced with kind permission of the Universitätsbibliothek in Leipzig)

associated with al-Battānī, deals with the five basic great circles on the heavenly sphere (giving the obliquity of the ecliptic as 23°35′), the division of the ecliptic into signs and degrees, the division of degrees into minutes, seconds, thirds, etc., and the results of multiplying or dividing such divisions. The contents is similar to Chapter 2 of al-Battānī's *Ṣābiᵓ Zīj*, but the wording clearly different. This section is followed by a chapter on laying out almanacs (fols. 128ʳ–129ᵛ), explaining how to make entries for consecutive days and hours, convert equal hours to seasonal hours and mean time to true time, and calculate the mean solar longitude for the resulting times. The tables used are said to be for Baghdad, with latitude 33°.

A chapter on the conditions (*aḥwāl*) of the planets on fols. 34ᵛ, 45–47ᵛ, immediately after the chronological section to which it also refers and preceding the main section on the determination of planetary positions, gives another general introduction to astronomy, discussing the shape of the earth and the heavens, the zodiac and its subdivisions (with a list of the Arabic and Greek names of the signs), and the planets and their motions and spheres (with mention of the Arabic, Persian and Greek names of the planets). This section ends by stating that observations of the planets had been performed over and over again up to the time of ᶜAḍud al-Dawla (936–983), under whose rule Ibn al-Aᶜlam corrected the motions of the Sun, Moon and planets. Since "the Mars of Ibn al-Aᶜlam ... is the best of the Marses in all zījes", it was included in "this corrected Maᵓmūnic zīj" (*hādha 'l-Zīj al-Maᵓmūni 'l-muṣlaḥ*; fol. 47ᵛ, line 2; see Figures 3a and 3b).

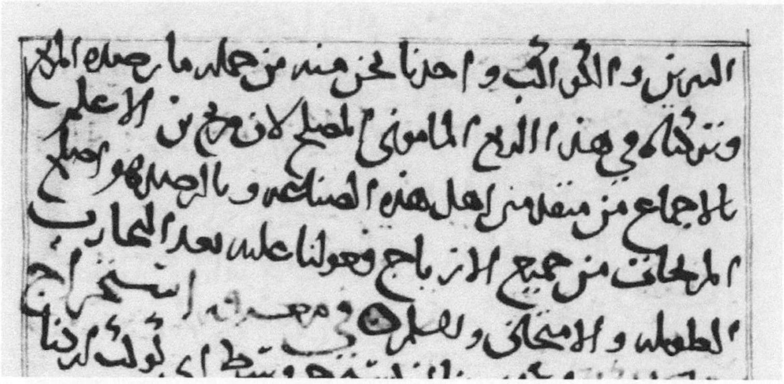

Figure 3b: Last part of section mentioning the "Maʾmūnic Zīj" (**L**, fol. $47^{v}$, top)
(reproduced with kind permission of the Universitätsbibliothek in Leipzig)

***Summary***

From the above overview of the contents of **L**, we can see that the following materials are explicitly attributed to early Muslim astronomers:

- al-Maʾmūn: eternal table for year transfers with instructions (fols. $24^{r}$–$25^{r}$), value for apogee motion underlying the table on fol. $22^{v}$;
- Yaḥyā ibn Abī Manṣūr: a simple method for lunar eclipses (fols. $57^{r}$), the *manzila* method for lunar eclipses (fols. $36^{r}$–$40^{v}$);
- Ḥabash al-Ḥāsib: the *manzila* method for lunar eclipses (fols. $36^{r}$–$40^{v}$);
- al-Battānī: general introduction on great circles, divisions of the ecliptic, and multiplication and division of parts of degrees (fols. $124^{v}$–$128^{r}$), section on chronology (folios $117^{r}$–$121^{v}$), value for apogee motion underlying the table on fol. $22^{v}$;
- Ibn al-Aᶜlam: value for apogee motion underlying the table on fol. $22^{r}$, list of mean motion parameters for Mars (fol. $61^{r}$), equations (and stations?) for Mars (fols. $87^{v}$–$90^{r}$), table for year transfers (fols. $5^{v}$–$6^{v}$, $7^{v}$–$8^{r}$).

Other materials in **L** that can be assumed to derive from the original *Mumtaḥan Zīj*, partially because they are also found in **E**, include the sections on chronology (fols. 113–$123^{v}$) and spherical astronomy (fols. 130, ..., 100–109, 4), the material on solar eclipses (fols. $40^{v}$–$43^{r}$), and the instructions for finding lunar latitudes (fols. $129^{v}$, $131^{r}$) and stations (fols. $50^{v}$, $35^{r}$) from tables. Judging from the underlying values for the obliquity of the ecliptic, also the tables for the tangent of the declination (fol. $27^{r}$), the normed right ascension (fols. $103^{r}$–$104^{r}$; **E**, pp. 93–94), the oblique ascension for Baghdad (fols. 104$^{v}$–106$^{r}$), and the oblique ascension for all latitudes (fol. 28$^{v}$) can be assumed to stem from the ninth century.

By a direct comparison, some more chapters in **L** can be seen to derive from the zījes of Ḥabash al-Ḥāsib and al-Battānī. The sections on the sine and versed sine (fols. 52$^{v}$–54$^{v}$) and on the lunar eclipse circle (fol. 152$^{v}$) are also contained in the Istanbul manuscript of the zīj of Ḥabash. The consecutive sections on spherical astronomy (fols. 133$^{r}$–139$^{r}$, 141$^{r}$–141$^{v}$), "applications" (*ittiṣālāt*, fols. 141$^{v}$–144$^{r}$), and the construction of a sundial for seasonal hours (fols. 144$^{r}$–147$^{r}$) are similar to chapters in al-Battānī's *Ṣābi᾽ Zīj*.[12] Furthermore, the table for the oblique ascension at Raqqa (fols. 106$^{v}$–108$^{r}$) and the geographical table (fols. 153$^{r}$–155$^{v}$, 152$^{r}$) are clearly related to the corresponding tables in the *Ṣābi᾽ Zīj*. As far as Ibn al-Aᶜlam is concerned, not only the equations but also the mean motions for Mars (fols. 85$^{v}$–87$^{r}$) can be attributed to him.

We have seen that the manuscripts **L** and **E** have the following materials in common:

- large part of the chronological section on fols. 113–123$^{v}$ (except for an extract from al-Battānī's *Ṣābi᾽ Zīj* on fols. 117$^{r}$–121$^{v}$);
- text and tables for finding the Christian feast and fasts;
- table for the 8-year cycle of the Hijra calendar;
- method for converting from the Arabic to the Byzantine calendar by means of the years of the *fatra* and the Seven Sleepers;
- a chapter on finding sines and arcsines from a table of sines;
- a cotangent table for gnomon length 12;
- largest part of the spherical astronomical section on fols. 130, 100–109, 4;
- spherical astronomical tables: normed right ascension, oblique ascension for Raqqa;
- tables for solar, lunar and planetary mean motions and equations, including the tables for Mars taken from Ibn al-Aᶜlam;
- a list of mean motion parameters for Mars associated with Ibn al-Aᶜlam;
- chapters on the true lunar and planetary positions;
- a listing of visibility limits for the lunar crescent;
- a chapter on solar and lunar conjunctions and oppositions;
- an isolated section on the hourly motion of Sun and Moon;

12 I have not compared these sections in detail with Nallino's edition; more often than not the texts appear to be so different that one might consider the possibility that they come from a different source in spite of obvious similarities. In particular, it might be possible that they stem from the earlier edition of al-Battānī's zīj which is mentioned in the *Fihrist* by Ibn al-Nadīm (AD 987).

• a simple method for lunar eclipses attributed to Yaḥyā ibn Abī Manṣūr;
• a brief section on the colour of lunar eclipses;
• text and tables on solar eclipses associated with Yaḥyā ibn Abī Manṣūr;
• a table for the latitude component of lunar parallax;
• a table with coordinates and other data for 18 stars for the year 380 Yazdigird;
• a chapter describing the *namūdhār* of Vettius Valens.

Both manuscripts also include tables for the solar declination / lunar latitude, for the ordinary (as opposed to: normed) right ascension, for the oblique ascension at Baghdad and at the locality of the compiler, and for planetary latitudes, but the parameters underlying these tables are different between the two sources.

Material found in **L** but not in **E** includes:

• two tables for the conversion of dates with explanatory text;
• universal oblique ascension table based on the Indian obliquity value 24°;
• accurate table for the tangent of the declination based on obliquity 23°33′;
• tables of apogee motions and positions for Mars according to Ibn al-Aᶜlam;
• separate tables for planetary stations, and a theoretical text on this topic;
• two methods for determining the visibility of the lunar crescent;
• tables for planetary visibility with instructions;
• *manzila* tables for lunar eclipses associated with Yaḥyā and Ḥabash;
• chapter on the time of the heliacal rising of Sirius;
• geographical table related to that of al-Battānī in the *Ṣābiʾ Zīj*;
• table for year transfers attributed to al-Maʾmūn;
• table for year transfers based on Ibn al-Aᶜlam.

Material found in **E** but not in **L** includes:

• tables of Abu ʾl-Wafāʾ and Kūshyār ibn Labbān;
• section on the Jewish calendar (discussed in Vernet 1954);
• tangent table with values to seconds for radius 60;
• further topics in spherical astronomy;
• three sections on the latitude of the visible climate, with tables;
• some chapters on the size of the planetary spheres;
• star table for AD 829 associated with the Mumtaḥan observations;
• different methods for the projection of the rays;
• various sections on astrological indicators.

A comparison with **L** makes it easier to restore the order of the first forty pages in **E** (the remainder of the manuscript appears be in correct order; note that each odd-numbered page in the facsimile is the obverse of the next *lower* even page, so that, for instance, pages 4–5 constitute one folio): 3–5, 20–21, 6–7, 8–9, 34–35, 10–31, 36–37, 32–33, 38–89 (pp. 65–66 contain spherical astronomical tables for latitude 36° that do not normally belong in between the planetary tables).

## Acknowledgements

I would like to express my gratitude to the editors for allowing me to submit this article at such short notice. The comments on an earlier version by Julio Samsó of the Departamento de Árabe of the University of Barcelona and David A. King of the Institut für Geschichte der Naturwissenschaften in Frankfurt am Main have been most useful. Mohamed Abu Zayed (Frankfurt) assisted me with some Arabic passages that were not immediately clear to me. I also much appreciated the help of Steffen Hoffmann during my visit to the manuscripts room of the Universitätsbibliothek in Leipzig and later in providing me with a microfilm copy of the manuscript Vollers 821. The research on the manuscript was carried out as part of a larger project on Islamic astronomical tables (Ki 418) funded by the German Research Foundation (DFG). Finally, it was David King's note "full of Ibn al-Aᶜlam", made some time back in the 1970s during a library research trip to the "DDR", that led me to look more carefully at what had been catalogued simply as a chronological text.

## Bibliography

Caussin de Perceval, Jean-Jacques-Antoine 1804. "Le livre de la grande table hakémite, Observée par le Sheikh, ..., ebn Iounis", *Notices et Extraits des Manuscrits de la Bibliothèque nationale* **7**, pp. 16–240 (references are to the separatum, which has page numbers 1–224).

van Dalen, Benno 1993. *Ancient and Mediaeval Astronomical tables: mathematical structure and parameter values* (doctoral thesis), Utrecht (Mathematisch Instituut).

—— 1998. "Taʾrīkh I-2. Era chronology in astronomical handbooks", in: *The Encyclopaedia of Islam. New edition*, Leiden (Brill), vol. 10, pp. 264–271.

Debarnot, Marie-Thérèse 1987. "The *Zīj* of Ḥabash al-Ḥāsib: A Survey of MS Istanbul Yeni Cami 784/2", in: King & Saliba 1987, pp. 35–69.

Girke, Dorothea 1988. *Drei Beiträge zu den frühesten islamischen Sternkatalogen. Mit besonderer Rücksicht auf Hilfsfunktionen für die Zeitrechnung bei Nacht*, Preprint Series No. 8, Frankfurt am Main (Institut für Geschichte der Naturwissenschaften).

Kennedy, Edward S. 1956a. "A Survey of Islamic Astronomical Tables", *Transactions of the American Philosophical Society, New Series* **46-2**, pp. 123–177. Reprinted in 1989 with page numbering 1–55.

—— 1956b. "Parallax Theory in Islamic Astronomy", *Isis* **47**, pp. 33–53. Reprinted in Kennedy *et al.* 1983, pp. 164–184.

—— 1977. "The Astronomical Tables of Ibn al-Aᶜlam", *Journal for the History of Arabic Science* **1**, pp. 13–23.

Kennedy *et al.* 1983. *Studies in the Islamic Exact Sciences by E.S. Kennedy, Colleagues and Former Students* (David A. King & Mary Helen Kennedy, eds.), Beirut (American University in Beirut).

Kennedy, Edward S. & Agha, Muhammad 1960. "Planetary Visibility Tables in Islamic Astronomy", *Centaurus* **7**, pp. 134–140. Reprinted in Kennedy *et al.* 1983, pp. 144–150.

Kennedy, Edward S. & Faris, Nazim 1970. "The Solar Eclipse Technique of Yaḥyā b. Abī Manṣūr", *Journal for the History of Astronomy* **1**, pp. 20–38. Reprinted in Kennedy *et al.* 1983, pp. 185–203.

Kennedy, Edward S. & Kennedy, Mary Helen 1987. *Geographical Coordinates of Localities from Islamic Sources*, Frankfurt am Main (Institut für Geschichte der Arabisch-Islamischen Wissenschaften).

King, David A. 1986. *Islamic Mathematical Astronomy*, London (Variorum). Second edition with corrections: Aldershot (Variorum) 1993.

—— 1987. "Some Early Islamic Tables for Determining Lunar Crescent Visibility", in: King & Saliba 1987, pp. 185–225. Reprinted in King 1993, II.

—— 1993. *Astronomy in the Service of Islam*, Aldershot (Variorum).

—— 2000. "Too Many Cooks ... A New Account of the Earliest Muslim Geodetic Measurements", *Suhayl* **1**, pp. 207–241.

—— 2004. *In Synchrony with the Heavens. Studies in Astronomical Timekeeping and Instrumentation in Medieval Islamic Civilization. Volume One. The Call of the Muezzin (Studies I–IX)*, Leiden (Brill).

King, David A. & Saliba, George (eds.) 1987. *From Deferent to Equant: A Volume of Studies in the History of Science in the Ancient and Medieval Near East in Honor of E.S. Kennedy*, New York (The New York Academy of Sciences).

King, David A. & Samsó, Julio 2001. "Astronomical Handbooks and Tables from the Islamic World (750–1900): an Interim Report" (with a contribution by Bernard J. Goldstein), *Suhayl* **2**, pp. 9–105.

Kunitzsch, Paul 1964. "Das Fixsternverzeichnis in der „Persischen Syntaxis" des Georgios Chrysokokkes", *Byzantinische Zeitschrift* **57**, pp. 382–411. Reprinted in Paul Kunitzsch, *The Arabs and the Stars*, Northampton (Variorum) 1989, II.

Mercier, Raymond P. 1989. "The Parameters of the Zīj of Ibn al-Aʿlam", *Archives Internationales d'Histoire des Sciences* **39**, pp. 22–50.

von Mžik, Hans 1926. *Das Kitāb Ṣūrat al-Arḍ des Abū Ǧaᶜfar Muḥammad ibn Mūsā al-Ḫuwārizmī*, Leipzig (Harrassowitz).

Nallino, Carlo Alfonso 1899–1907. *al-Battānī sive Albatenii. Opus astronomicum*, 3 vols., Milan (Hoepli). Reprinted: Frankfurt am Main (Institut für Geschichte der Arabisch-Islamischen Wissenschaften) 1997.

Neugebauer, Otto 1962. *The Astronomical Tables of al-Khwārizmī*, Copenhagen (Det Kongelige Danske Videnskabernes Selskab / Munksgaard).

—— 1975. *A History of Ancient Mathematical Astronomy*, 3 vols., Berlin (Springer).

Rosenfeld, Boris A. & İhsanoğlu, Ekmeleddin 2003. *Mathematicians, Astronomers, and Other Scholars of Islamic Civilization and Their Works (7th–19th c.)*, Istanbul (IRCICA).

Sachau, C. Edward 1879. *The Chronology of Ancient Nations*, London (Oriental Translation Fund of Great Britain and Ireland / Allen). Reprinted: Frankfurt am Main (Institut für Geschichte der Arabisch-Islamischen Wissenschaften) 1998.

Saliba, George 1970. "Easter Computation in Medieval Astronomical Handbooks", *Al-Abhath* **23**, pp. 179–212. Reprinted in Kennedy *et al.* 1983, pp. 677–709.

Sezgin, Fuat 1978. *Geschichte des arabischen Schrifttums. Band VI: Astronomie bis ca. 430 H.*, Leiden (Brill).

—— 1979. *Geschichte des arabischen Schrifttums. Band VII: Astrologie – Meteorologie und Verwandtes bis ca. 430 H.*, Leiden (Brill).

—— (ed.) 1986. *The Verified Astronomical Tables for the Caliph al-Maʾmūn. Al-Zīj al-Maʾmūnī al-mumtaḥan by Yaḥyā ibn Abī Manṣūr. Reproduced from MS árabe 927, Escorial Library* (with an introduction by E. S. Kennedy), Frankfurt am Main (Institut für Geschichte der Arabisch-Islamischen Wissenschaften).

—— 2000. *Geschichte des arabischen Schrifttums. Band X–XII: Mathematische Geographie und Kartographie im Islam und ihr Fortleben im Abendland*, Frankfurt am Main (Institut für Geschichte der Arabisch-Islamischen Wissenschaften). An English translation is in preparation.

Suter, Heinrich 1914. *Die astronomischen Tafeln des Muḥammed ibn Mūsā al-Khwārizmī in der Bearbeitung des Maslama ibn Aḥmed al-Madjrīṭī und der latein. Uebersetzung des Athelhard von Bath*, Copenhagen (Det Kongelige Danske Videnskabernes Selskab / Høst).

Tihon, Anne 1989. "Sur l'identité de l'astronome Alim", *Archives Internationales d'Histoire des Sciences* **39**, pp. 3–21.

Van Brummelen, Glen R. 1998. "Mathematical Methods in the Tables of Planetary Motion in Kūshyār ibn Labbān's *Jāmiʿ Zīj*", *Historia Mathematica* **25**, pp. 265–280.

Vernet, Juan 1954. "Un antiguo tratado sobre el calendario judio en las «Tabulae Probatae»", *Sefarad* **14**, pp. 59–78. Reprinted in Vernet 1979, pp. 213–232.

—— 1956. "Las «Tabulae probatae»", in: *Homenaje a Millás-Vallicrosa*, Barcelona (Consejo Superior de Investigaciones Científicas), vol. II, pp. 501–522. Reprinted in Vernet 1979, pp. 191–212.

—— 1979. *Estudios sobre historia de la ciencia medieval*, Barcelona (Bellaterra).

Viladrich, Mercè 1988. "The Planetary Latitude Tables in the *Mumtaḥan Zīj*", *Journal for the History of Astronomy* **19**, pp. 257–268.

—— 1996. "The *Mumtaḥan* tradition in al-Andalus. Analysis of data from the *Calendar of Cordova* related to the entrance of the sun in the zodiacal signs", in: *From Baghdad to Barcelona. Studies in the Islamic Exact Sciences in Honour of Prof. Juan Vernet* (Josep Casulleras & Julio Samsó, eds.), Barcelona (Instituto "Millás Vallicrosa" de Historia de la Ciencia Arabe), vol. 1, pp. 253–265.

Vollers, Karl 1906. *Katalog der islamischen, christlich-orientalischen, jüdischen und samaritanischen Handschriften der Universitäts-Bibliothek zu Leipzig* (mit einem Beitrag von Johannes Leipoldt), Leipzig (Harrassowitz). Reprinted: Osnabrück (Biblio Verlag) 1975.

## Appendix: Complete table of contents of MS Leipzig, Universitätsbibliothek, Vollers 821

The following table of contents covers the text as well as the tables in the Leipzig manuscript. For each chapter, the folio number and line at which it starts is indicated. For each smaller table, the position on the folio is indicated by one of the abbreviations t (top), m (middle), b (bottom), l (left) and r (right). For tables longer than a single page the range of folios is given. My own additions to the titles are given between square brackets. To make clear which entries are texts and which are tables, I have inserted *[Fī]* before titles of chapters that start with "*Jadwal*" and *[Jadwal]* before titles of tables that do not start with "*Jadwal*". Tables whose function is described in English do not have a heading. A longer horizontal line between two folios denotes a break in the manuscript. A shorter horizontal line preceded by a question mark indicates a situation where the previous page ends with a section or table and the following page starts with a new one, but the topics are so different that a break seems possible. Parts in between any two consecutive lines occur in the manuscript in the correct order. After each entry, a capital letter between parentheses refers to the paragraph of Section 2 in which the topic is discussed.

$1^{r}$: Second half of the table of numbers of days between epochs (continued from fol. $151^{v}$, A).

$1^{v}$–$2^{v}$: *Jadwal al-tawārīkh al-muṣaḥḥaḥ li-istikhrāj baʿḍihā min baʿḍ wa-hiya al-majmūʿa / Jadwal al-sinīn al-mabsūṭa li-l-tawārīkh* (continued on fol. $34^{r}$, A).

---

$3^{r}$:1 Continuation of a text on the determination of the seasonal hours and equal hours that have passed of the night (but not that on fol. $4^{v}$, C).

$3^{r}$:14 *Maʿrifat maṭraḥ shuʿāʿāt al-kawākib* (according to the method of Drīnūsh, presumably Dorotheus; K).

$3^{v}$:1 *Fī maʿrifat ruʾūs al-ahilla bi-l-jadwal* (A).

$3^{v}$:b *Jadwal maʿrifat ruʾūs al-ahilla al-ʿarabiyya* (A).

---

$4^{r}$: Last quarter of the oblique ascension table for latitude 37°30′ (continued from fols. $108^{v}$–$109^{v}$, C).

$4^{v}$:1 *Maʿrifat al-ẓill min qibal al-irtifāʿ bi-l-jadwal* (B, C).

4ᵛ:15 *Maᶜrifat mā maḍā min al-layl min sāᶜa bi-qiyās al-kawākib al-thābita* (continuation missing from the manuscript, C).

---

5ʳ:1 Last words of an unidentified section: *wa-kawākibihi wa-l-salām.*

5ʳ:1 *Maᶜrifat taḥāwīl sini 'l-ᶜālam* (K).

5ᵛ:8 *Maᶜrifat taḥwīl sini 'l-ᶜālam wa-l-mawālīd bi-l-jadwal* (K).

6ᵛ:11 *Maᶜrifat al-samt min qibal al-irtifāᶜ wa-l-ṭāliᶜ* (K).

7ʳ:6 *Fī maᶜrifat ṭāliᶜ al-ᶜālam* (K).

7ʳ:10 *Fī maᶜrifat taḥwīl sana al-ᶜālam* (K).

7ᵛ–8ʳ: *Jadwal taḥwīl sini 'l-ᶜālam wa-l-mawālīd li-tisᶜīn sana ᶜalā mā raṣadahu al-sharīf Ibn al-Aᶜlam raḥimahu Allāh* (K).

8ᵛ:1 *Fī maᶜrifat al-dā'ir min al-falak* (C).

?---

9ʳ:1 *Maᶜrifat istikhrāj ṣawm al-naṣārā wa-mā yataᶜallaqu bi-hi* (A).

10ʳ: Table for Christian feasts and fasts to be entered with the remainders of the divisions of the Byzantine year minus 1204 by 28 and 19 (A).

10ᵛ–12ᵛ: Table for Christian feasts and fasts to be entered with a number from 1 to 70 found from the previous table (continued on fol. 29ʳ, A).

---

13ʳ: Continuation from fol. 30ᵛ of a text concerning the calculation of lunar eclipses (H).

15ᵛ:4–20ᵛ *Fī maᶜrifat kusūf al-shams* (text: *al-qamar*) *wa-taᶜdīl azmānihi wa-ṭāliᶜ kull zaman minhā* (continuation missing from the manuscript, H).

---

21ʳ: Continuation from fol. 112ᵛ of a text on the correction of the ascendant (K).

21ᵛ:8 *Maᶜrifat ṭāliᶜ al-qubba* (K).

21ᵛ:1 *[Jadwal] awj al-mirrīkh* (D).

22ʳ: *[Jadwal] ḥarakāt al-kawākib fi 'l-awjāt ᶜalā ra'y Ibn al-Aᶜlam wa-huwa li-kull 70 sana daraja* (D).

22ᵛ: *[Jadwal] ḥarakāt awjāt al-kawākib fi 'l-sinīn wa-l-shuhūr wa-l-ayyām bi-l-miḥna wa-l-raṣad wa-huwa fī kull 66 sana daraja wāḥid ᶜalā ra'y al-Battānī wa-l-Ma'mūn wa-hādhihi ḥarakāt al-kawākib al-thābita ᶜalā mā wujidu* (D).

---

23ʳ:1 *Basmala*, followed by a text on a simplified method for calculating oblique ascensions, with two small tables (C).

23ᵛ: Table of numbers of days of progressions, retrogressions and stations of the five planets (F).

---

24ʳ:1 *[Fī] jadwal taḥāwīl al-sinīn li-l-Ma'mūn li-l-dahr* (K).

24ʳ:12 *Maᶜrifat taḥwīl sanat al-ᶜālam* (K).

24ᵛ–25ʳ: *Jadwal taḥāwīl al-sinīn li-l-dahr li-l-Ma'mūn* (K).

25ᵛ–26ᵛ: *Jadwal maṭāliᶜ al-burūj fi 'l-falak al-mustaqīm* (C).

27ʳ: *Jadwal fuḍūl al-maṭāliᶜ li-ᶜarḍ kullihā* (C).

27ᵛ:1 *Maᶜrifat al-ijtimāᶜ wa-l-istiqbāl li-l-shams wa-l-qamar* (H).

28ᵛ: Universal table of oblique ascensions, only latitudes 1 to 33° (C).

---

$29^r$: Last page of the table of Christian feasts and fasts (continued from fols. $10^v$–$12^v$, A).

$29^v$:1–$30^v$ *Maᶜrifat kusūf al-qamar wa-taᶜdīl azmānihi wa-taᶜdīl kull zaman minhu* (continued on fol. $13^r$, H).

$31^r$: *[Jadwal] al-ẓill* (B).

$31^v$: *[Jadwal] mawāḍiᶜ al-kawākib al-thābita min falak al-burūj wa-mamarrihā bi-dāʾirat niṣf al-nahār wa-aḥwālihā li-sanat 380 li-taʾrīkh Yazdigird* (continued on fol. $153^r$, I).

$32^r$: *Jadwal ikhtilāf ᶜarḍ al-qamar li-l-ruʾya* (H).

$32^v$–$33^r$: Lunar eclipse tables for nearest and furthest distance (H).

$33^v$: *Jadwal taᶜdīl al-kusūfayn* (H).

$34^r$: *Jadwal tamām al-sinīn al-mabsūṭa li-l-tawārīkh* (continued from fol. $2^v$, A).

$34^v$: *Fī dhikr aḥwāl al-burūj wa-l-kawākib* (continued on fol. $45^r$, L).

$35^r$:1 Continuation from fol. $50^v$ of a standard text on planetary stations (F).

$35^r$:12 *Maᶜrifat ᶜarḍ al-qamar* (E).

$35^v$:2 *Fī maᶜrifat taqwīm al-jawzahar* (E).

$35^v$:5 *Maᶜrifat ẓuhūr al-kawākib wa-ikhtifāʾihā* (continuation missing from the manuscript, G).

$36^r$:1 Continuation from fol. $110^v$ of a text on lunar crescent visibility (G).

$36^r$:10 *Maᶜrifat kusūf al-qamar bi-l-manāzil* (H).

$36^v$:11 *Maᶜrifat alwān al-kusūf* (H).

$37^r$: *Jadwal istikhrāj al-manāzil li-kusūf al-qamar* (H).

$37^v$–$40^r$: *[Jadwal] al-manzila al-ūla li-kusūf al-qamar*, up to *al-manzila al-sādisa* (H).

$40^v$:1 *Fī maᶜrifat kusūf al-shams* (H).

$41^v$–$42^r$: *Jadwal al-samt li-ᶜilm kusūf al-shams* (H).

$42^v$: *Jadwal ᶜarḍ al-shams* (H).

$43^r$: *Jadwal yuᶜrafu minhu kusūf al-shams ṣaḥīḥ mujarrab idhā uḥsina al-ᶜamal bihi* (H).

$43^v$:1 *Maᶜrifat kusūf al-qamar bi-l-jadwal muqarrab* (H).

$43^v$:13 *Maᶜrifat sini 'l-Iskandar [min sini 'l-hijra]* (A).

$44^r$: Table: *Fī maᶜrifat al-kusūf bi-hādha 'l-jadwal* (H).

$44^v$:1 *Maᶜrifat taḥāwīl al-sinīn bi-ziyādāt al-adwār* (K).

$44^v$:10 *Maᶜrifat nuzūl al-shams ruʾūs* (text: *ruʾūs al-shams*) *al-burūj wa-masīrihā fī kull burj* (K).

$44^v$:15 *Maᶜrifat maṭlaᶜ al-shiᶜrā* (continued on fol. $112^r$, I).

$45^r$: Continuation from fol. $34^v$ of an introduction to astronomy (L).

$47^v$:5 *Fī maᶜrifat istikhrāj awsāṭ al-kawākib* (D).

$49^r$:4 *Fī maᶜrifat taqwīm al-shams* (D).

$49^r$:12 *Fī maᶜrifat taqwīm al-qamar* (D).

$49^v$:15 *Bāb fī maᶜrifat taqwīm zuḥal [wa-mushtarī wa-mirrīkh]* (D).

50$^{v}$:9 *wa-ammā taqwīm al-zuhara wa-l-ʿuṭārid ...* (D).
50$^{v}$:12 *Maʿrifat rujūʿ al-kawākib wa-istiqāmatihā* (continued on fol. 35$^{r}$, F).

---

51$^{r}$: Continuation of a text on the calculation of planetary visibility (G).
51$^{v}$–52$^{r}$: *Jadwal ruʾyat al-kawākib al-khamsa wa-ikhtifāʾihā* (G).
52$^{v}$:1 *Fī maʿrifat al-jayb* (B).
53$^{v}$:3 *Fī maʿrifat ʿamal al-juyūb wa-l-qisī bi-l-jadwal* (B).
53$^{v}$:18 *Fī maʿrifat taqwīs al-jayb* (B).
54$^{r}$:8 *Maʿrifat ʿamal al-jayb al-maʿkūs* (B).
54$^{r}$:16 *Fī maʿrifat taqwīs al-jayb maʿkūsan* (B).
54$^{v}$:5 *Maʿrifat al-mayl al-aʿẓam wa-ʿilm ḥisābihi* (C).
55$^{r}$:11 *Maʿrifat ḥisāb al-mayl li-ajzāʾ falak al-burūj* (C).
55$^{v}$:4 *Maʿrifat ḥisāb ʿarḍ al-qamar* (E).
55$^{v}$:13 *Maʿrifat mayl al-shams bi-l-jadwal wa-huwa buʿduhā ʿan dāʾirat muʿaddil al-nahār* (C).
56$^{r}$:13–56$^{v}$ *Fī maʿrifat ʿarḍ al-qamar bi-l-jadwal* (E).

?———

57$^{r}$: *Fī maʿrifat kusūf al-qamar bi-l-jadwal li-Yaḥyā ibn Abī Manṣūr ṣaḥīḥ*, with three small tables (H).
57$^{v}$–60$^{r}$: *Jadwal maqāmāt al-kawākib* (F).
60$^{v}$:1 *Fī maʿrifat ʿurūḍ al-kawākib ʿalā madhhab Baṭlamiyūs muqarrab* (E).
60$^{v}$:6 *Fī maʿrifat jawzahar al-kawākib* (E).
60$^{v}$:11 *Maʿrifat mawāḍiʿ jawzaharāt al-kawākib* (E).
60$^{v}$:15 *Awjāt al-kawākib* (D).
61$^{r}$:6 List of mean motion parameters for Mars by Ibn al-Aʿlam (D).
61$^{v}$–63$^{v}$: *Jadwal ʿarḍ zuḥal / mushtarī / mirrīkh / zuhara / ʿuṭārid* (E).
64$^{r}$:1 *Fī maʿrifat al-mayl wa-ʿarḍ al-qamar* (C, E).
64$^{v}$–65$^{v}$: *[Jadwal] al-mayl [wa-]ʿarḍ al-qamar* (C, E).

?———

66$^{r}$–66$^{v}$: *[Jadwal] wasaṭ al-shams fi 'l-sinīn al-majmūʿa wa-l-mabsūṭa wa-l-shuhūr / wasaṭ al-shams fi 'l-ayyām wa-l-sāʿāt wa-l-kusūr* (D).
67$^{r}$–68$^{r}$: *Jadwal taʿdīl al-shams al-awj 82 39* (D).
68$^{v}$–69$^{r}$: *Jadwal masīr wasaṭ al-qamar* (double elongation, D).
69$^{v}$–70$^{r}$: *Jadwal masīr tadwīr al-qamar* (mean longitude, D).
70$^{v}$–71$^{r}$: *Jadwal masīr khāṣṣat al-qamar* (mean anomaly, D).
71$^{v}$–72$^{r}$: *Jadwal wasaṭ al-jawzahar* (D).
72$^{v}$–75$^{r}$: *Jadāwil taʿdīl al-qamar* (D, E).
75$^{v}$–76$^{r}$: *Jadwal masīr wasaṭ zuḥal* (D).
76$^{v}$–77$^{r}$: *Jadwal masīr khāṣṣat zuḥal* (D).
77$^{v}$–80$^{r}$: *Jadāwil taʿādīl zuḥal al-awj 8 4 30* (D, F).
80$^{v}$–81$^{r}$: *Jadwal masīr wasaṭ al-mushtarī* (D).
81$^{v}$–82$^{r}$: *Jadwal masīr khāṣṣat al-mushtarī* (D).
82$^{v}$–85$^{r}$: *Jadāwil taʿādīl al-mushtarī awjuhu 172 32* (D, F).
85$^{v}$–86$^{r}$: *[Jadwal] wasaṭ al-mirrīkh* (D).

86ᵛ–87ʳ: *[Jadwal] wasaṭ masīr al-mirrīkh* (corrected in the same hand to: *Jadwal masīr khāṣṣat al-mirrīkh)* (D).
87ᵛ–90ʳ: *Jadāwil taʿādil al-mirrīkh wa-hādha 'l-mirrīkh huwa mirrīkh Ibn al-Aʿlam wa-huwa ajwaduhā 128 50* (D, F).
90ᵛ–91ʳ: *Jadwal masīr khāṣṣat al-zuhara* (D).
91ᵛ–94ʳ: *Jadāwil taʿādil al-zuhara al-awj ...* (D, F).
94ᵛ–95ʳ: *Jadwal masīr khāṣṣat al-ʿuṭārid* (D).
95ᵛ–98ʳ: *Jadāwil taʿādil ʿuṭārid al-awj 1 0 2 0 (?) / 244 30* (D, F).
98ᵛ:1 *Fī maʿrifat al-jayb bi-l-jadwal* (B).
99ᵛ: *Jadwal al-jayb wa-l-qaws* (B).

?———

100ʳ:1 *Maʿrifat niṣf qaws al-nahār wa-ajzāʾ sāʿāt al-nahār wa-l-layl* (continued from fol. 130ᵛ with one folio missing, C).
100ʳ:11 *Maʿrifat sāʿāt al-nahār al-mustawiya* (C).
100ʳ:16 *Maʿrifat ajzāʾ sāʿāt al-nahār* (C).
100ᵛ:3 *Maʿrifat taḥwīl al-sāʿāt al-mustawiya ila 'l-muʿwajja* (C).
100ᵛ:10 *Fī maʿrifat jayb al-nahār wa-l-layl* (C).
100ᵛ:18 *Fī maʿrifat kam maḍā min al-nahār min sāʿat bi-qiyās al-shams wa-l-ʿamal bihi* (C).
101ʳ:10 *Fī maʿrifat irtifāʿ niṣf al-nahār* (C).
101ʳ:15 *Maʿrifat taḥwīl daraj al-sawā ilā daraj al-maṭāliʿ* (and vice versa, C).
101ᵛ:5 *Maʿrifat saʿat al-mashāriq, wa-nabdaʾu bi-l-shams* (C).
101ᵛ:15 *Maʿrifat saʿat al-mashāriq* (C).
102ʳ:5 *Fī maʿrifat iqāmat al-ṭāliʿ wa-l-buyūt al-ithnā ʿashara* (C).
103ʳ–104ʳ: *Jadwal maṭāliʿ al-falak al-mustaqīm* (C).
104ᵛ–106ʳ: *Jadwal maṭāliʿ al-burūj li-ʿarḍ Baghdād wa-ḥayth al-ʿarḍ 33 0* (C).
106ᵛ–108ʳ: *Jadwal maṭāliʿ al-burūj li-ʿarḍ al-Raqqa wa-ḥayth al-ʿarḍ 36* (C).
108ᵛ–109ᵛ: *Jadwal maṭāliʿ al-burūj li-ʿarḍ Mayyāfāriqīn wa-Āmid wa-Arzan wa-Badlīs wa-Khilāṭ wa-ḥayth al-ʿarḍ 37 30* (continued on fol. 4ʳ, C).

---

110ʳ:1 *Maʿrifat masīr al-nayyirayn li-sāʿa mustawiya* (H).
110ʳ:7 *Maʿrifat ruʾyat al-hilāl* (G).
110ᵛ:15 *Wajh ākhar fi 'l-ruʾya* (continued on fol. 36ʳ, G).

---

111ʳ: *Jadwal ikhtilāf al-manẓar* (H).
111ᵛ: *Jadwal taʿdīl al-kusūfāt al-shamsiyya* (H).

---

112ʳ:1 Continuation from fol. 44ᵛ of a text on the heliacal rising of Sirius (I).
112ʳ:8 *Maʿrifat sinī Yazdigird al-fārisī [min sini 'l-Iskandar]* (A).
112ʳ:18 *Maʿrifat ruʾyat al-ahilla* (G).
112ᵛ:7 *Bāb al-namūdhārāt* (continued on fol. 21ʳ, K).

---

113ʳ:1 Continuation of a text on the determination of the weekday of the beginning of a month (in an undetermined calendar, A).
113ʳ:3 *Fī maʿrifat al-aṣl al-rūmī* (A).

113$^{v}$:5 *Maʿrifat ʿalāmat sanat al-rūm wa-madākhil shuhūrihim fi 'l-ayyām al-sabʿa* (A).

114$^{r}$:3 *Fī istikhrāj al-tawārīkh baʿḍihā min baʿḍ* (deals in fact with the beginnings of the Arabic months, A).

114$^{r}$:17 *Maʿrifat istikrāj al-ʿarabī min al-fārisī* (A).

114$^{v}$:6 *Fī maʿrifat taʾrīkh Dhi 'l-qarnayn min taʾrīkh al-furs* (A).

114$^{v}$:15 *Maʿrifat taʾrīkh al-furs min taʾrīkh Dhi 'l-qarnayn* (A).

115$^{r}$:6 *Fī maʿrifat istikhrāj al-yūnānī min al-ʿarabī* (A).

115$^{v}$:1 *Fī maʿrifat istikhrāj sini 'l-furs min al-ʿarabī* (A).

115$^{v}$:10 *Fī maʿrifat istikhrāj al-qibṭī min al-ʿarabī* (A).

116$^{r}$:3 *Fī maʿrifat istikhrāj al-ʿarabī min al-yūnānī* (A).

116$^{r}$:15 *Fī maʿrifat ruʾūs shuhūr al-furs* (A).

116$^{v}$:8 *Fī maʿrifat sini 'l-rūm wa-ruʾūs shūhūrihim* (A).

117$^{r}$:1 *Fī maʿrifat raʾs sanat al-qibṭ wa-shuhūrihim* (A).

117$^{r}$:9 *Fī maʿrifat ruʾūs al-shuhūr al-ʿarabiyya ʿalā raʾy al-Battānī* (A).

121$^{v}$:10 *Fī maʿrifat al-sana al-kabīsa* (in the Byzantine calendar, A).

121$^{v}$:14 *Fī maʿrifat dukhūl shuhūr al-ahilla bi-l-jadwal* (A).

122$^{r}$:4 *Fī maʿrifat shuhūr al-yūnāniyyīn wa-l-sana al-kabīsa bi-l-jadwal* (A).

122$^{r}$:11 *Maʿrifat shuhūr al-furs wa-raʾs sinīhim bi-l-jadwal* (A).

122$^{v}$: *al-Jadwal al-mujarrad li-l-Hijra* (A).

123$^{r}$: *al-Jadwal al-mujarrad li-Dhi 'l-qarnayn al-Iskandar* (A).

123$^{v}$:t *al-Jadwal al-mujarrad li-istikhrāj raʾs sanat al-furs* (A).

123$^{v}$:1 *Fī maʿrifat taʿdīl daqāʾiq al-ḥiṣṣa* (L).

123$^{v}$:8 *Fī maʿrifat al-kawkab rājiʿ huwa aw mustaqīm* (F).

124$^{r}$:16 *Fī maʿrifat ḥiṣaṣ al-kawkab li-yawm* (D).

124$^{r}$:18 *Maʿrifat masīr khāṣṣat al-kawkab li-yawm* (D).

124$^{v}$:5 *Wajh muqarrab malīḥ fī taʿdīl al-ayyām* (D).

124$^{v}$:10 *wa-hādha 'lladhī yaʾtī min al-Battānī* ... (on the heavenly sphere, divisions of the ecliptic, multiplication and division; L).

128$^{r}$:9 *Fī maʿrifat sāʿāt al-taqwīm fī kull balad dhi 'l-sāʿāt al-muʿtadila al-wusṭā* (L).

129$^{v}$:12 *Fī maʿrifat ʿarḍ al-qamar wa-jihatihi* (continued on fol. 131$^{r}$, E).

---

130$^{r}$:1 Continuation of a text on the hours of day and night (only the title is missing, which is given as *Maʿrifat al-sāʿāt al-muʿwajja wa-tusammā ajzāʾ al-sāʿāt* in **F**, p. 162, line 3; C).

130$^{r}$:9 *Wajh ākhar fī ʿilm azmān al-sāʿāt* (C).

130$^{r}$:13 *Maʿrifat sāʿāt al-layl* (C).

130$^{v}$:5 *Fī maʿrifat maṭāliʿ al-burūj fi 'l-buldān* (one folio missing, then continued on fol. 101$^{r}$, C).

---

131$^{r}$:1 Continuation from fol. 129$^{v}$ of a text on lunar latitude (E).

131$^{r}$:12 *Fī ṭulūʿ al-khamsa al-mutaḥayyira wa-ghurūbihā* (G).

133$^{r}$:8 *Fī maʿrifat miqdār mayl falak al-burūj ʿan falak muʿaddil al-nahār* (C).

133ᵛ:14 *Maᶜrifat maṭāliᶜ al-burūj fi 'l-falak al-mustaqīm* (C).
134ᵛ:4 *Fī maᶜrifat irtifāᶜ al-quṭb min qibal ziyādat al-nahār al-aṭwal* (C).
135ʳ:1 *Fī maᶜrifat ziyāda al-nahār al-aṭwal wa-mā dūnahu min ziyādāt al-nahār* (C).
135ᵛ:3 *Fī maᶜrifat al-irtifāᶜ wa-l-ẓill aḥadihimā min qibal al-ākhar bi-l-ḥisāb wa-l-jadwal* (B, C).
135ᵛ:18 *Fī maᶜrifat samt al-irtifāᶜ wa-l-ẓill min dāʾirat al-ufq fī kull balad wa-fī kull waqt* (C).
136ᵛ:6 *Fī maᶜrifat khaṭṭ niṣf al-nahār fī kull balad* (C).
137ᵛ:17 *Fī maᶜrifat maṭāliᶜ al-burūj fī kull balad* (C).
139ʳ:5 *Taṣnīf al-ᶜamal bi-l-maṭāliᶜ* (C).
141ʳ:3 *Fī maᶜrifat ᶜurūḍ al-buldān* (C).
141ʳ:13 *Fī maᶜrifat irtifāᶜ al-shams fī waqt aṣnāf al-nahār min kull yawm* (C).
141ᵛ:10 *Fī taḥqīq aqdār al-ittiṣālāt* (K).
144ʳ:9 *Fī ṣifat al-rukhāma li-maᶜrifat al-sāᶜāt al-zamāniyya al-māḍiya min al-nahār bi-l-shams fī kull balad turīd* (L).
147ʳ:4 *Fī maᶜrifat ᶜamal al-tawārīkh baᶜḍihā min baᶜḍ min qibal al-jadwal* (A).
147ᵛ:4 *Tafṣīl al-tawārīkh min qibal al-jadwal* (A).
147ᵛ:17 *Maᶜrifat taʾrīkh al-rūm wa-Diqliṭiyānūs min qibal al-jadwal* (A).
149ʳ:1 *Fī maᶜrifat taʾrīkh Diqliṭiyānūs bi-l-jadwal* (A).
149ʳ:7 *Maᶜrifat taʾrīkh al-hijra min qibal al-jadwal* (A).
149ᵛ:17 *Maᶜrifat al-tawārīkh baᶜḍihā min baᶜḍ bi-l-jadwal* (A).
150ᵛ:1 *Maᶜrifat taʾrīkh al-ṭūfān min taʾrīkh Dhi 'l-qarnāyn* (A).
151ᵛ: Table without title: number of days between epochs (continued on fol. 1ʳ, A).

---

152ʳ: Last page of the geographical table (continued from fol. 155ᵛ, J).
152ᵛ:1 *Fī maᶜrifat mawḍiᶜ al-raʾs wa-l-dhanab* (E).
152ᵛ:6 *Fī dhikr kusūf al-qamar bi-dāʾirat al-kusūf* (continuation missing from the manuscript, H).

---

153ʳ: Second page of the star table for 380 Yazdigird (continued from fol. 31ᵛ, I).
153ᵛ–155ᵛ: *Awsāṭ al-buldān wa-ismāʾihā / Aṭwāl al-buldān wa-ᶜurūḍihā ᶜalā mā fī Kitāb Ṣūrat al-arḍ* (continued on fol. 152ʳ, J).

---

156ʳ-157ᵛ: Text and tables on chronology in a different hand.

# VIII

## Re-editing the tables in the *Ṣābi' Zīj* by al-Battānī

Benno VAN DALEN, München, and Fritz S. PEDERSEN, Copenhagen

**Summary**

Between 1899 and 1909, Carlo Alfonso NALLINO published his monumental study of the *Ṣābi' Zīj*, an important Arabic astronomical handbook with tables compiled around AD 900 by Abū 'Abd Allāh Muḥammad ibn Jābir ibn Sinān AL-BATTĀNĪ. NALLINO'S work has remained one of the most important publications on Islamic astronomy up to this time. However, his transliterations of AL-BATTĀNĪ'S tables turn out to be frequently unfaithful to the only source that was available to him, the unique complete manuscript of the *Ṣābi' Zīj* in the Escorial library. In this article we explore the reasons for these deviations and present a critical edition of some of AL-BATTĀNĪ'S tables based on a range of Arabic, Castilian and Latin sources.

### 1. Introduction

Abū 'Abd Allāh Muḥammad ibn Jābir ibn Sinān AL-BATTĀNĪ (d. 929) was famous for the accuracy of the astronomical observations that he carried out in Raqqa in present-day Syria between 877 and 918.[1] Besides an astrological history, his most important extant work is an astronomical handbook with tables,[2] often referred to as the *Ṣābi' Zīj* after the Sabian religion to which his ancestors adhered. AL-BATTĀNĪ'S *zīj* is the earliest surviving one that is almost purely based on Ptolemaic astronomy, since the *Mumtaḥan Zīj* by YAḤYĀ IBN ABĪ MANṢŪR (d. c. 830) and the *Damascene Zīj* by ḤABASH AL-ḤĀSIB (c. 870) still show much clearer traces of Indian material. The *Ṣābi' Zīj* was influential in the Islamic East, where its parameters were adopted by various later astronomers such as KŪSHYĀR IBN LABBĀN (fl. 1025) and Abū Ja'far Muḥammad AL-ṬABARĪ (fl. 1100), as well as in the West, where it was translated into Castilian and Latin and many of its tables were widely distributed as part of the *Toledan Tables*.

A full study of the only surviving complete Arabic manuscript of AL-BATTĀNĪ'S *zīj*, kept in the Escorial Library near Madrid, was published by the Italian Arabist Carlo Alfonso NALLINO (1872–1938) as (NALLINO 1899–1907). This monumental work includes a Latin translation of the text of the *zīj* in vol. I, a transcription of the tables in

---

1 More information on AL-BATTĀNĪ'S life and works can be found in (HARTNER 1970, VAN DALEN 2007).

2 The standard work on Islamic astronomical handbooks with tables, in Arabic and Persian called *zījes*, is still (KENNEDY 1956). A new survey of Islamic astronomical tables is currently being prepared by the first author. This will also include information on almost 100 works that have become known only during the last fifty years.

vol. II, and an edition of the Arabic text in vol. III. Half of vols. I and II is taken up by extensive commentaries that explain the numerous technical issues connected with the text and the tables of the *zīj* and relate these to various other sources that NALLINO had available.

Having been given such excellent scholarly treatment at a rather early time, AL-BATTĀNĪ'S *zīj*, and in particular the Escorial manuscript, may in more recent times not have received the attention they deserve. Parts of the *zīj* have been used for topical studies (see, for instance, SWERDLOW 1973, KUNITZSCH 1974, RAGEP 1996), but no new critical assessment has been made of the text and the tables as a whole. Since NALLINO had only the Escorial manuscript at his disposal, and the criteria for establishing the tabular values in his transcription are partly obscure and appear to be inconsistent, a fresh look at the tables in the *Ṣābi' Zīj* is in fact highly desirable. In some cases NALLINO'S transcription turns out to differ so much from the manuscript tradition, and therewith in all probability from AL-BATTĀNĪ'S original table, that it cannot be reliably used to either establish direct interrelations with other sources or carry out a mathematical analysis of the underlying parameters and methods of computation.

The purpose of this article is to make an inventory of the problems attached to NALLINO'S transcription of AL-BATTĀNĪ'S tables and to pave the way for a possible new, critical edition that takes into account all relevant additional sources, in particular some later Arabic and Persian *zījes* that adopted tables from AL-BATTĀNĪ, the Castilian translation of the *Ṣābi' Zīj* prepared around 1260 for ALFONSO X (1221–1284), and the Latin tradition of the *Toledan Tables*. For this purpose we shall present an edition of AL-BATTĀNĪ'S table of the solar declination and full apparatuses for a number of other tables, each with its own peculiar features. For each of the tables we shall discuss the types of errors that they contain, specify our own criteria for establishing the tabular values, and comment on the policy followed by NALLINO for his transcription.

Since medieval Latin manuscripts turn out to play an important role in our re-edition of some of AL-BATTĀNĪ'S tables, it is a great pleasure for us to be able to present this research in a *Festschrift* for Prof. Menso FOLKERTS, who has dedicated most of his career to such manuscripts. We would like to acknowledge the kind assistance of the curators of the Escorial library and the other libraries listed below in supplying us with microfilms of manuscripts from their collections.

### *1.1 Sources*

*Edition*

N = (NALLINO 1899–1907, vol. II), NALLINO'S transcription of the tables from AL-BATTĀNĪ'S *zīj*, which is the only available publication of the tables. This is not a critical edition in the modern sense of the word. It rests on one Arabic manuscript, the unique complete copy of the *zīj* from the Escorial Library (E below), which NALLINO, often tacitly, corrected in various ways, in particular by means of recomputation and by adopting values from PTOLEMY'S *Almagest* and *Handy Tables* where these can be seen to have been AL-BATTĀNĪ'S ultimate source (cf. vol. I, p. vi; all such references without explicit indication of a publication will be to NALLINO 1899–1907). Only for tables that cannot be recomputed, such as the geographical table and the star catalogue, does NALLINO promise to cite the Arabic manuscript consistently (*ibid.*). Also for the Ptolemaic tables for planetary visibility he gives complete variant readings from a number of sources.

*Manuscripts*

For our critical edition and apparatuses of sample tables we have used the following manuscripts or parallel sources:

E = Escorial, ms. árabe 908. Described in (NALLINO 1899–1907, vol. I, p. lxi). In Western Arabic script from the late 11th or early 12th century. This is the only complete Arabic manuscript of AL-BATTĀNĪ's work, and the only manuscript used by NALLINO throughout.

A = Paris, Arsenal, ms. 8322. Described in (NALLINO 1899–1907, vol. I, pp. lvii–lx; vol. II, pp. vii–viii). In Castilian, writing of the late 13th century, prepared for king ALFONSO X, incomplete. Translation of a version of the text and tables that is quite similar to E. This manuscript did not become available to NALLINO until his transcription of the tables had been printed (vol. II, p. vii). He excerpted it to some extent for his commentary on the tables, though not for the tables to be discussed here. The text of the Castilian translation of the *Ṣābi' Zīj* was published in (BOSSONG 1978).

O = Oxford, Bodleian Library, ms. Savile 22. In Latin, mid 13th century.

C = Cambridge, University Library, ms. Kk.I.1 (1935). In Latin, mid 13th century.
O and C represent a version ("class {k}") of the *Toledan Tables* that preserves Battanian readings much more faithfully than the mainstream versions of these tables (cf. PEDERSEN 2002, pp. 19 and 858; TOOMER 1968, pp. 12–13). O and C are closely related to each other and seem largely independent of E and A. In evaluating the evidence for readings in AL-BATTĀNĪ's tables we shall normally count O and C as a single source.

T = *Toledan Tables*. Unless otherwise specified, the readings given by us are the mainstream ones, established from manuscripts representing the "archaic" and "early vulgar" classes ({a0} and {a1}, see PEDERSEN 2002, p. 868). This excludes readings peculiar to mss. O and C above. As a result, some of the values here cited for T are not the same as those printed in the edition in (PEDERSEN 2002).

We have not included tables derived from PTOLEMY'S *Almagest* or *Handy Tables*, since NALLINO'S transcriptions of these usually do not present many problems. For the lunar equation of anomaly we have adduced Arabic and Persian *zījes* that adopted tables from AL-BATTĀNĪ, in particular the *Mufrad Zīj* by Abū Ja'far Muḥammad AL-ṬABARĪ and the *Īlkhānī Zīj* by Naṣīr al-Dīn AL-ṬŪSĪ. Details of the manuscripts used for these works can be found in Section 3.5. Note that the 12th-century Latin translation of AL-BATTĀNĪ'S *zīj* by PLATO OF TIVOLI (12th c.), which was printed twice in the 16th and 17th centuries, included only the text.

*Constructed values*

We have adduced three types of "constructed values" for judging the plausibility of manuscript readings. These comprise:

r = Modern recomputation for the standard Ptolemaic formula and the parameter(s) indicated. Although the exact historical method of computing any one table is not generally known in detail, a modern recomputation can in most cases be assumed to give a fair approximation to AL-BATTĀNĪ'S original tabular values. Various of the tables discussed here include parts with systematic larger deviations from the recomputation, which may, for instance, be due to the use of interpolation; this is indicated for each individual table.

d = A value that makes the surrounding pattern of first-order differences highly regular, i. e., either constant or alternating between two adjacent values in a regular fashion (for example, 2, 2, 3, 2, 2, 3, 2, 2, 3). This type of constructed value will be applied only to tables of linear functions (such as the sub-tables of mean motion tables except those for extended years and months), and to tables that were clearly computed by means of linear interpolation between certain nodes.

s = A value that can be derived in a trivial way (e.g., identity, linear transformation) from a value elsewhere in the table and can be expected to be the same as the present value by virtue of some symmetry or antisymmetry, as described in each case. This type of constructed value will only be applied to cases where the symmetry was obviously used by the compiler of the table.

### *1.2 Apparatuses*

For each of the tables discussed in Sections 2 and 3 we present an apparatus of variants that is intended to illustrate the manuscript evidence for each reading, to justify our choice of values, and to assess the rationality of NALLINO'S choices when editing from E alone. In establishing the tabular values, we shall give preference to E and A, since these are the only sources actually ascribed to AL-BATTĀNĪ, and primarily to E, since A seems to share some readings with the mainstream *Toledan Tables*. Generally, an apparatus does not list all variants in all manuscripts; however, the readings from E and A and the readings accepted by us are listed wherever they differ from N, so that E and A can be reconstructed by the reader who compares NALLINO'S table to the apparatus.

*Scribal errors and easy corrections*

Nearly every source for every single table discussed in this article is distorted by a large number of scribal errors. Many of these are caused by similarities between the Arabic letters used in the *abjad* notation with which most numbers in Arabic mathematical and astronomical tables are written, or between the Roman or Hindu-Arabic numerals as used in medieval Latin manuscripts. Which mistakes are particularly likely in a given source depends to a large extent on the type of writing of the original. In particular, due to a number of differences in the *abjad* notation between Eastern and Western Arabic writing, some mistakes are likely to occur only with an Eastern original and others only with a Western original. Most of the scribal errors in the Castilian and Latin manuscripts that we have used can be seen to derive from their Arabic originals (cf. NALLINO'S discussion of scribal errors in vol. II, pp. v–vi).

Some of the most frequent obvious scribal errors are the following (here "t" denotes any number of tens and "u" any number of units unequal to zero; "etc." indicates that also compounds starting with the same numerals may be confused):[3]

*Arabic*: 0 ↔ 5, t2 ↔ t4, t2 ↔ t7, t3 ↔ t4, t3 ↔ t8, t4 ↔ t5, t4 ↔ t6, t4 ↔ t7, t6 ↔ t7, 7 ↔ 50, 9 ↔ 20, 1u ↔ 3u, 1u ↔ 5u, 17 ↔ 40, 3u ↔ 5u, 4u ↔ 5u, 40 ↔ 47, 50 ↔ 55, various other confusions of the form $t_1$u ↔ $t_2$u (also with $t_1 = 0$), 80 ↔ 100 etc. Only for Eastern Arabic sources: 60 ↔ 300 etc.
Only for Western Arabic sources: 60 ↔ 90 etc.

*Latin*: Roman notation: I (1) ↔ L (50), X (10) ↔ Ẋ (40); N (9) ↔ II (2); miscounting of strokes, e.g. IIII (4) ↔ III (3). Hindu-Arabic notation: no typical errors.

---

3 Unlike the practice in most other Arabic and Persian *zījes*, in the Escorial manuscript of the *Ṣābi' Zīj* angles that occur as arguments of planetary equations and other functions are always written as total numbers of degrees from 0 to 360 rather than as zodiacal signs plus a number of degrees from 0 to 29. This leads to different patterns of scribal errors from what is found in many other sources.

Our sources also frequently contain other types of scribal mistakes, such as slides of large parts of a column, a number of consecutive tabular values whose degrees have all been set to a particular number, mistaken digits which can only be explained by assuming that they were copied from an erroneous location, etc. When establishing our accepted readings we shall consider all plausible corrections of obvious scribal errors (to which we shall refer as "easy corrections") for tabular values with apparent mistakes. For instance, in a table with deviations from the recomputation of generally less than 1″, we shall consider the possibility that the reading 7″ is a scribal mistake for 50″ if the other witnesses attest to 49, 50 or 51″. In particular in the Escorial manuscript, readings are not always easy to distinguish on the microfilm that we have used (nor possibly in the manuscript itself). In cases where the reading is ambiguous (e. g., if the dots on an Arabic letter are not strictly speaking correct for any possible number), we shall often give the manuscript the benefit of the doubt and shall not indicate a variant reading in the apparatus.

*Apparatus entries*
An apparatus entry has the following form:

**69** 56;0,54 [*6° O; 5′ E]: AOC; 56″ E, 53″ TNrd.

Here 69 (in boldface) is the argument of the tabular value concerned, and 56;0,54 is our accepted value. This notation is shorthand for 56°0′54″, where the degree sign stands for units, also in cases where the variable is not an arc, and ′ and ″ for minutes and seconds. In this case the accepted value is found in the manuscripts AOC,[4] whereas we have rejected two other readings for the seconds, namely 56″ in E and 53″ in T. The latter reading is also given by NALLINO following the recomputation and/or a construction on the basis of tabular differences. The variant readings in the degrees and minutes printed between square brackets are considered irrelevant in this connection; they indicate that the value actually found in O is *6;0,54, where the asterisk stands for an illegible digit, and that the value in E is 56;5,56. Note that for the present purpose we ignore the exact layout of the tables and scribal mistakes in the arguments, and limit ourselves to variant readings in the tabular values as such.

*Classes of apparatus entries*
In each apparatus there are four basic classes of entries, namely, *Scattered evidence*, *Outliers*, *(Practically) certain values*, and *Remaining scribal errors in E and/or A*. These classes may have somewhat varying titles due to the particular circumstances of each table, and they may be supplemented by further classes to illustrate specific points, mostly related to the way NALLINO dealt with certain tabular values.

The basic classes are specified in terms of "likely values" or "likely readings". These are defined as readings that fit the available constructed values (r, d, s) satisfactorily according to the circumstances of each case. Depending on the overall accuracy of a table, we may allow deviations from recomputed values of 1, 2 or 3 units in the final sexagesimal position. For tables with systematic larger deviations on certain intervals, we shall say that a value is likely if it "fits in its surroundings", i. e., if it differs from the recomputation by roughly the same amount as its neighbours. For example, in Table 4 the value 5;0,45 for argument 97 with an error of −11″ fits in its surroundings because the neigh-

4 Only in exceptional cases will there be no actual witness for the value that we accept. Occasionally the only witness may be a recomputed ("r") or symmetrical ("s") value.

bouring errors are −12″ and −10″. The decision which likely value we accept in each case depends on the particular circumstances of the table and will be detailed under the heading *Criteria*. In the case of close calls between likely values we generally give priority to E and then to A.

In general, whenever N differs from our accepted value, we shall include a full entry in the apparatus. Not included in the apparatus at all will be cases where there is only one likely reading, and EAN (*before* any easy correction of scribal errors) agree about it. If any one present of (OC) or T has a variant reading in such cases, or if they have variant readings that differ among themselves and appear paleographically unrelated to each other, such readings are rejected tacitly.

The four main classes of apparatus entries are specified in the following general terms:

*Scattered evidence*: Cases where there is evidence for several likely readings.
*Outliers*: Cases where there is no evidence for any likely reading, even when applying easy corrections of obvious scribal errors as explained above.
*(Practically) certain values*: Cases where there is only one likely reading but a minority of the witnesses, after the correction of obvious scribal errors, attests to a different value.
*Remaining scribal errors in E and/or A*: All remaining cases where EAN differ among each other. In practice these are cases where we have accepted the value from N, and E and A differ from N only due to obvious scribal mistakes (in other words, NALLINO has rightly corrected the scribal errors in E). For this class we use a briefer notation indicating only the deviations from N (or, equivalently, from our accepted value). For each entry included, incidental variant readings from any other witnesses are also recorded. Cases where both E and A have a scribal error in the final sexagesimal position will be listed in a separate class *Values with scribal errors in the seconds in both E and A*, or else under *(Practically) certain values*. A combined entry in the section *Remaining scribal errors ...* occasionally summarizes errors in a range of consecutive tabular values (e. g., "**122–126** 27–25–22–19–17′ (minutes all one too large) A"). Any further entries for arguments within such a range are preceded by a plus sign in order to facilitate the construction of the complete tabular values as they are found in the manuscripts.

*Interrelationship of manuscripts*
It may be said that we cannot determine the overall interrelationship of our manuscripts in any reliable way. The two AL-BATTĀNĪ witnesses E and A are each quite faulty, and they also have many errors in common. O and C are closely related. For certain variant readings, E may join OC whereas A joins T, such that neither of the readings in question is an obvious error. It is thus unwise to posit a common archetype, not to speak of postulating that this archetype reflects AL-BATTĀNĪ'S intention. Probably the Battanian manuscript tradition is a continuum generated by piecemeal corrections and cross-copying (for a possible instance of a table that occurs in more than one version, see Section 3.2).

*Terminology*
We shall use the following terms without further explanation. An *error* (other than a scribal error) is a deviation from the recomputation as specified for each table. Whenever a signed numerical value for an error is given, it is calculated as "tabular value minus recomputation". An exact value is a value without error, i. e., equal to the recomputation. Due to lack of space we cannot include full explanations of all astronomical functions and

other basic concepts of Ptolemaic astronomy discussed in this article. For further information we refer the reader to (KENNEDY 1956, PEDERSEN 1974, NEUGEBAUER 1975, VAN DALEN 1993).

### 2. Re-edition of the table for the solar declination

AL-BATTĀNĪ based his table for the solar declination on the highly accurate obliquity value 23°35′, which had already been observed earlier in the 9th century and was confirmed by his own observations. The table is essentially different from any other declination table in early Islamic sources.[5] It contains a large number of errors of two seconds and more that cannot be explained as scribal errors. Since these errors show only very few regular patterns, most of them cannot be due to, for example, the use of linear or quadratic interpolation and they most plausibly stem from inaccuracies in the process of calculation. NALLINO states in his commentary (vol. II, p. 221) that, besides correcting obvious scribal errors into plausible values, he corrected every declination value with an error of two seconds or more into a value recomputed by the Italian astronomer Giovanni Virginio SCHIAPARELLI (1835–1910) (in some cases SCHIAPARELLI'S correction differs from our own recomputation by a second, probably due to small computational mistakes or intermediate rounding). As a result, the table that NALLINO prints is very different from what we may assume about AL-BATTĀNĪ'S original table. In Table 1 we provide our own edition of the declination table, in which we have tried to stay as close as possible to the tables in our only two witnesses, the Arabic and Castilian versions of the *Ṣābi' Zīj*.

*Apparatus*
*Edition*: N pp. 57–58. *Manuscripts*: E ff. 177v–178r; A ff. 43r–43v. *Constructed values*: r = recomputation for an obliquity of the ecliptic of 23°35′.

*Criteria*: A likely value is any value differing from the recomputation by at most 3″. Whenever E and A, or any easy corrections of them, agree on a likely value, we accept it. If either E or A attests to a likely value, we likewise accept it. If E and A attest to two different likely values, we tend to accept the value from E.

*Commentary*: Where E and A agree, if necessary after correcting obvious scribal errors, they show numerous deviations of 2″ and more from the recomputation. Only some of these errors occur together in smaller groups and may hence be related, for example due to the use of interpolation. Five scattered errors larger than 8″ appear to be completely unrelated to any others and are here listed separately under *Outliers*. As NALLINO states in his commentary (vol. II, p. 221), he corrected all values in E that differ from SCHIAPARELLI'S recomputed values by 2″ or more (in fact, occasionally by only 1″) to those recomputed values. It is unclear to which extent he actually used paleographical considerations in

5 The Escorial manuscript of the *Mumtaḥan Zīj* by YAḤYĀ IBN ABĪ MANṢŪR contains a declination table for the obliquity value 23°33′ observed under AL-MA'MŪN, whereas the tradition of the *Toledan Tables* includes AL-KHWĀRIZMĪ'S table for the Ptolemaic value 23°51′ and a table for the typical Zarqalian value 23°33′30″. The declination tables in the Leipzig manuscript of the *Mumtaḥan Zīj* (see VAN DALEN 2004) and in the Istanbul and Berlin versions of the *zīj* of ḤABASH AL-ḤĀSIB are for obliquity 23°35′, but were all calculated by different methods, partly involving a heavy use of linear interpolation.

| Solar declination | | | | | | | |
|---|---|---|---|---|---|---|---|
| arg. | accepted | error | Nallino | arg. | accepted | error | Nallino |
| 1 | 0;24, 0 | | | 46 | 16;43,33 | –1 | |
| 2 | 0;48, 0 | | | 47 | 17; 0,47 | –3 | 17; 0,50 |
| 3 | 1;11,59 | | | 48 | 17;18, 0 | +12 | 17;17,48 |
| 4 | 1;35,57 | | | 49 | 17;34,30 | +2 | 17;34,29 |
| 5 | 1;59,54 | | | 50 | 17;50,52 | +2 | 17;50,51 |
| 6 | 2;23,49 | | | 51 | 18; 6,55 | +2 | |
| 7 | 2;47,41 | | | 52 | 18;22,36 | –1 | |
| 8 | 3;11,31 | | | 53 | 18;38, 5 | +3 | 18;38, 2 |
| 9 | 3;35,15 | –3 | 3;35,17 | 54 | 18;53, 7 | | |
| 10 | 3;59, 1 | –1 | | 55 | 19; 7,52 | +1 | |
| 11 | 4;22,42 | +1 | | 56 | 19;22,18 | +2 | 19;22,16 |
| 12 | 4;46,18 | +1 | 4;46,17 | 57 | 19;36,19 | | |
| 13 | 5; 9,49 | | | 58 | 19;50, 2 | +1 | 19;50, 1 |
| 14 | 5;33,16 | +1 | | 59 | 20; 3,21 | | |
| 15 | 5;56,37 | | | 60 | 20;16,20 | | |
| 16 | 6;19,53 | | | 61 | 20;28,41 | –16 | 20;28,57 |
| 17 | 6;43, 4 | +1 | | 62 | 20;41,11 | | |
| 18 | 7; 6, 7 | +1 | | 63 | 20;52,54 | –8 | 20;53, 2 |
| 19 | 7;29, 4 | +1 | 7;29, 3 | 64 | 21; 4,27 | –3 | 21; 4,30 |
| 20 | 7;51,53 | | | 65 | 21;15,36 | +1 | |
| 21 | 8;14,36 | | | 66 | 21;26, 9 | –8 | 21;26,17 |
| 22 | 8;37,12 | +1 | | 67 | 21;36,34 | | |
| 23 | 8;59,37 | | | 68 | 21;46,28 | +1 | |
| 24 | 9;21,56 | +1 | | 69 | 21;55,58 | +2 | 21;55,56 |
| 25 | 9;44, 5 | +1 | | 70 | 22; 5, 0 | | |
| 26 | 10; 6, 5 | +1 | | 71 | 22;13,41 | +2 | 22;13,39 |
| 27 | 10;27,55 | +1 | 10;27,54 | 72 | 22;21,45 | –8 | 22;21,52 |
| 28 | 10;49,34 | | | 73 | 22;29,42 | | |
| 29 | 11;11, 3 | | | 74 | 22;37, 3 | –1 | 22;37, 5 |
| 30 | 11;32,22 | | | 75 | 22;44, 3 | +1 | 22;44, 2 |
| 31 | 11;53,29 | | | 76 | 22;50,33 | | |
| 32 | 12;14,25 | | | 77 | 22;56,38 | | |
| 33 | 12;35,10 | +1 | | 78 | 23; 2,16 | | |
| 34 | 12;55,42 | +2 | 12;55,40 | 79 | 23; 7,29 | | |
| 35 | 13;15,59 | | | 80 | 23;12,15 | +1 | |
| 36 | 13;36, 9 | +5 | 13;36, 4 | 81 | 23;16,34 | +1 | |
| 37 | 13;55,59 | +3 | 13;55,56 | 82 | 23;20,24 | –1 | |
| 38 | 14;15,37 | +3 | 14;15,35 | 83 | 23;23,50 | +1 | |
| 39 | 14;35, 2 | +4 | 14;34,58 | 84 | 23;26,48 | +1 | |
| 40 | 14;54, 8 | | | 85 | 23;29,18 | +1 | |
| 41 | 15;13, 2 | | | 86 | 23;31,19 | –2 | 23;31,21 |
| 42 | 15;31,41 | +1 | | 87 | 23;32,55 | –2 | 23;32,56 |
| 43 | 15;50, 3 | | | 88 | 23;34, 2 | –3 | 23;34, 5 |
| 44 | 16; 8, 9 | –1 | | 89 | 23;34,43 | –3 | 23;34,46 |
| 45 | 16;26, 0 | | | 90 | 23;35, 0 | | |

**Table 1:** Table of the solar declination, with the values accepted by us on the basis of manuscripts E and A, differences from the recomputation, and NALLINO's values if different from ours.

those cases where he corrected scribal errors into paleographically likely readings, because in a majority of the relevant cases the paleographically likely readings are equal to recomputed values.[6]

*Scattered evidence* (E and A, or any easy corrections of them, attest to two different likely values): **19** 7;29,4: A; 6″ E, 3″ Nr. **21** 8;14,36: ENr; 37″ A. **29** 11;11,3: ENr; 2″ A. **47** 17;0,47: A; 8″ E, 50″ Nr. **53** 18;38,5: E; 4″ A, 2″ Nr. **82** 23;20,24: EN; 25″ Ar. **87** 23;32,55: 35″ E, 54″ A, 56″ N, 57″ r.

*Outliers* (scattered errors of 8″ and more that do not fit into any patterns and were corrected by NALLINO into recomputed values):[7] **48** 17;18,0 [16′ E, 58′ A]: EA; 17′48″ Nr. **61** 20;28,41: A; 21″ E, 57″ Nr. **63** 20;52,54: A; 14″ E, 53′2″ Nr. **66** 21;26,9: EA; 17″ Nr. **72** 22;21,45: EA; 52″ N, 53″ r.

*Other corrections by* NALLINO *into recomputed values* (E, or any easy correction of it, differs by 2″ or more (in three cases by only 1″) from values recomputed by SCHIAPARELLI): **9** 3;35,15: A; 55″ E, 17″ N, 18″ r. **12** 4;46,18 [36′ E]: 5″ E, 58″ A, 17″ Nr. **27** 10;27,55: A; 19″ E, 54″ Nr. **34** 12;55,42: EA; 40″ Nr. **36** 13;36,9 [56′ A]: EA; 4″ Nr. **37** 13;55,59: EA; 56″ Nr. **38** 14;15,37: A; 17″ E, 35″ N, 34″ r. **39** 14;35,2: EA; 34′58″ Nr. **49** 17;34,30 [36′ E]: EA; 29″ N, 28″ r. **50** 17;50,52: EA; 51″ N, 50″ r. **56** 19;22,18: E; 58″ A, 16″ Nr. **58** 19;50,2: A; 50″ E, 1″ Nr. **64** 21;4,27: EA; 30″ Nr. **69** 21;55,58: A; 18″ E, 56″ Nr. **71** 22;13,41 [2*° E]: EA; 39″ Nr. **74** 22;37,3: EA; 5″ N, 4″ r. **75** 22;44,3: EA; 2″ Nr. **86** 23;31,19: EA; 21″ Nr. **88** 23;34,2: EA; 5″ Nr. **89** 23;34,43: E; 13″ A, 46″ Nr.

*Values with scribal errors in the seconds in both E and A*: **14** 5;33,16: N; 6″ E, 56″ A, 15″ r. **16** 6;19,53 [59′ A]: Nr; 33″ EA. **24** 9;21,56: N; 16″ EA, 55″ r. **52** 18;22,36: N; 6″ EA, 37″ r. **57** 19;36,19: Nr; 43″ E, 59″ A. **59** 20;3,21: Nr; 11″ EA. **62** 20;41,11: Nr; 31″ E, 0″ A.[8] **78** 23;2,16: Nr; 4″ E, 56″ A. **79** 23;7,29: Nr; 9″ E, 39″ A. **85** 23;29,18: N; 48″ E, 38″ A, 17″ r.

*Remaining scribal errors in E and/or A*: **1** 31° A; 31″ E. **7** 46′ E. **10** 54′ E. **11** 32′ A. **15** 50″ E. **17** 54″ E. **20** 43″ E. **22** 17′ A; 52″ E. **31** 25″ E. **32** 54′ A. **35** 55′ A; 15″ E. **41** 3′ E; 53′ A. **43** 55′ A. **46** 53′ E. **60** 56′ A. **65** 55′ A. **80** 55″ A. **84** 27′ E.

---

6 For six of the ten values with scribal errors in the seconds in both E and A, and for five of the eight values with scribal errors in the seconds in either E or A, the value accepted by us on paleographical grounds is equal to the recomputation; in all other cases it differs from it by only 1″.

7 E and A are in agreement for three of the five values listed here; only for arguments 61 and 63 is the accepted value somewhat uncertain.

8 Note that the error in A may be dependent on that in E: read 0 [illegible] for 31 لا. In fact, the abjad forms for 0 and 31 in the Escorial manuscript are so close that confusion is certainly possible. NALLINO (vol. II, pp. vi and 271) interprets the abjad numeral 31 found instead of zero on seven occasions in the Escorial manuscript as the Arabic negation *lā*.

### 3. Apparatus for selected other tables

#### *3.1 Sine*

AL-BATTĀNĪ'S sine table has an unusual format, displaying values for every 30′ of arc rather than for every 15′, such as ḤABASH'S table derived from PTOLEMY'S *Almagest*, or 1° as most other early Islamic sine tables. Apparently it thus followed the format of PTOLEMY'S table of chords, which likewise presents values for every 30′ of arc. The values of AL-BATTĀNĪ'S sine table are not related to those in any other Islamic *zījes* that we have checked. The only witnesses for this table are therefore the Arabic and Castilian versions of the *Ṣābi' Zīj* and the manuscript tradition of the *Toledan Tables*. Since our reconstruction of the archetype shows only 32 deviations from the recomputation (29 of which are +1″ or −1″), NALLINO'S strategy to replace doubtful values by recomputed ones (vol. II, pp. 220–221) in this case leads to a generally correct recovery of the original table.

*Apparatus*
*Edition*: N pp. 55–56. *Manuscripts*: E ff. 176v–177r; A ff. 28r–28v; O ff. 18r–19r;[9] C ff. 164r–164v.[10] *Constructed values*: r = recomputation for radius of the base circle 60. The readings in the mainstream *Toledan Tables* (see PEDERSEN 2002, pp. 957–959) differ so frequently from the Battanian tradition that we do not consider their inclusion to be useful.

*Criteria*: A likely value is any value that lies within 1″ of the recomputation. Whenever E and A, or any easy correction of them, agree on a likely value, we accept this value. If E, A and at least one of OC agree on any value, we likewise accept it. If E and A attest to two different values (not necessarily likely) in the neighbourhood of the recomputation, none of which is supported by O or C, we accept the value that is closer to the recomputation; if both values are equally close, we accept the value from E.

*Commentary*: AL-BATTĀNĪ'S sine table as we have reconstructed it shows 32 deviations from the recomputation, in only three cases larger than 1″. Of the latter the error 44″ (for 49″) at argument 88 is the only one attested by all three manuscript sources. Our accepted values differ from NALLINO at arguments 4, 11½, 51½, 69, 87, and 87½. NALLINO corrected 55 values from E, mostly for obvious scribal errors, on the basis of PTOLEMY'S table of chords, Abū 'Alī al-Ḥasan AL-MARRĀKUSHĪ'S sine table for every 15′ of arc, and a recomputation (vol. II, pp. 220–221). Only 14 of these corrections were *not* into exact sine values, twelve of which obviously for paleographical reasons (the only two exceptions being the outliers 51½ and 88).

*Scattered evidence* (E, A and (OC), or any easy corrections of them, attest to two or three different likely values): **7** 7;18,44: OCNr; 32″ corrected to 42″ (?) E, 45″ A. **12½**

---

[9] In O the seconds for arguments 45½ to 60 were mistakenly copied into the values for arguments 15½ to 30. As a result, readings for the seconds for arguments 15½ to 30 are to be considered missing.

[10] In O and C the minutes of the values for arguments 51–56½ have slid upwards by three rows. As a result, the readings for 51, 51½ and 52 are to be considered missing.

12;59,11: AOCNr; 10″ E. **24½** 24;52,54: EOCNr; 55′ A. **51** 46;37,44 [(no minutes) OC]: OCNr; 46″ E, 45″ A. **52½** 47;36,4 [46′ E, 33′ O]: EANr; 5″ OC. **60½** 52;13,17 [33′ A]: EOCNr; 16″ A. **73½** 57;31,45 [56° A]: AOCNr; 44′ E. **76** 58;13,4: EOCNr; 3″ A. **77** 58;27,44: ENr; 45″ A, 14″ OC. **78½** 58;47,44: EOCNr; 45″ A. **87** 59;55,4 [54′ A]: AOCr; 3″ EN.

*Outliers* (three scattered values differing from the recomputation by 2″ or more): **29½** 29;32,45: EN; 44″ A, 43″ Cr. **51½** 46;57,25 [(minutes missing) OC]: EA; 24″ OCN, 23″ r. **88** 59;57,44: EAC; 49″ ON, 48″ r.

*Reasonably certain values* (two or three of E, A and (OC), or any easy corrections of them, agree about a likely value, whereas any alternative readings are not likely): **4** 4;11,8: EAOC; 7″ Nr. **4½** 4;42,27: AOCNr; 25″ E. **11** 11;26,55 [36′ E]: ANr; 35″ E, 14″ O, 9″ C. **11½** 11;57,43 [*6′ E]: AOCr; 55″ E, 44″ N. **25½** 25;49,50: ENr; 7″ A, 51″ C. **39** 37;45,34: EAN; 32″ OC, 33″ r. **39½** 38;9,53: EONr; 13″ A, 54″ C. **44½** 42;3,16 [4′ E]: ENr; 56″ A, 17″ OC. **47** 43;52,53: EON; 33″ A, 54″ C, 52″ r. **69** 56;0,54: AOC; 56″ E, 53″ Nr. **86½** 59;53,17: ECNr; 54″ A, 18″ O.

*Values with scribal errors in the seconds in both E and A*: **24** 24;24,15 [26′ E]: CNr; 35″ E, 25″ A. **42½** 40;32,8: CN; 28″ EA, 4″ O, 7″ r. **48** 44;35,19: OCNr; 49″ E, (seconds blank) A. **80** 59;5,19 [0′ A]: OCN; 59″ A, 39″ E; 18″ r. **87½** 59;56,35: OCN; 55″ EA; 34″ r.

*Remaining scribal errors in E and/or A*: **1** 1′ A. **1½** 35′54″ A. **2** 6′ EA, *8″ E. **5** (26″ corrected into *6″) E. **5½** 8″ A. **6** 36′38″ E. **9½** 42″ E. **10** 35′ E. **10½** 36′ E; 4″ OC. **12** 37′0″ E. **14** 15° EA. **14½** 12″ E. **15** 41″ E. **16½** 12′ E. **18** 24″ E. **18½** 38″ A. **20** 36″ E. **21** 55″ A. **21½** 55″ E. **26** 38′ E, 28′ A. **27** 34′ E, 44′ A. **27½** 38″ E. **28** 32′ E. **29** 0′ A. **33½** 18″ E. **35½** 55′ A. **36** 56′ A. **36½** 32″ E. **40** 35′ E. **43** 52″ A. **46** 58″ E. **46½** 0″ A. **47½–48½** 45° A. **+47½** 54′52″ A. **49** 56′ A; 18″ EOC. **49½** 23″ E. **50½** 57′ A. **52** 56′55″ A, 50″ OC. **53** 14′ E; 0″ A. **53½** 33′ A. **54** 12′ E. **54½** 55′ A. **55½** 27′31″ E, 36′ C. **56** 52″ E. **56½–58** 55° A. **+57** 59′ A. **+57½** 46′52″ E. **+58** 19″ E. **59** 35′ A. **61** 18′ OC; 18″ E. **63** 26′ EOC. **64** 54′ E. **64½–66** 55° A. **+65½** 34′ E; 12″ OC. **+66** 16″ E. **67** 33′ A. **69½** 30″ E. **72–75** 56° A. **+73½**: see under *Scattered evidence*. **75½** 0′ A. **77½** 35′ EA. **79** 31″ A. **79½** 53″ E. **81½** 38″ E. **82** 35′ A; 18″ EOC. **85** 49″ E, 59″ OC. **86** 54″ A.

### *3.2 Cotangent for a gnomon length of 12 parts*

AL-BATTĀNĪ's shadow table with values to minutes for a gnomon length of 12 parts is different from similar tables in other early Islamic works such as the *zījes* of Abū 'Abd Allāh Muḥammad ibn Mūsā AL-KHWĀRIZMĪ (first half of 9th c.) and KŪSHYĀR and the Almanac of AZARQUIEL (late 11th c.). It shares with these sources two typical errors in the cotangents of 1 and 2°, but it has its own characteristic errors for arguments 18 and 36. Ten errors of ±1′ in the Escorial manuscript of the *Ṣābi' Zīj* are not found in the other witnesses for AL-BATTĀNĪ'S cotangent table, namely, the Castilian version of the *zīj* and the tradition of the *Toledan Tables*. This may point to the existence of two different versions of this table.

*Apparatus*

*Edition*: N p. 60. *Manuscripts*: E fol. 179r; A fol. 44v; O fol. 19v; C fol. 163v;[11] T pp. 991–993.[12] *Constructed values*: r = recomputation for gnomon length 12.

*Criteria*: A likely value is any value within 1′ of the recomputation. In nearly every single case a clear majority of the witnesses attests to one particular value, likely or not, which we consequently accept (only for argument 36 do two or more witnesses give values rejected by us). However, for a group of apparently interrelated errors of ±1′ in E, which (with two minor exceptions) do not occur in any of the other sources, we accept the value from E (cf. below). We include in the apparatus full entries for all values corrected by NALLINO and, under the heading *Possible indications ...*, for the group of apparently interrelated errors in E. As elsewhere, OC have quite a large number of scribal errors in common that do not appear in any of the other sources; these are not listed here, but can be found in (PEDERSEN 2002, pp. 991–993).

*Commentary*: AL-BATTĀNĪ's cotangent table for gnomon length 12 as we have restored it from the available sources has 19 deviations from the recomputation, 16 of which are only ±1′. NALLINO corrects eight of the errors to exact values, including all three larger ones. He mentions the typical errors in the cotangents of 1° and 2° in his commentary (vol. II, p. 221) and explains them from the use of sine values rounded to three sexagesimal places (12 · 59;59,27 / 1,2,50 = 687;25,41,13,... and 12 · 59;57,48 / 2;5,38 = 343;38,51,33,...). All sources for AL-BATTĀNĪ's cotangent table have two characteristic errors in common, namely those for arguments 18 (−2′) and 36 (−3′). The Escorial manuscript has ten errors of a single minute on the interval 71 to 89°, where, with two minor exceptions, the Castilian translation and the tradition of the *Toledan Tables* have exact values. This makes it plausible that two versions of AL-BATTĀNĪ's cotangent table existed in the western Islamic world (cf. the somewhat similar situation for the excess of half daylight in Section 3.3, for which only the *Toledan Tables* include more accurate values for two out of three intervals with larger deviations from the recomputation).

*Corrections by* NALLINO *not supported by EA(OCT)*: **1** 687;26 [387° A]: EAOCT; 29′ Nr. **2** 343;39 [693° E, 383° A]: EAOCT; 38′ Nr. **15** 44;46: EAOCT; 47′ Nr. **18** 36;54: EAT; 14′ OC, 56′ Nr. **36** 16;28: EA; 38′ OC, 30′ T, 31′ Nr. **46** 11;34: EAOC; 35′ TNr. **67** 5;6: AOCTNr; 3′ E.

*Corrections by* NALLINO *supported by A(OCT)*: **60** 6;56: AOCTNr; 57′ E. **62** 6;23′ ATNr; 24′ E, (minutes missing) OC. See also **71** and **72** under the category *Possible indications ....*

*Possible indications for the existence of a revised Battanian table* (ten values for arguments on the interval 71–87 for which E has deviations from the recomputation of +1′ (twice) or −1′ (eight times), whereas A (with one exception) and OCT (with one exception

[11] In OC the minutes of the values for arguments 61–87 have slid upwards by three rows; as a result, readings of the minutes for arguments 61–63 are to be considered missing.

[12] The mainstream version of T is very close to AL-BATTĀNĪ and includes his two characteristic errors for arguments 18 and 36 (but not the ten errors towards the end of the table that are not contained in AOC either).

for T) give exact cotangent values; of these ten values, NALLINO corrected only those at 71 and 72 into recomputed values): **71** 4;9: E; 8′ AOCTNr. **72** 3;55: ET; 54′ AOCNr. **76** 2;59 [3° E]: EN; 3°0′ AOCTr. **78** 2;32: EN; 33′ AOCTr. **79** 2;19: EN; 20′ AOCTr. **80** 2;6: EN; 7′ AOCTr. **84** 1;15: N; 35′ E, 16′ AOCTr. **85** 1;2: EN; 3′ AOCTr. **87** 0;37: EN; 38′ AOTr, 48′ C. **89** 0;12: EAN; 13′ OCTr.

*Remaining scribal errors in E and/or A*: **5** 136° E, 134° OC. **13** 19′ E. **17** 55′ EA. **21** 56′ EA. **23** 56′ E. **40** 58′ A. **43** 32′ E. **48** 28′ E. **59** 53′ A. **61** 29′ E, (minutes missing) OC. **63** 50′ E, (minutes missing) OC.

### *3.3 Half excess of longest daylight as a function of geographical latitude*

This table displays, for every half degree of geographical latitude up to 60, the maximum equation of daylight, i. e., half the difference between the length of the longest day and 12 hours, expressed in equatorial degrees. It can be used for the calculation of oblique ascensions at arbitrary latitudes as explained in Chapter 13 of the *Ṣābi' Zīj* (vol. I, pp. 27 (translation) and 187–189 (commentary); vol. III, p. 39 (Arabic)). The standard method by which the half excess of longest daylight can be computed is explained in Chapter 9 of the *Ṣābi' Zīj*, although the present table is not explicitly mentioned. We have not found other copies of this table in early Arabic *zījes*, so that the only available witnesses are the Arabic and Castilian versions of AL-BATTĀNĪ's *zīj* and a branch of the tradition of the *Toledan Tables*. Whereas more than half of the tabular values that we accept are highly accurate, three intervals show groups of larger systematic errors, two of which were adjusted to more or less exact values in the *Toledan Tables*.

*Apparatus*
*Edition*: N p. 59. *Manuscripts*: E fol. 178v;[13] A fol. 44r; T pp. 1128–1129.[14] *Constructed values*: r = recomputation for an obliquity of the ecliptic of 23°35′; d: as explained in the introduction.

*Criteria*: A likely value lies within 1′ of the recomputation or, in the case of values within one of the three intervals with systematic deviations from the recomputation, fits in its surroundings, if possible in such a way that the tabular differences become regular.[15] If E and A, or any easy corrections of them, agree on a likely value, we accept it. If, on the third interval of inaccurate values, neither E nor A has a likely value (e. g., for arguments 52, 54 and 57½), we tend to accept E.

---

13 In E the arguments and degrees for arguments 2½–11½ have slid downwards by one row, making the degrees for arguments 2½, 5, 7, 9, and 11½ one too small.

14 This table is not innate in the mainstream tradition of the *Toledan Tables*; it occurs only in manuscript Oo and class {d} (cf. PEDERSEN 2002, p. 1128) and may be an accidental loan. It differs systematically from the archetype of EA on the following intervals: 5½–10, 48½–49½ and 51½–55½, 58–59½. Consequently, on these intervals T has not been given the same weight as a witness for the Battanian table, and its values are cited between parentheses.

15 The second part of this definition applies only to argument 6½, since no further problematic values appear in the first two intervals with systematic deviations, and the errors on the third interval are so irregular that all values for which E and A differ fall under the category *Scattered evidence*.

| Half excess of longest daylight | | | | | | | |
|---|---|---|---|---|---|---|---|
| lat. | accepted | error | diff. | lat. | accepted | error | diff. |
| 15 | 6;43 | | 0;14 | 23 | 10;43 | +2 | 0;16 |
| 15½ | 6;57 | | 0;14 | 23½ | 10;59 | +2 | 0;17 |
| 16 | 7;11 | | 0;14 | 24 | 11;16 | +4 | 0;17 |
| 16½ | 7;25 | −1 | 0;15 | 24½ | 11;33 | +4 | 0;18 |
| 17 | 7;40 | | 0;14 | 25 | 11;51 | +6 | 0;17 |
| 17½ | 7;54 | −1 | 0;15 | 25½ | 12; 8 | +7 | 0;16 |
| 18 | 8; 9 | | 0;14 | 26 | 12;24 | +6 | 0;17 |
| 18½ | 8;23 | −1 | 0;15 | 26½ | 12;41 | +7 | 0;16 |
| 19 | 8;38 | −1 | 0;15 | 27 | 12;57 | +6 | 0;17 |
| 19½ | 8;53 | −1 | 0;15 | 27½ | 13;14 | +6 | 0;16 |
| 20 | 9; 8 | −1 | 0;15 | 28 | 13;30 | +5 | 0;17 |
| 20½ | 9;23 | −1 | 0;16 | 28½ | 13;47 | +4 | 0;16 |
| 21 | 9;39 | | 0;16 | 29 | 14; 3 | +3 | 0;17 |
| 21½ | 9;55 | +1 | 0;16 | 29½ | 14;20 | +2 | 0;16 |
| 22 | 10;11 | +1 | 0;16 | 30 | 14;36 | | |
| 22½ | 10;27 | +2 | 0;16 | | | | |

**Table 2:** Fragment of the table of the half excess of longest daylight as a function of geographical latitude, with the values accepted by us, differences from the recomputation, and first-order tabular differences.

*Commentary*: AL-BATTĀNĪ'S table deviates in a systematic fashion from the recomputation on three intervals: for arguments 5–10½ with a maximum error of +5′; for arguments 21½–29½ with a maximum error of +7′; and for arguments 52–55½ with a maximum error of −5′. The very regular pattern of tabular differences up to argument 45 points to a heavy use of linear interpolation. The first two "dents" can thus be explained by only two incorrect values at interpolation nodes.[16] Of the three "dents", the version of this table in the manuscript tradition of the *Toledan Tables* has only the second one; instead of the first and third dent it has (almost) exact values. Since three quarters of the tabular values in the *Toledan Tables* agree with AL-BATTĀNĪ, we may nevertheless assume that the two tables are interdependent. It cannot at present be decided with certainty whether the version in the *Toledan Tables* corrected the dents resulting from incorrect linear interpolation in

16 The first-order differences within the first two intervals with systematic deviations from the recomputation are highly regular between the nodes 4–8–11 and 21–25–30 (to explain the table fully from linear interpolation between these nodes, we need to assume that a minor mistake was made between arguments 21 and 25, resulting into tabular differences 16, 16, 16, 16, 16, 17, 17, 18). As a result, if linear interpolation is carried out using the incorrect tabular values for arguments 8 and 25 (for the latter, cf. Table 2), the Battanian table can be reproduced almost exactly.

AL-BATTĀNĪ'S archetype (this possibility may be more probable), or the version extant in the Arabic and Castilian versions of the *Ṣābi' Zīj* incorrectly patched some more or less exact parts of an archetype surviving in the *Toledan Tables*. NALLINO (vol. II, p. 224) states that his transcription of this table was emended by SCHIAPARELLI. It turns out that, at the first and second dents with their very regular tabular differences, he only corrected scribal errors. At the third, much more irregular, dent he changed various values into recomputed ones without any obvious system.

*Scattered evidence* (E and A, or any easy corrections of them, attest to two different likely values, or neither E nor A has a likely value): **4** 1;45: ATNrd; 44′ E. **6½** 2;54: ANd; 55′ E, (52′ T), 51′ r. **14** 6;15: ATNrd; 16′ E. **31½** 15;31: ATrd; 32′ EN. **47** 27;55: ETNr; 54′ A. **52** 33;56: E; 54′ A, 57′ N, (34°1′ T), 58′ r. **54** 36;53: EN; 57′ A, (55′ T), 56′ r.

*Reasonably certain values* (E and A, or any easy corrections of them, agree on a likely value, or either E or A agrees with T on a likely value whereas the third witness does *not* provide a likely value): **11** 4;51: EA; 53′ N, 52′ Tr. **39** 20;42: ATrd; 45′ E, 40′ N. **39½** 21;5: ATrd; 13′ E, 3′ N. **50** 31;22: ATN; 24′ E, 21′ r. **53** 35;22: EA; (26′ T), 24′ Nr. **53½** 36;7: EA; (10′ T), 9′ Nr. **54½** 37;39: EA; (42′ T), 43′ N, 44′ r. **55** 38;29: EA; (32′ T), 34′ Nr. **55½** 39;22: EA; (25′ T), 26′ Nr. **57** 42;16: AT; 56′ E, 14′ Nr. **57½** 43;17: ET; 14′ A, 15′ Nr. **59½** 47;45: EA; 49′ (T)N, 50′ r.

*Remaining scribal errors in E and/or A*: **3** 39′ E. **7½** 32′ E, (18′ T). **13** 40′ E. **14½** 25′ E. **15** **′ E. **15½** 54′ A. **16** 51′ A. **18½** 28′ A. **19** 33′ A. **19½** 58′ E. **21** 19′ E. **22** 51′ A. **24** 56′ A. **25½** 81′ (sic!) A. **27½** 54′ A. **30½** 24′ E, 55′ T. **31** 53′ A. **32** 55′ A. **34** 4′ E. **36½** 11′ E. **37** 53′ A. **37½** 3*′ E. **38** 37′ E. **38½** 59′ A. **46** *2′ E. **51½** 38′ A, (20′ T). **56½** **′ E. **58** 59′ E, (20′ T). **60** 50′ A.

### *3.4 Mean motion of the northern lunar node for the Arabic calendar*

AL-BATTĀNĪ includes in his *zīj* two sets of planetary mean motion tables, one for the Arabic calendar, the common calendar in the Islamic world, and one for the Byzantine or Syrian calendar, which was more frequently used in the region where AL-BATTĀNĪ lived and worked. Both sets cover a range of many centuries and are based on new parameters which were apparently derived from the observations that AL-BATTĀNĪ made in Raqqa between 877 and 918. Although the mean motion tables present straightforward linear functions, they are not usually error-free. Occasionally the sub-tables contain systematic errors that make it impossible to reliably determine the underlying parameter for the daily mean motion and to decide between readings differing by one or two seconds. We present a full apparatus for the table of the motion of the ascending lunar node in the Arabic calendar and reproduce in its entirety the particularly problematic sub-table for extended years from this table. We have determined the underlying values for the daily mean motion by means of the least number of errors criterion,[17] omitting from the analysis all uncertain values as listed under the category *Nontrivial variants*.

---

[17] For a discussion of the least number of errors criterion, see (VAN DALEN 1993, 60–62). A fully explained application of the criterion can be found in (VAN DALEN 2000).

*Apparatus*
*Edition*: N pp. 19–23. *Manuscripts*: E ff. 164v–166v; A ff. 36v–38r. *Constructed values*: r = recomputation for a daily mean motion of 0;3,10,37,24,5,57 °/day; d: as explained in the introduction.

*Criteria*: In general, a likely value is a value that differs by at most 1″ from our recomputation. For inaccurate and irregular sub-tables (in particular, the extended years and days) a likely value is a value that produces a roughly regular pattern of tabular differences. We shall usually accept any likely value on which E and A, or any easy corrections of them, agree. In cases where E and A attest to two different likely values, we accept the value which is closest to our recomputation or, in the case of less accurate sub-tables, fits best in its surroundings. Values constructed on the basis of tabular differences (indicated by "d") will be included in the apparatus only for purely linear sub-tables, i. e., not for the extended years and months.

*Commentary*: NALLINO's values for collected years are all correct for the indicated parameter. However, in the manuscripts the degrees have suffered from great distortions: in E those for the last ten values are 40° too low, and in A most of those for arguments 331–871 are 20° too high. On the other hand, the extended years, reproduced in Table 3, cannot be satisfactorily recomputed for any parameter value (the minimum number of errors in this sub-table is 15) and cannot be plausibly restored either. Also the sub-table for days has a large number of deviations from the recomputation due to an extremely irregular pattern of first-order differences.[18]

*Nontrivial variants*: **Collected years** (1, 31, 61, …, 871 Hijra): **241** 57;7,34 [4′ A]: ENrd; 35″ A. **841** 155;33,13 [115° E, 175° A]: Nrd; 48″ E, 10″ A. **Extended years**: **3** 56;17,12 [57′ E]: N; 52″ E, 13″ Ar. **6** 112;34,25: Ar; (seconds illegible: 1 or 30?) E, 24″ N. **7** 131;22,17: Ar; 15″ EN. **9** 168;51,37: A; 34″ E, 36″ N, 38″ r. **10** 187;39,29: Ar; 27″ EN. **23** 71;32,57 [38′ E, 33′ A]: N; 17″ E, 14″ A, 33′0″ r. **26** 127;53,20 [58′ E]: (perhaps most likely on paleographical grounds[19]); 9″ EAN, 24″ r. **29** 184;10,34 [174° A; **′ (21′ ?) E]: N; 24″ EA, 36″ r. **30** 202;55,14: AN; **″ E, 17″ r. **Months**[20]: Shawwāl 15;37,14: Ar; 54″ E, 15″ N. Dhu 'l-Ḥijja 18;44,41: ANr; 45′15″ E. **Days**: **21** 1;6,44: A; 45″ EN, 43″ r.[21] **Hours**: -.

*Remaining scribal errors in E and/or A*: **Collected years**: **1** 213° E. **31** 47′ A. **151** 48° E. **301** 152° A. **331–571** and **661–871** (all degrees 20 too high except where noted) A. **601–811** (all degrees 40 too low except where noted) E. **+631** 165° A; 56″ E. **+691** 172° E, 241°57′ A. **+781** 120° A; 50″ E. **871** 218°…50″ E. **Extended years**: **4** 12″ E. **5** 48″

18 As for all mean motions, there are only minor differences between the sub-tables for days and hours in AL-BATTĀNĪ'S tables for the Arabic calendar and those in his tables for the Byzantine calendar.

19 A sound correction on the basis of a recomputation or tabular differences is here basically impossible. Acceptable alternatives are also 19″ and 21″.

20 To the sub-tables for months of all mean motion tables for the Arabic calendar NALLINO has added an extra calculated entry for Dhu 'l-Ḥijja in a leap year, which is absent from the manuscripts; cf. (vol. II, p. 204, note to pag. 21).

21 The correction into 44″ is mentioned by NALLINO in his commentary (vol. II, p. 204). The basically identical sub-table for days in the table for Byzantine years shows the corrected value.

| Mean motion of the northern lunar node in extended Arabic years | | | | | | | |
|---|---|---|---|---|---|---|---|
| arg. | accepted | error | Nallino | arg. | accepted | error | Nallino |
| 1 | 18;44,41 | | | 16 | 300;13,53 | –1 | |
| 2 | 37;32,32 | | | 17 | 318;58,34 | –1 | |
| 3 | 56;17,12 | –1 | | 18 | 337;46,25 | –1 | |
| 4 | 75; 1,52 | –1 | | 19 | 356;31, 6 | –1 | |
| 5 | 93;49,43 | –2 | | 20 | 15;15,46 | –2 | |
| 6 | 112;34,25 | | 112;34,24 | 21 | 34; 3,36 | –3 | |
| 7 | 131;22,17 | | 131;22,15 | 22 | 52;48,17 | –3 | |
| 8 | 150; 6,56 | –1 | | 23 | 71;32,57 | –3 | |
| 9 | 168;51,37 | –1 | 168;51,36 | 24 | 90;20,48 | –4 | |
| 10 | 187;39,27 | –2 | | 25 | 109; 5,28 | –4 | |
| 11 | 206;24, 8 | –2 | | 26 | 127;53,20 | –4 | 127;53, 9 |
| 12 | 225; 8,49 | –2 | | 27 | 146;38, 2 | –2 | |
| 13 | 243;56,39 | –3 | | 28 | 165;22,43 | –2 | |
| 14 | 262;41,20 | –2 | | 29 | 184;10,34 | –2 | |
| 15 | 281;26, 2 | –1 | | 30 | 202;55,14 | –3 | |

**Table 3:** Sub-table for extended Arabic years from the table of the motion of the lunar node, with the values accepted by us, differences from a recomputation for a daily mean motion of 0;3,10,37,24,5,57 °/day, and NALLINO's values if different from ours.

E. **12** 59″ E. **13** 248° E. **15** 25′ EA. **16** 58′28″ E, 60°…23″ A. **17** 328°…24″ A. **19** 1′ E. **20** 35′ E. **22** 57″ A. **25** 0′ A. **27** 30′ E, 8′ A. **28** 48″ E. **Months**: Muḥarram 49″ E. Ṣafar 30′ E, 4′ A. Rabīʿ II 34′ E. Jumādā I 33′ E. Jumādā II 51″ E. Dhu 'l-Qaʿda 32′53″ E. **Days: 4** 13′ E. **11** 35′50″ E. **13** 55″ E. **15** 60″ (sic!) A. **17** 34′ E. **18** 37′2″ E. **19** 43″ E. **25** 20″ A. **28** 19″ E. **30** 49″ E. **Hours: 7** 1′ A. **16** 50″ E. **18** 33″ A.

### *3.5 Lunar equation of anomaly at apogee*

Of the tables in the *Ṣābi' Zīj* for the solar, lunar and planetary equations, AL-BATTĀNĪ can be assumed to have computed only very few. He calculated the solar equation on the basis of his new solar eccentricity 2;4,45ᴾ, whose determination he describes in Chapter 28 of the *zīj* (cf. NALLINO 1899–1907, vol. I, pp. 44 (translation) and 213 (commentary); vol. III, p. 223 (Arabic)). He computed the lunar equation of anomaly at apogee anew for the Ptolemaic epicycle radius but with values to seconds, and may have carried out his own calculation of the equation of centrum for Venus on the basis of the new maximum of 1°59′ which had been observed by YAḤYĀ IBN ABĪ MANṢŪR and the other astronomers involved in the compilation of the *Mumtaḥan Zīj* under the Abbasid caliph AL-MA'MŪN (c. 830).[22] Basically all other tables for the planetary equations in the *Ṣābi' Zīj* were adopted from PTOLEMY'S *Handy Tables*, possibly through the intermediary of the earliest Arabic *zījes* with Ptolemaic planetary tables, of which the *Mumtaḥan Zīj* and the *Zīj* by ḤABASH AL-ḤĀSIB (c. 870) have survived.

For the tables that AL-BATTĀNĪ adopted from the *Handy Tables* we thus also have a number of earlier sources available. For the tables that were originally calculated by him, our main witnesses besides the Arabic and Castilian versions of the *Ṣābi' Zīj* are the tradition of the *Toledan Tables* and incidental Arabic and Persian *zījes* that made use of AL-BATTĀNĪ'S work. For the lunar equation of anomaly, one of the tables for which NALLINO'S edition is furthest removed from what we may assume about AL-BATTĀNĪ'S original table, the witnesses include the *Mufrad Zīj* by Abū Ja'far Muḥammad AL-ṬABARĪ (Persian, c. 1100, for Amul to the south of the Caspian Sea), and the *Īlkhānī Zīj* by Naṣīr al-Dīn AL-ṬŪSĪ (Persian, c. 1270, for Maragha).

*Apparatus*
*Edition*: N pp. 78–83. *Manuscripts*: E ff. 189v–192r; A ff. 52r–54v;[23] O ff. 48r–50v; C ff. 153r–154r; T pp. 1253–1258.[24] Also collated: F = *Mufrad Zīj*, ms. Cambridge University Library, Browne O.1, f. 105v; K = *Īlkhānī Zīj*, ms. Paris Bibliothèque Nationale de France, persane 163, ff. 34v–36r.[25] *Constructed values*: r = recomputation for epicycle radius 5;15$^{p}$.

*Criteria*: A likely value is a value whose error fits within its surroundings. In the parts of the table that were clearly computed by means of some form of interpolation (see below), this may include values with deviations of more than 10″ from the recomputation. Except in six cases of scattered evidence, a clear majority of the sources always favours a particular likely value, which we consequently accept.

*Commentary*: This table has numerous groups of five to ten consecutive errors of the same sign, in some parts of the table as large as 15″, which point to a heavy use of linear and/or quadratic interpolation. AL-BATTĀNĪ apparently adjusted the table around its maximum in order to obtain exactly the Ptolemaic maximum of 5;1,0°, whereas a correct calculation for the used epicycle radius 5;15$^{p}$ would have led to a maximum of 5;1,11°. NALLINO states in his commentary (vol. II, p. 226) that he compared AL-BATTĀNĪ'S table to values recomputed by means of logarithms for every 6° and by means of interpolation in between. He notes some of the larger errors, in particular those of −12″ at argument 96 and of −11″ at argument 114, but, unlike his policy for some of the other tables here discussed, he refrains

22 Alternatively, the equation of centrum for Venus may simply have been rounded from the solar equation (cf. GOLDSTEIN and SAWYER 1967, pp. 167–168; TOOMER 1968, pp. 65–66), but there are two groups of deviations that do not support this assumption.

23 In A the minutes of the values for arguments 71–87 and 104–107 have slid upwards by one row, so that readings of the minutes for arguments 71 and 107 are to be considered missing.

24 On some intervals (28–40, 65–71, 88–89, 103, 130–145) T deviates significantly from the other sources, which may point to a systematic attempt to improve the Battanian table in parts where it has rather large errors. Accordingly, on these intervals T has been given less weight as a witness and is quoted between parentheses.

25 The errors in F and K do not seem to depend on the errors in the other witnesses, nor to be correlated with the recomputed values.

| Lunar equation of anomaly | | | | | | | |
|---|---|---|---|---|---|---|---|
| arg. | accepted | error | Nallino | arg. | accepted | error | Nallino |
| 91 | 5; 0,26 | −1 | | 106 | 4;55,38 | +4 | |
| 92 | 5; 0,44 | −2 | | 107 | 4;54,33 | +4 | 4;54,34 |
| 93 | 5; 0,55 | −5 | | 108 | 4;53,20 | +1 | 4;53,30 |
| 94 | 5; 0,59 | −9 | | 109 | 4;52, 0 | −3 | 4;52,10 |
| 95 | 5; 1, 0 | −11 | | 110 | 4;50,33 | −8 | 4;50,43 |
| 96 | 5; 0,57 | −12 | | 111 | 4;49, 2 | −12 | 4;49,12 |
| 97 | 5; 0,49 | −11 | 5; 0,45 | 112 | 4;47,26 | −15 | 4;47,32 |
| 98 | 5; 0,37 | −10 | | 113 | 4;45,47 | −15 | |
| 99 | 5; 0,21 | −6 | | 114 | 4;44, 6 | −11 | |
| 100 | 5; 0, 1 | −1 | | 115 | 4;42,23 | −4 | 4;42,24 |
| 101 | 4;59,32 | | | 116 | 4;40,31 | | |
| 102 | 4;58,55 | −1 | | 117 | 4;38,31 | +1 | |
| 103 | 4;58,11 | −3 | | 118 | 4;36,23 | +1 | |
| 104 | 4;57,24 | −2 | | 119 | 4;34, 9 | −1 | |
| 105 | 4;56,33 | | | 120 | 4;31,50 | −1 | |

**Table 4:** Fragment of the lunar equation of anomaly at the farthest distance of the epicycle from the earth, with our accepted values, differences from the recomputation, and NALLINO's values if different from ours.

from correcting these errors in his transcription. In seven cases, here listed in a separate category, he omits to correct some rather obvious scribal errors, although he mentions three of these in his commentary. Between arguments 107 and 115, in the part of the table with the largest errors, four scribal errors in E caused him to introduce a number of corrections that can now easily be seen to be unjustified. Table 4 shows this most erroneous part of the table with the errors in AL-BATTĀNĪ'S tabular values and NALLINO'S adjustments.

*Scattered evidence* (two likely readings each supported by two or more witnesses): **28** 2;11,4: E; 5″ ATN, 6″ OCK, 7″ F, 2″ r. **34** 2;36,43: EAKNr; 44″ OCF, 42″ T. **93** 5;0,55: ETFKN; 54″ AOC, 1′0″ r. **103** 4;58,11: AFKN; 51″ E, 12″ OCT, 14″ r. **122** 4;26,56 [27′ A]: EATN; 57″ OCFK, 58″ r. **167** 1;13,56 [14′ OC]: EKN; 57″ ATF, 53″ OC, 58″ r.

*Practically certain values* (a majority of the witnesses, or any easy corrections of them, agrees on a likely reading and at least one of the alternative readings is also likely but none of them is supported by two or more witnesses; this category also includes values for which the seconds in both E and A contain obvious scribal mistakes): **5** 0;24,7: AOCTKNr; 5″ E, 4″ F. **7** 0;33,44: AOCTFKr; 45″ EN. **25** 1;57,48: ETFKN; 4″ A, 47″ OC, 44″ r. **29** 2;15,26: OCFKN; 46″ E, 25″ A(T), 24″ r. **40** 3;1,4: EAOCFN; 3″ TK; 2″ r. **54** 3;51,7:

AOCTK; 30″ E, 4″ F, 6″ Nr. **55** 3;54,17: AOCTFK; 16″ ENr. **57** 4;0,24: AOCTFKr; 25″ EN. **62** 4;14,37: AOCTK; 36″ EN, 18″ F, 39″ r. **65** 4;22,24: EOCFN; 25″ A(T), 23″ Kr. **67** 4;27,7: EOCKN; 12″ A(T)r, 4″ F. **70** 4;33,54 [34′ OC]: AOCFK; 55″ EN, (51″ T), 52″ r. **87** 4;58,16: OCTFKNr; 56″ E, 26″ A. **97** 5;0,49: AOCTFK; 45″ EN, 1′0″ r. **126** 4;16,5 [17′ A]: AOCTFK; 6″ EN, 4″ r. **140** 3;26,57: OCFK; 7″ E, 56″ A, (52″ T), 58″ N, 27′0″ r. **158** 2;2,37: EOCKN; 36″ AT, 27″ F, 35″ r. **161** 1;46,44: AOTFKr; 45″ EN, 42″ C.

*Scribal errors not corrected by* NALLINO:[26] **35** 2;40,52: AOCT; 12″ EN, 50″ F, 51″ K, 53″ r. **36** 2;44,57: ACK; 45′17″ EN, 58″ O(T), 50″ F, 45′0″ r. **37** 2;49,2 [48′ C]: AOC(T)FK; 42″ EN, 4″ r. **41** 3;4,57: AOCTFK; 17″ EN, 56″ r. **61** 4;11,53: AOCTK; 33″ EN, 13″ F, 56″ r. **63** 4;17,18: AOCTFKr; 8″ EN. **73** 4;39,52: AOCTFKr; 38′ EN.

*Other mistaken readings and corrections by* NALLINO:[27] **19** 1;30,25 [26″ Fr]: EAOCTK-Fr; 31′ N. **107** 4;54,33 [(minutes missing) A]: AOCTFK; 24″ E, 34″ N, 29″ r. **108** 4;53,20: EAOCTK; 30″ N, 17″ F, 19″ r. **109** 4;52,0: EAOCTFK; 10″ N, 3″ r. **110** 4;50,33: AOCTFK; 13″ E, 43″ N, 41″ r. **111** 4;49,2: AOCTFK; 12″ EN, 14″ r. **112** 4;47,26: EAOCTK; 32″ N, 25″ F, 41″ r. **115** 4;42,23: AOCTFK; 34″ E, 24″ N, 27″ r.

*Remaining scribal errors in E and/or A*: **1** 55″ A. **13** 0′ A. **31** 26′ E; 9″ A, 30″ O, (1″ T). **45** 4″ A. **51** 56″ E. **52** 45′ E, 43′ C. **66** 25′ A; (51″ T). **69** 48″ A, (41″ T). **71** 34′ E, **′ A; (57″ T). **72** 36′ A. **82** 56″ E. **85** 59″ E. **101** 52″ A. **102** 59′ OC; 35″ E. **120** 7″ A. **121** 25′ E. **122–126** 27–25–22–19–17′ (minutes all one too large) A. **+122**: see under *Scattered evidence*. **+126**: see under *Practically certain values*. **139** 16″ E, (52″ T), 57″ F. **146** 53″ E. **151** 38′ OC; 39″ A. **154** 7″ E. **160** 51′ EOC. **163** 16″ E. **177** 4″, 13″ C.

### *3.6 Lunar latitude*

AL-BATTĀNĪ used PTOLEMY'S standard method for calculating the lunar latitude based on the assumption that the Moon moves on a circle inclined to the plane of the ecliptic. He also used the Ptolemaic value for the maximum latitude, namely 5°. In some other Arabic and Persian *zījes* we find different methods of calculating the lunar latitude, such as the so-called method of sines, as well as different values for the maximum latitude, such as the *Mumtaḥan* value of 4°46′. Of the numerous Islamic tables for a maximum lunar latitude of 5°, none appear to be directly related to the table in the *Ṣābi' Zīj*. KŪSHYĀR displays values to minutes that were not simply rounded from those of AL-BATTĀNĪ, and the *Mufrad Zīj* includes a table whose values were computed by a different method and differ by up to 28″ from AL-BATTĀNĪ'S values for arguments around 40°. Our only other witnesses for this table are therefore found in the Castilian translation of the *Ṣābi' Zīj* and in the manuscript tradition of the *Toledan Tables*.

---

26 NALLINO mentions the errors for arguments 35–37 in his commentary (vol. II, p. 226), but not the others here listed. Only for arguments 35 and 36 do some of the witnesses attest to other likely values.

27 The mistake at argument 19 is a misprint. NALLINO reports the mistakes for arguments 108–109 in his commentary as if they were readings in E (vol. II, p. 226). For arguments 107–115, see the sample values in Table 4.

*Apparatus*

*Edition*: N pp. 78–83. *Manuscripts*: E ff. 189v–192r; A ff. 52r–54v; O ff. 48r–50v; C ff. 153r–154r;[28] T pp. 1253–1258.[29] *Constructed values*: r = recomputation according to the standard formula $\beta(a) = \arcsin(\sin\beta_{max} \cdot \sin a)$ for $\beta_{max} = 5°$; s (for an argument *a*) = the symmetrical value (as accepted by us) for argument $180° - a$.

*Criteria*: A likely value is a value with a deviation of at most 2″ from the recomputation or, for parts of the table with larger errors, a value whose error fits within its surroundings. Except for seven values with scattered evidence, a majority of the witnesses for every value (and nearly always also the symmetrical value) agree on a likely reading, which we consequently accept.

*Commentary*: AL-BATTĀNĪ'S lunar latitude table as we have restored it is rather accurate for arguments 0–10, 70–110 and 170–180, but shows groups of between five and ten consecutive errors of at most 5″ in all other parts of the table. Since consecutive errors usually have the same sign, it is plausible that these errors result from some kind of (quadratic) interpolation. The table has only three divergences from the symmetry $\beta(\lambda) = \beta(180° - \lambda)$, namely for arguments 15/165, 56/124, and 60/120; in two of these cases it concerns the highly common confusion of 6 and 7, in the third case the only slightly less common confusion of 4 and 6. For one pair of arguments, namely 61/119, the deviations in A and T are themselves symmetrical. NALLINO states that he checked AL-BATTĀNĪ'S table against a recomputation by SCHIAPARELLI (vol. II, p. 227). For his transcription he mainly corrected the obvious scribal errors in E, erring in six cases listed under *Practically certain values*. He somewhat unsuccesfully tried to improve the values for arguments 88–92 and those on the interval 131–138, but followed E faithfully in most other cases.

*Scattered evidence*: **61** 4;22,20 [21′ E]: EOCNs; 22″ AT, 18″ r.[30] **63** 4;27,14 [26′ EN]: TNrs; 54″ E, 19″ A, 15″ O, 16″ C. **89** 4;59,57: AOCrs; 54″ E, 58″ T, 56″ N. **91** 4;59,57: Ars; 54″ E, 52″ OC, 58″ T, 56″ N. **105** 4;49,44: ETNs; 45″ Ar, 42″ OC. **119** 4;22,20: EOCNs; 22″ AT, 18″ r.[31] **124** 4;8,36: EOCN; 37″ ATrs.

*Rather certain values* (a majority of the witnesses supports one likely reading, but one or more other readings are also likely): **36** 2;56,10: AOCTs; 11″ ENr. **60** 4;19,47: EOCTN; 46″ As, 44″ r. **149** 2;34,24 [35′ E]: AOCTs; 25″ EN, 22″ r. **158** 1;52,17: EOCTNs; 16″ Ar. **165** 1;17,34: EOCTN; 35″ A, 33″ r, 36″ s.

*Practically certain values*:[32] (the majority of the witnesses supports one likely reading, and any other readings can be explained as straightforward scribal mistakes): **15** 1;17,36: EAOCT; 33″ Nr, 34″ s. **25** 2;6,40: AOCTs; 17″ E, 37″ N, 39″ r. **28** 2;20,40: AOCTs; 5″

28 In OC the seconds of the values for arguments 41–54 have slid upwards by one row, so that a reading of the seconds for argument 41 is to be considered missing.

29 The mainstream version of the *Toledan Tables* does not have significant systematic deviations from the version in the *Ṣābi' Zīj*.

30 A and T are symmetrical for arguments 61 and 119 in spite of these deviations; cf. argument 119.

31 See the previous footnote.

32 These are mostly scribal errors in E which were incorrectly restored by NALLINO.

E, 39″ N, 42″ r. **88** 4;59,50: EOCTs; 5*″ A, 47″ N, 49″ r. **92** 4;59,50: EOCTs; 30″ A, 47″ N, 49″ r. **101** 4;54,28: AOCTrs; 23″ E, 27″ N. **103** 4;52,17: AOCTs; 40″ E, 16″ N, 18″ r. **164** 1;22,35: AOCTs; 30″ E, 34″ N, 36″ r.

*Corrections by* NALLINO *on the interval 131–138:*[33] **131** 3;46,17: AOCTrs; 40″ E, 20″ N. **132** 3;42,49: EAOCTrs; 51″ N. **133** 3;39,17: AOCTrs; 40″ E, 20″ N. **134** 3;35,41 [34′ E]: EAOCTs; 51″ N, 40″ r. **135** 3;32,0: EAOCTrs; 10″ N. **136** 3;28,15: AOCTs; 55″ E, 35″ N, 16″ r. **137** 3;24,27 [25′ E]: EOCs; 26″ AT, 47″ N, 28″ r. **138** 3;20,35: AOCs; 30″ E, 36″ Tr, 55″ N.

*Values with scribal errors or illegible digits in the seconds in both E and A*: **29** 2;25,17: OCTNs; 57″ E, 14″ A, 18″ r. **115** 4;31,49: OCTNs; 4*″ E, 59″ A, 50″ r. **116** 4;29,34: OTNrs; **″ E, 59″ A, 39″ C. **117** 4;27,14: OCTNrs; **″ E, 19″ A.

*Remaining scribal errors in E and/or A*: **1** 53″ A, 14″ C. **5** 25′50″ E. **12** 36″ E. **19** 34′ E; 26″ O, 36″ C. **27** 12″ E. **30** 32″ A. **31** 35′ A. **32** 12″ A. **33** 57″ OCT. **35** 16″ A, 57″ T. **43** 25′ A; 26″ T. **44** 55″ E. **52** 56″ E. **53** 58″ E. **56** 9′ E. **58** 15′ E. **64** 30″ E. **65** 44″ E. **72** 44′ A. **79** 48″ E, 39″ OC. **81** 14″ E, 27″ OC. **82** 7″ E. **87** 19′ A. **99** 57″ E. **110** **′ E. **113** 37′ E. **121** 30″ A. **143** 2°58′ (sic!) E. **144** 3° A. **146** 46′ E. **147** 48′ A. **159** 46′ E. **161** 36′ E. **175** **″ E, 6″ T. **179** 53″ E.

## 4. Conclusion

Our sample edition and apparatuses for a number of tables from AL-BATTĀNĪ'S *Ṣābi' Zīj* have shown a wide variety of tabular characteristics and editorial difficulties, and a corresponding variety of policies by NALLINO to deal with these in his extensive study (1899–1907).

The tables discussed include highly accurate ones, such as the sine (Section 3.1) and the cotangent (3.2); generally accurate ones with incidental systematic deviations which may be due to erroneous interpolation, such as the half excess of longest daylight (3.3); rather inaccurate tables with irregular errors, such as the solar declination (Section 2); tables that are accurate in many parts but have irregular groups of errors in other parts, such as the lunar equation of anomaly (3.5) and the lunar latitude (3.6); and a sub-table of the table for the motion of the lunar node with relatively large but more or less regular errors (3.4).

Having to deal with these characteristics on the basis of the Escorial manuscript only, NALLINO resorted to various strategies. Sometimes he explains these in his commentary, but in other cases we need to infer them from the values in his transcription of the tables. He appears to have had two basic policies for dealing with tabular values with larger errors, whether computational or scribal: correction into values recomputed by SCHIAPARELLI, and correction into paleographically likely values. He applies corrections into recomputed values in particular in the tables for the solar declination (in which every error of 2″ or more is emended), the sine, and the cotangent. On the other hand, he does not correct the

[33] NALLINO introduces additional errors in an attempt to correct four scribal errors on the interval 131–138. Note that the symmetrical values for arguments 42 to 49 are basically correct.

larger errors in the tables of the lunar equation of anomaly and the lunar latitude. Especially in the table of the solar declination NALLINO mixes his two approaches: he corrects into paleographically likely values if these are at most 1″ removed from the recomputation, otherwise into recomputed values. In the lunar tables in particular, NALLINO introduces additional errors in an attempt to correct some unclear scribal errors in the Escorial manuscript. Also in various other cases a proper correction was simply impossible for him on the basis of the single source that he had available.

Even though the tables discussed here reflect some of the most obvious problems in NALLINO'S transcription, caution is required for many other tables as well. We may conclude that a new, critical edition of the tables from the *Ṣābi' Zīj*, taking into account in particular the Castilian translation of the *zīj* and the tradition of the *Toledan Tables*, is highly desirable. Until this is available, researchers interested in a more detailed comparison of AL-BATTĀNĪ'S tables with other sources or in an analysis of their mathematical properties are recommended to consult the manuscripts of the Arabic and Castilian versions of the *zīj* and the second author's edition of the *Toledan Tables* besides NALLINO'S monumental study.

## Bibliography


BOSSONG, Georg: Los canones de Albateni. Herausgegeben sowie mit Einleitung, Anmerkungen und Glossar versehen. Tübingen: Niemeyer, 1978.

VAN DALEN, Benno: Ancient and Mediaeval Astronomical Tables: mathematical structure and parameter values (doctoral dissertation). University of Utrecht, Mathematical Institute, 1993.

– : The Origin of the Mean Motion Tables of Jai Singh. Indian Journal of History of Science *35* (2000) 41–66.

– : A Second Manuscript of the *Mumtaḥan Zīj*. Suhayl *4* (2004) 9–44.

– : Battānī. In: HOCKEY, Thomas; RAGEP, F. Jamil; et al. (Eds.): The Bibliographical Encyclopedia of Astronomers. Berlin: Springer, 2007, vol. 1, pp. 101–103.

GOLDSTEIN, Bernard R.; SAWYER, Frederick W.: Remarks on Ptolemy's Equant Model in Islamic Astronomy. In: MAEYAMA, Yasukatsu; SALTZER, Walter G. (Eds.): Prismata: Naturwissenschaftliche Studien. Festschrift für Willy Hartner. Wiesbaden: Steiner, 1977, pp. 165–181. Reprinted in *idem*: Theory and Observation in Ancient and Medieval Astronomy. London: Variorum reprints, 1985, ch. VII.

HARTNER, Willy: al-Battānī. In: GILLISPIE, Charles C. (Ed.): Dictionary of Scientific Biography. New York: Scribner's Sons, 1970–1980, vol. 1, pp. 507–516.

KENNEDY, Edward S.: A Survey of Islamic Astronomical Tables. Transactions of the American Philosophical Society *N.S. 46–2* (1956) 123–177. Reprinted in 1989.

KUNITZSCH, Paul: New Light on al-Battānī's *Zīj*. Centaurus *18* (1974) 270–274. Reprinted in *idem*: The Arabs and the Stars. Texts and Traditions on the Fixed Stars, and their Influence in Medieval Europe. Northampton (UK): Variorum, 1989, ch. V.

NALLINO, Carlo Alfonso: al-Battānī sive Albatenii opus astronomicum. Ad fidem codicis Escurialensis arabice editum, Latine versum, adnotationibus instructum, 3 vols., Milan: Ulrico Hoepli, 1899–1907. Reprinted Frankfurt: Minerva, 1969 (vols. I and II) and Hildesheim: Olms, 1977 (vol. III). Again reprinted as Islamic Mathematics and Astronomy, vols. 11–13. Frankfurt: Institut für Geschichte der Arabisch-Islamischen Wissenschaften, 1997.

NEUGEBAUER, Otto: A History of Ancient Mathematical Astronomy, 3 vols. Berlin: Springer, 1975. Reprinted in 2004.

PEDERSEN, Olaf: A Survey of the Almagest. Odense: Odense University Press, 1974.

PEDERSEN, Fritz S.: The Toledan Tables. A Review of the Manuscripts and the Textual Versions with an Edition, 4 vols. Copenhagen: Reitzel, for Det Kongelige Danske Videnskabernes Selskab, 2002.

RAGEP, F. Jamil: al-Battānī, Cosmology and the History of Trepidation in Islam. In: CASULLERAS, Josep; SAMSÓ, Julio (Eds.): From Baghdad to Barcelona. Studies in the Islamic Exact Sciences in Honour of Prof. Juan Vernet. Barcelona: Instituto "Millás Vallicrosa" de Historia de la Ciencia Arabe, 1996, vol. 1, pp. 267–298.

SWERDLOW, Noel M.: Al-Battānī's Determination of the Solar Distance. Centaurus *17* (1973) 97–105.

TOOMER, Gerald J.: A Survey of the Toledan Tables. Osiris *15* (1968) 5–174.

# Dates and Eras in the Islamic World: Era Chronology in Astronomical Handbooks *

Practically all *zījes*, mediaeval Islamic astronomical handbooks with tables and explanatory text, include an extensive chapter on chronology. This chapter, usually to be found at the very beginning of the *zīj*, contains the chronological information needed for the calculation and evaluation of planetary positions and other astronomical data from the whole mediaeval period. Thus various calendars and eras are described and methods for converting dates from one calendar into another explained. For each calendar at least the following are given: a list of the names and lengths of the months, a method to compute from a given date the number of days elapsed since the epoch and vice versa, and a rule for calculating the week day (*madkhal*). A set of numerical tables facilitates the calculations involved. In addition, many *zījes* contain lists of festivals in various calendars, mathematical tables for determining the dates of the movable Christian feasts, and regnal lists of caliphs, kings and emperors.

This article describes the aspects of era chronology that are typically found in *zījes* and explains the conversion of dates from one calendar into another step by step. More general aspects of calendars used in the Islamic world, including their historical development and the names and lengths of the months, are discussed in the article TAʾRĪKH 1 in the *Encyclopaedia of Islam, New Edition* ($EI^2$). Depending on the context, the word *taʾrīkh* as it is used in *zījes* can be translated into English as "era", "epoch", "calendar" or "date".

## a. Calendars and eras

Table 1 shows which calendars are described in detail in a number of important *zījes*. For more information about the authors of these works,

---

*First published as TAʾRĪKH I.2 in *The Encyclopaedia of Islam, New Edition*, vol. 10 (2000), pp. 264–271. The original page numbers (with "a" indicating the first column on each page and "b" the second column) are shown in square brackets within the text. The transcription of Arabic words was changed to the more common one used in all other articles in this volume.

TABLE 1

*Calendars described in detail in some important zījes.*

| *Author* | *Title* | *Place* | *Year*[a] | Arabic[b] | Byzantine[c] | Persian[d] | Coptic[e] | Jewish | Jalālī | Uighur | various[f] |
|---|---|---|---|---|---|---|---|---|---|---|---|
| Yaḥyā b. Abī Manṣūr | al-Zīj al-Mumtaḥan | Baghdad | 215/830 | A | A | E | D | ×[1] | | | CF[1] |
| al-Khwārizmī[2] | al-Zīj al-Sindhind | Baghdad | 215/830 | A | A[3] | E | D | | | | |
| Ḥabash al-Ḥāsib | al-Zīj al-Mumtaḥan (?) | Baghdad | 225/840 | C[4] | A | E | P/A/D | | | | EK |
| al-Battānī | al-Zīj al-Ṣābiʾ | Raqqa | 285/900 | A | A[5] | E | P/A | | | | K |
| Kūshyār b. Labbān | al-Zīj al-Jāmiʿ | Persia | 390/1000 | A | A | E/L | | | | | CF |
| Ibn Yūnus | al-Zīj al-Ḥākimī | Cairo | 395/1005 | A | A | E | D | | | | C |
| al-Bīrūnī | al-Qānūn al-Masʿūdī | Ghazna | 420/1030 | C | A | E | A/D | × | | | CEFIKMS |
| al-Khāzinī | al-Zīj al-Sanjarī | Marw | 515/1120 | C | A | E/L | | | | | CFK |
| al-Ṭūsī | Zīj-i Īlkhānī | Maragha | 660/1262 | A | A | L | | × | × | × | CF |
| al-Baghdādī | al-Zīj al-Waqibiya (?) | Baghdad | 685/1285 | C | A | E | A/D | × | × | | CEFKM |
| al-Kāshī | Zīj-i Khāqānī | Samarqand | 825/1420 | A | A | L | | | × | × | F |
| Ulugh Beg | Zīj-i Sulṭānī | Samarqand | 850/1445 | A | A | L | | | × | × | F |

[a]Since most of the *zījes* listed here cannot be dated precisely, this column contains approximate dates of compilation. [b]A: astronomical (epoch Thursday 15 July 622), C: civil (epoch Friday 16 July 622) [c]A: Alexander (Seleucid era) [d]E: early variant (extra days inserted after Ābān), L: late variant (extra days inserted at the end of the year) [e]P: Philip, A: Augustus, D: Diocletian [f]This column contains the following abbreviations: C: material on the calculation of Christian feasts (Easter or the Great Lent), E: Egyptian calendar, F: list of feasts and festivals, I: Indian calendar, K: regnal list of kings and caliphs, M: Muʿtaḍid calendar, S: Soghdian calendar

[1]This material may not have been part of the original work. [2]Based on the Latin translation of the revision by Maslama al-Majrīṭī. [3]The leap day is inserted at the end of December. [4]Based on an accurate value for the length of a lunation and an intercalation scheme very different from the usual ones. [5]The year starts with Aylūl (September), the tables with Ādhār (March). [6]Jamāl al-Dīn Abi 'l-Qāsim b. Maḥfūẓ al-Munajjim al-Baghdādī reworked in his *zīj* material of early Muslim astronomers such as Ḥabash al-Ḥāsib, Kūshyār b. Labbān, and Abu 'l-Wafāʾ al-Būzjānī. It seems plausible that for his extensive chapter on chronology he borrowed from al-Bīrūnī, but this has not yet been investigated.

TABLE 2

*Epochs occurring in zījes and the differences between them in days.*

| Epoch | Date in Julian calendar | Differences between the epochs in days (sexagesimal/decimal) | | | | | | | | |
|---|---|---|---|---|---|---|---|---|---|---|
| Flood | Fr 18 Feb. 3102 B.C. | **Flood** | 860 172 | 1 014 932 | 1 019 273 | 1 122 241 | 1 236 564 | 1 359 973 | 1 363 597 | 1 526 770 |
| Nabonassar | We 26 Feb. 747 B.C. | 3,58,56,12 | **Nabonassar** | 154 760 | 159 101 | 262 069 | 376 392 | 499 801 | 503 425 | 666 598 |
| Philip | Su 12 Nov. 324 B.C. | 4,41,55,32 | 42,59,20 | **Philip** | 4 341 | 107 309 | 221 632 | 345 041 | 348 665 | 511 838 |
| Alexander | Mo 1 Oct. 312 B.C. | 4,43, 7,53 | 44,11,41 | 1,12,21 | **Alexander** | 102 968 | 217 291 | 340 700 | 344 324 | 507 497 |
| Augustus | Sa 30 Aug. 30 B.C. | 5,11,44, 1 | 1,12,47,49 | 29,48,29 | 28,35, 8 | **Augustus** | 114 323 | 237 732 | 241 356 | 404 529 |
| Diocletian | Fr 29 Aug. 284 A.D. | 5,43,29,24 | 1,44,33,12 | 1, 1,33,52 | 1, 0,21,31 | 31,45,23 | **Diocletian** | 123 409 | 127 033 | 290 206 |
| Hijra (A) | Th 15 Jul. 622 A.D. | 6,17,46,13 | 2,18,50, 1 | 1,35,50,41 | 1,34,38,20 | 1, 6, 2,12 | 34,16,49 | **Hijra (A)** | 3 624 | 166 797 |
| Yazdigird | Tu 16 Jun. 632 A.D. | 6,18,46,37 | 2,19,50,25 | 1,36,51, 5 | 1,35,38,44 | 1, 7, 2,36 | 35,17,13 | 1, 0,24 | **Yazdigird** | 163 173 |
| Malik-Shāh | Fr 15 Mar. 1079 A.D. | 7, 4, 6,10 | 3, 5, 9,58 | 2,22,10,38 | 2,20,58,17 | 1,52,22, 9 | 1,20,36,46 | 46,19,57 | 45,19,33 | **Malik-Shāh** |

Hijra (A) indicates the "astronomical" Hijra epoch. In the sexagesimal numbers under the diagonal, the digits are separated by commas, e.g., 42,59,20 denotes $42 \times 60^2 + 59 \times 60 + 20$.

Most important variants: The above date for the epoch of Augustus, signifying his year of accession, is only found in the *zīj* of Ḥabash al-Ḥāsib. Kūshyār b. Labbān, Ibn Yūnus, and al-Baghdādī give the epoch as 13 November 30 B.C., based on the assumption that New Year in the ancient Egyptian and the Coptic calendar coincided in the time of Philip instead of Augustus. al-Battānī uses 29 Augustus 25 B.C., possibly indicating the year of introduction of the calendar. The data in al-Bīrūnī's *al-Qānūn al-Masʿūdī* are inconsistent and point to both the earlier epoch and the later one.

The above date for the epoch of Diocletian is given by Yaḥyā b. Abī Manṣūr and Ibn Yūnus, whereas the Berlin manuscript of the *zīj* of Ḥabash al-Ḥāsib and Kūshyār b. Labbān use 12 November 284 A.D. A Philip epoch for the Coptic calendar is employed by Ḥabash al-Ḥāsib and al-Battānī.

the reader is referred to the respective articles in the *EI*$^2$ or the *Dictionary of Scientific Biography*, New York 1970-80. It can be noted that three calendars occur in practically all *zījes*: the purely lunar Muslim ("Arabic" or "Hijra") calendar, the Byzantine (*rūmī*) calendar which is essentially equivalent to the Julian, and the Persian Yazdigird calendar with a constant year length of 365 days. The use of most other calendars discussed below is restricted to particular geographical regions or historical periods.

Table 2 displays the precise dates of the most important epochs used in *zījes* as well as the differences in days between these epochs, which are needed for the conversion of dates from one calendar into another as described below. Note that, except in the **[col. 267a]** examples of date conversions, the Hijra dates given in this article are the common "civil" dates rather than the "astronomical" ones applied in most *zījes*. The names of calendars and epochs used in this article are literal translations of the names occurring in *zījes*, which in some cases may not be the historically or linguistically most correct ones.

*Arabic or Hijra calendar.* Whereas in civil usage the beginnings of the Arabic months are determined by actual observations of the lunar crescent after new moon (see the articles TAʾRĪKH 1 and RUʾYAT AL-HILĀL in the *EI*$^2$), most mediaeval astronomers used the schematic calendar with nineteen ordinary years of 354 days and eleven leap years of 355 days in a thirty-year cycle. The number of days in the first *n* years of this cycle is determined by multiplying *n* by the average year length of $354\frac{11}{30}$ days and rounding the result to the nearest whole number. If for $n = 15$ the resulting half of a day is truncated, this leads to the set of leap years 2, 5, 7, 10, 13, 16, 18, 21, 24, 26 and 29. If it is rounded upwards, the fifteenth year of each cycle becomes a leap year rather than the sixteenth. The first variant seems to be more common in early *zījes*, the second one in later Persian *zījes*.

Whereas most Muslim astronomers used the "astronomical" Hijra era based on the mean new moon of Thursday, 15 July 622 A.D., al-Bīrūnī and some others adopted the "civil" epoch determined by the first visibility of the lunar crescent on Friday, 16 July (see the article HIDJRA in the *EI*$^2$). The early Abbasid astronomer Ḥabash al-Ḥāsib based his rules and tables for the Hijra calendar on the actual time of the above-mentioned new moon and the traditional Babylonian value for the length of a lunation, which is slightly different from that implicit in the schematic lunar calendar. Thus he arrived at a highly unusual intercalation scheme, in which the years 3, 6, 9, 11, 14, 17, 19, 22, 25, 28, and 30 in the thirty-year cycle were leap years.

*Persian or Yazdigird calendar.* Because of its constant year length of 365 days, the Persian calendar is very convenient for calculating planetary motions during long periods of time. For this reason it was adopted as the base calendar in many *zījes*. The five extra days of the Persian year are usually called *andarja* (from the Persian *andar gāh(ān)*, "intermediate times"), *mustaraqa* ("stolen days"), or *lawāḥiq* ("appendages"). Various *zījes* describe the type of intercalation supposedly carried out in the ancient Persian calendar (see the article TAʾRĪKH 1, section vii, in the *EI*$^2$ and F. de Blois, *The Persian calendar*, in *Iran*, xxxiv [1996], 39-54), as a result of which the extra days were inserted after the eighth month, Ābān, until approximately the year 397/1007. Starting with Kūshyār b. Labbān, more and more Muslim astronomers moved the extra days back to the end of the year (cf. Table 1). In all *zījes* the epoch of the Persian calendar is the beginning of the year of accession of the last Persian king Yazdigird III; in only a few cases are the Sogdian and Khwarazmian forms of the Persian calendar described. In most *zījes* the names of the 30 days of the Persian months and the five extra days are listed.

*Byzantine (*rūmī*) or Syrian calendar.* This calendar is essentially the same as the Julian, but counts the years according to the Seleucid era, which is mistakenly named after Alexander the Great (see the article AL-ISKANDAR in the *EI*$^2$). In *zījes* it usually employs the Syro-Macedonian month names listed in the article TAʾRĪKH 1, section vii, in the *EI*$^2$ and has Tishrīn I (October) as the first month of the year. Once every four years a leap day is inserted at the end of Shubāṭ (February), leap years being those Alexander years which leave a remainder equal to three when divided **[col. 267b]** by four. Most deviations from these basic characteristics were incidental, e.g. al-Battānī started the year with Aylūl (September), the Latin version of the *Zīj* of al-Khwārizmī inserted the leap day at the end of Kānūn II (December), and in some tables the months were reckoned from Ādhār (March). Occasionally transliterations of the Latin month names were given.

*Egyptian or Coptic calendar* (taʾrīkh al-qibṭ). In some *zījes* this designation stands for the ancient Egyptian calendar as used by Ptolemy, which occurs with the Nabonassar (*Bukhtanaṣṣar*), Philip (*bylbs*, usually unvocalised in the manuscripts) and Antoninus (*ʾntnns*) epochs. Mostly, however, it is used for the Alexandrian or Coptic variant of this calendar introduced by the Roman emperor Augustus (*ʾghsts*), which runs parallel to the Julian calendar. The five extra days of the Coptic year (six in a leap year) are called *nasī* or epagomenai (*ʾbwghmnʾ*). As epochs we find the beginning of the year of accession of Augustus and that of Diocletian (*dqltyʾnws*). In the first case leap years are those which leave a remainder zero after division by four, in the second case those which leave a remainder three. Al-Battānī used a Philip era for the Coptic calendar (Tuesday, 29 August 324 B.C.), which he mistook for the epoch of Ptolemy's *Handy Tables*.

It may be noted that New Year in the ancient Egyptian calendar coincides with that in the Coptic calendar around the time of introduction of the latter, 25 B.C. However, in a number of *zījes* we find rules based on the assumption that New Year in both calendars coincides at the time of the Philip epoch, 324 B.C. As a result, the Augustus and Diocletian epochs in these *zījes* are 75 days later than the dates shown in Table 2.

*Jalālī calendar.* Whereas Islamic religious life was governed by the purely lunar Hijra calendar, for agricultural and taxation purposes a true solar calendar was found to be indispensable. In *zījes* the most extensive descriptions of such calendars concern the Jalālī (also called Malikī; see the article D̲JALĀLĪ in the *EI*²), which was introduced in the late 5th/11th century by the Great Saljūq Sultan Jalāl al-Dawla Malik-Shāh I. New Year in this calendar is defined as the day (from noon till noon) on which the sun reaches the vernal equinoctial point, to be determined by astronomical calculation rather than by a straightforward intercalation scheme. Here the method varies from *zīj* to *zīj* depending on solar theory and parameter values. In most cases the times of successive vernal equinoxes are computed by repeatedly adding a constant length of the solar year to the time of the equinox of the epoch. A more accurate theory is contained in the Khāqānī Zīj by the 9th/15th-century computational genius al-Kāshī, who took into account the influence of the motion of the solar apogee on the solar equation and hence correctly obtained varying time spans between consecutive equinoxes. Al-Kāshī constructed his sophisticated table for the computation of vernal equinoxes on the basis of intervals of 33 Jalālī years, practically equal to an integer number of days, and used parabolic interpolation to calculate the equinoxes within these intervals.

Independent of the precise method of computing the vernal equinox, in all *zījes* the Jalālī year ordinarily has 365 days, every fourth or incidentally fifth year being a leap year of 366 days. In some *zījes* the leap year following four ordinary years occurs alternately after 25 or 29 years, in others after 29 or 33 years. The beginnings of the Jalālī months are said to be defined by the entry of the sun into the **[col. 268a]** zodiacal signs, but in practice the traditional Persian months of 30 days are used, "*jalālī*" being appended to the original names and a sixth extra day being added in a leap year. The epoch of the Jalālī calendar generally is the vernal equinox of the year 471/1079. Other epochs in use for true solar calendars based on the vernal equinox are the Ilkhān era of Ghāzān Khān mentioned by al-Kāshī (Tuesday, 12 Rajab 701/13 March 1302) and the era of Chinggis Khan used in the *zīj* of al-Sanjufīnī, written in Tibet in 1366 (Wednesday, 12 Shaᶜbān 603/14 March 1207).

*Chinese-Uighur (Turco-Mongolian) calendar.* Under the Mongol Īlkhān dynasty the so-called Chinese-Uighur calendar was the commonly used one in Persia. It is prominently described in a number of *zījes*, mostly in Persian, the earliest of which is the Īlkhānī Zīj by al-Ṭūsī. The calendar is a luni-solar one and is very similar to the *Revised Ta Ming li*, the calendar of the Chinese Chin Dynasty, which was adopted by Chinggis Khan after the Mongol conquest of northern China in 1215. With the *Ta Ming li*, the Chinese-Uighur calendar shares the underlying solar and lunar parameters, the era of the creation, and the general rules for determining the beginnings of years and months and the position of the leap month. Different from official Chinese calendars, in the Uighur calendar the solar and lunar equations, required for the computation of the true new moons, are calculated as parabolic functions. The determination of the lunar equation involves a period relation of ultimately Babylonian origin, which equates nine lunar anomalistic cycles to 248 days.

In the Chinese-Uighur calendar, days are counted either from the beginning of the current month or in the Chinese sexagesimal cycle. Years are grouped into three consecutive cycles of 60 years which are called by

Chinese words Shang-yüan, Chung-yüan, and Hsia-yüan. In the Īlkhānī Zīj, the epoch of the Chinese-Uighur calendar is Thursday, 30 Rabīᶜ I 662/31 January 1264.

(Various technical, philological and historical aspects of the Chinese-Uighur calendar have been discussed in E.S. Kennedy, *The Chinese-Uighur calendar as described in the Islamic sources*, in *Isis*, lv [1964], 435-43*; R.P. Mercier, *The Greek 'Persian Syntaxis' and the Zīj-i Il-khānī*, in *Archives internationales d'histoire des sciences*, xxxiv [1984], 35-60; and C. Melville, *The Chinese Uighur animal calendar in Persian historiography of the Mongol period*, in *Iran*, xxxii [1994], 83-98. The complete method of computing Uighur dates as found in the Īlkhānī Zīj has been laid out in B. van Dalen, E.S. Kennedy and Mustafa K. Saiyid, *The Chinese-Uighur calendar in Ṭūsī's Zīj-i Īlkhānī*, in *Zeitschrift für Geschichte der arabisch-islamischen Wissenschaften*, xi [1997], 111-52.)

*Others.* A number of *zījes* contain descriptions of the Jewish luni-solar calendar. These range from simple listings of month names and elementary properties to complete sets of rules and tables for the computation of the Tishri new moon, which determines the beginning of the Jewish year. (The material on the Jewish calendar in the unique manuscript copy of the Mumtaḥan Zīj has been published in: J. Vernet, *Un antiguo tratado sobre el calendario judío en las 'Tabulae Probatae'*, in *Sefarad*, xiv [1954], 59-78, repr. in: Vernet, *Estudios sobre historia de la ciencia medieval*, Barcelona-Bellaterra 1979, 213-32. A treatise on the Jewish calendar by al-Khwārizmī has been analyzed in: E.S. Kennedy, *Al-Khwārizmī on the Jewish Calendar*, in *Scripta Mathematica*, xxvii [1964], 55-59*.)

A couple of *zījes* mention the calendar of the Abbasid caliph al-Muᶜtaḍid (r. 892-902), a modification of the **[col. 268b]** Persian calendar which fixes New Year in June and inserts a leap day once every four years. Its epoch is Wednesday, 13 Rabīᶜ II 282/11 June 895. Like his monumental *al-Āthār al-bāqiya* (tr. C.E. Sachau, *The chronology of ancient nations*, London 1879, repr. Frankfurt 1969), al-Bīrūnī's *al-Qānūn al-Masᶜūdī* contains a wealth of chronological material, which has not yet been properly investigated (cf. Kennedy, *Al-Bīrūnī's Masudic Canon*, in *al-Abḥāth*, xxiv [1971], 59-81*). Besides the common calendars discussed above, al-Bīrūnī includes details of the calendar reform by al-Muʿtaḍid and of the Jewish, Sogdian/Khwarazmian, and Indian calendars (on the last, see Kennedy, S. Engle and J. Wamstad, *The Hindu calendar as described in Al-Bīrūnī's Masudic Canon*, in *Journal of Near Eastern Studies*, xxiv [1965], 274-84*).

## b. Days since the epoch (*aṣl*)

In order to convert dates from one calendar into another, it is necessary to calculate the number of days between the epoch and a given date. This is done by summing the days of the completed (*tāmm*) years, the days of the completed months of the current (*nāqiṣ* "incomplete" or *munkasir* "broken") year, and the day of the current month. Conversely, it will be necessary to convert a given number of days since the epoch into the corresponding date, which is done by first computing the number of completed years and then distributing the remaining days over the months of the current year.

In many *zījes*, the number of days in a given number of completed years is called the *aṣl* ("basis"). The computation of the *aṣl* and, conversely, of the number of completed years from a given number of days since the epoch, are discussed below, examples being given in section c., *Conversions*. Whereas in *zījes* all calculations are written out in words, here the following modern notation will be used: $a$ denotes the *aṣl*, $c$ the number of days that have elapsed in the current year (including the current day), $d$ the total number of days that have elapsed since the epoch (including the current day), and $C$ the number of completed years. The following subscripts will be appended to these symbols: H for Hijra, A for Alexander (Seleucid era), and Y for Yazdigird. All divisions operate with whole numbers, usually discarding a possible remainder. When the remainder is used it will be denoted by $r$.

Now, for a given number of completed years $C$, the *aṣl* $a$ is calculated in the following way:

| | | |
|---|---|---|
| Arabic calendar: | $a_{\mathrm{H}} = (C_{\mathrm{H}} \times 10631 + 14)/30$ | (1) |
| Persian calendar: | $a_{\mathrm{Y}} = C_{\mathrm{Y}} \times 365$ | (2) |
| Byzantine calendar: | $a_{\mathrm{A}} = (C_{\mathrm{A}} \times 1461 + 1)/4$ | (3) |

It can be noted that all three operations are roughly a multiplication by the average year length. The constants 14 and 1 which are added in the calculations for the Arabic and Byzantine calendar are required in order to have the result increase by an extra day precisely when a leap year is encountered. To obtain the variant of the Arabic calendar in which the fifteenth year of each cycle is a leap year instead of the sixteenth, the constant 14 should be replaced by 15. Note that, instead of adding a constant and then discarding the remainder of the division, one may define a rounding rule for the remainder. For instance, equivalent to the above, the Arabic *aṣl* can be found as $(C_H \times 10631)/30$, where a remainder from 1 to 15 (14 if the fifteenth year of each cycle is a leap year) is discarded and a remainder from 16 (15) to 29 is made into an extra day. For the Byzantine calendar, **[col. 269a]** the remainder of the division $(C_H \times 1461)/4$ should be discarded if it is equal to 1 or 2 and made into an extra day if it is 3.

In most *zījes* the rules for the calculation of the *aṣl* are not exact. Often the addition of a constant for the Arabic and Byzantine calendar is disregarded or the rounding rule presented is ambiguous, producing results which may be off by a day. Of the *zījes* listed in Table 1 only those by Ḥabash al-Ḥāsib and al-Bīrūnī take extra care to present exact rules.

After the *aṣl* has been determined, the total number of days since the epoch, $d$, can be computed by adding the lengths of the respective completed months of the current year and the day of the current month.

Conversely, in order to calculate the number $C$ of years that have been completed on day $d$ reckoned from the epoch, the following rules can be used:

| | | |
|---|---|---|
| Arabic calendar: | $C_{\mathrm{H}} = (30 \times d_{\mathrm{H}} - 15)/10631$ | (4) |
| Persian calendar: | $C_{\mathrm{Y}} = (d_{\mathrm{Y}} - 1)/365$ | (5) |
| Byzantine calendar: | $C_{\mathrm{A}} = (4 \times d_{\mathrm{A}} - 2)/1461$ | (6) |

In this case the operations are roughly a division by the average year length and constants are subtracted in all three calculations in order to obtain the correct answer also for the first and last day of every year. For the variant of the Arabic calendar in which the fifteenth year of each cycle is a leap year instead of the sixteenth, 16 should be subtracted instead of 15. The remaining days of the current year, $c$, can be found from the remainder $r$ of the above divisions in the following way:

$$\text{Arabic calendar:} \quad c_{\mathrm{H}} = r_{\mathrm{H}}/30+1 \tag{7}$$

$$\text{Persian calendar:} \quad c_{\mathrm{Y}} = r_{\mathrm{Y}}+1 \tag{8}$$

$$\text{Byzantine calendar:} \quad c_{\mathrm{A}} = r_{\mathrm{A}}/4+1 \tag{9}$$

(the remainders of the divisions by 30 and 4 are discarded). Note that the results are equal to $d-a$ if $a$ denotes the *aṣl* corresponding to the number of completed years $C$ obtained above. The rules found in *zījes* for the calculation of the number of completed years and the day of the current year from the number of days since the epoch generally appear to be even less exact than those for the *aṣl*.

## c. Conversions

The explanation in *zījes* of the conversion of dates from one calendar into another varies from convenient shortcuts for particular problems, especially in early *zījes*, via extensive theoretical expositions as found, for instance, with Ibn Yūnus, to brief general explanations supplementing the rules for using the tables, in particular in later Persian *zījes*.

Using the rules presented in the previous section the conversion of dates can be performed in a general and straightforward way. First calculate the number of days from the epoch of the given calendar to the given date. Then add or subtract the number of days between the given epoch and the desired epoch (Table 2) to obtain the number of days from the epoch of the desired calendar to the given date. Finally transform these into completed years, completed months and the day of the current month in the desired calendar.

*Example*: Convert 24 Ramaḍān 254 Hijra into the corresponding Byzantine date.

[1] According to formula (1), the *aṣl* of 253 completed Arabic years is $(253\times10631+14)/30 = 89655$. (As usual, the remainder of the division is discarded. The same result can be obtained by taking $8 \times 10631$ days for the eight completed cycles of thirty Arabic years and then adding $13\times 354$ for the remaining thirteen completed years plus 5 for the leap days accumulated **[col. 269b]** during these thirteen years.) Since Ramaḍān

is the ninth month of the Arabic year and each two months number 59 days, the given date is the ($4 \times 59 + 24 =$) 260th day of the year 254 Hijra. Adding this number to the *aṣl*, we obtain 89915 for $d_H$, the total number of days since the Hijra epoch (including the given day).

[2] Now $d_A$, the total number of days from the Alexander epoch up to (and including) the given date, is determined by adding the difference in days between the two epochs as found in Table 2 to $d_H$: $d_A = 89915 + 340700 = 430615$. (We thus assume that the given Hijra date is astronomical; in practice this will have to be verified on the basis of the week day.)

[3] The number of completed Byzantine years is obtained according to formula (6): $C_A = (4 \times 430615 - 2)/1461 = 1178$ ($r = 1400$), and the remaining days of the current year 1179 Alexander using formula (9): $c_A = r/4 + 1 = 351$. Because 1179 leaves a remainder three when divided by four, the current year is a leap year and Shubāṭ has 29 days. By noting that the first eleven months of a Byzantine leap year number 336 days (simply add the numbers of days or subtract the length of the last month Aylūl from 366), we find that the desired date is the (351 − 336)th day of the twelfth month, i.e. 15 Aylūl 1179 Alexander.

In early *zījes* in particular, actual date conversions were not usually carried out by the above general rule, which may involve very large numbers, but by shortcuts based on expressions for the differences between the epochs in years and days. Examples of such expressions are:

– The (astronomical) Hijra epoch falls 932 Byzantine years and 287 days after the Alexander epoch.
– The Yazdigird epoch occurs 259 days after the beginning of 943 Alexander or 989 days after the beginning of 941 Alexander. Since reckoning from 943 Alexander the 1st, 5th, 9th, ... years are leap years instead of the 3rd, 7th, 11th, ..., the constant in formula (6) has to be changed to −4 in order to produce the correct number of completed Byzantine years since the beginning of 943 Alexander. When the beginning of the year 941 is used, the formula need not be changed.

*Example*: Convert 2 Kānūn II 1168 Alexander into the corresponding Hijra date.

[1] The Byzantine New Year just preceding the Hijra epoch is that of the year 933 Alexander. According to formula (3), the number of

days of the Byzantine years completed since the beginning of this year is $(1168 - 933) \times 1461 + 1)/4 = 85834$. Adding to this the number of days in the current year, $31 + 30 + 31 + 2 = 94$, we obtain 85928 days from the beginning of 933 Alexander until the given date.

[2] The number of days from the (astronomical) Hijra epoch to the given date is 287 less, i.e. 85641.

[3] The number of completed Arabic years is now found by formula (4): $C_{\mathrm{H}} = (30 \times 85641 - 15)/10631 = 241$ ($r = 7144$). The days of the current year are then obtained using formula (7): $c_{\mathrm{H}} = r/30 + 1 = 239$. These fill up 8 completed months ($4 \times 59 = 236$ days), leaving 3 days in the ninth month. Thus the desired Arabic date is 3 Ramaḍān 242 Hijra.

As we have seen above, the rules used for date conversions in *zījes* are often not exactly formulated. Therefore it is necessary to check the result by means of the week day of the given and the calculated date (see below, d.). Note that, in order to facilitate the conversion of historical dates, many *zījes* contain extensive regnal lists of caliphs and other rulers indicating the beginning and duration of their reigns. **[col. 270a]**

## d. Madākhil

A *madkhal* (pl. *madākhil*, literally "entrance", translated as *feria*, *nota* or *signum*) is the week day of the first day of a year or month or of a particular date, represented by a number from 1 (Sunday), 2 (Monday), till 7 (Saturday). In some *zījes* the number is given a separate name, *ʿalāma* ("indicator"). Most *zījes* contain both rules and tables for the calculation of *madākhil*, which, as we have seen, are particularly important for checking the results of date conversions.

Most simply, the *madkhal* of a given year is calculated by adding the *aṣl* of the completed years $C$ to the *madkhal* of the epoch (cf. Table 2) and casting off multiples of seven. For example, for the Hijra year 254 we add the *aṣl* of the completed years found above, 89655, to 5 (Thursday), the *madkhal* of the astronomical epoch. Discarding multiples of seven we obtain 4, signifying that 1 Muḥarram 254 Hijra was a Wednesday.

For the Persian and Byzantine calendars we find direct methods for calculating the *madkhal* based on the fact that the number of days of an (ordinary) year is $52 \times 7 + 1$. Thus the *madkhal* of a Persian year is simply found as $(3+C_Y)$ mod 7 ($m$ mod $n$ denotes the remainder of the division $m/n$; note that 3 is the *madkhal* of the Yazdigird epoch). Similarly, the *madkhal* of a Byzantine year is obtained as $(2+C_A+(C_A+1)/4)$ mod 7, where the remainder of the division by 4 is discarded.

In order to obtain the *madākhil* of the following months, two and one are added alternately in the case of the Arabic calendar, two for each month and five for the extra days in the case of the Persian calendar, and the respective month lengths minus 28 in the case of the Byzantine calendar. To obtain the *madkhal* of the current date, the day of the current month minus one should be added to the *madkhal* of the current month. After each addition multiples of seven are discarded.

## e. Tables

In practically all *zījes*, the chronological chapter contains a set of mathematical tables which facilitate the calculations described above. Among these tables the following appear to be standard:

(1) *Tables for the calculation of the aṣl.* Such tables display in three subtables, either in sexagesimal or in decimal notation, the number of days corresponding to groups of years (*al-sinūn al-majmūʿa*), individual years (*al-sinūn al-mabsūṭa*), and months. For instance, for the Arabic calendar the numbers of days in multiples of 30 years and in 1, 2, . . . , 30 single years will be given. Some authors of *zījes* combine more than one calendar in a single table by using a sexagesimal set-up with groups of 60 years. In each case, the number of days from the epoch to a given date is obtained by adding the day of the current month to the sum of the appropriate values from the three subtables.

(2) *Direct conversion tables.* From this type of table the dates in one or more calendars corresponding to a set of (often equidistant) beginnings of years in a base calendar can be read off directly. Intermediate dates can then be found readily by adding the years and days in the desired calendar corresponding to the remaining single years and months in the base calendar, which are tabulated in separate subtables, as well as the elapsed days of the current month. For instance, when converting 24 Ramaḍān 254 Hijra into the corresponding Byzantine date, one may read directly from the table that 1 Muḥarram 241 Hijra corresponds to

the 233rd day of the year 1166 Alexander. Furthermore one finds that the first thirteen **[col. 270b]** years of the cycle of thirty Arabic years are equal to 12 Byzantine years plus 224 days, and the Arabic months before Ramaḍān to 236 days. Together with the elapsed days of the current month (24 – 1) this yields the (233 + 224 + 236 + 23 =) 716th day since the beginning of the year 1178 Alexander, i.e. the 351st day or 15 Aylūl of 1179 Alexander.

(3) *Tables for the calculation of madākhil.* Tables for *madākhil* make use of the fact that the same dates recur on the same week days after 210 years in the Arabic calendar, 7 in the Persian, and 28 in the Byzantine. The first step in the determination of a *madkhal* is therefore to cast off multiples of the length of the cycle concerned from the given year. For the Persian and Byzantine calendars, most *zījes* contain a double-argument table from which the initial week day can be read off directly for every month of every year within the respective cycles. In the case of the Arabic calendar the initial week days of years 1, 2, ..., 210 are displayed together with a constant for every month which should be added to the initial week day of the given year. In early *zījes*, this type of table is often found under the name *al-jadwal al-mujarrad*.

Besides the types discussed here, many *zījes* contain special tables for more complicated calendars such as the Jalālī and the Chinese-Uighur, and tables of feasts and fasts in various calendars. Islamic tables for the determination of the Great Lent and Easter have been analysed in G.A. Saliba, *Easter Computation in Medieval Astronomical Handbooks*, in *al-Abḥāth*, xxiii (1970), 179-212*.

## f. Various

In early *zījes* in particular, we find approximate methods for chronological calculations, such as tables for Arabic *madākhil* based on a cycle of eight years. In the manuscript of the *zīj* of Ḥabash al-Ḥāsib extant in Berlin, some problems are discussed of the types "Find a date for which the Alexander year is equal to the Hijra year" or "Find an Alexander year in which Farwardīn occurs twice", whereas al-Bīrūnī presents solutions for problems of "mixed dates", in which day, month, and year are given in three different calendars.

## Bibliography

Much of the information contained in this article has been obtained from an inspection of primary sources and has not previously been published. The publications already mentioned deal with specific topics related to era chronology in *zījes*. General literature concerning calendars in the Islamic world has been listed at the end of the article TAʾRĪKH 1 in the *EI*²; the bibliography below presents some additions and various works particularly useful in relation to era chronology in *zījes*. Entries with an asterisk were reprinted in E.S. Kennedy *et al.*, *Studies in the Islamic exact sciences* (*SIES*), Beirut 1983.

For more information about *zījes* the reader is referred to the article ZĪDJ in the *EI*² and to Kennedy, *A survey of Islamic astronomical tables*, in *Transactions of the American Philosophical Society*, N.S. lxvi/2 (1956), 123-77 (repr. 1989). Many important investigations of parts of *zījes* are contained in *SIES*.

The standard work on general mathematical chronology is still F.K. Ginzel, *Handbuch der mathematischen und technischen Chronologie*, 3 vols., Leipzig 1906-14, repr. Leipzig 1958. A valuable overview of era chronology in the Islamic world is presented in S.H. Taqizadeh, *Various eras and calendars used in the countries of Islam*, in *Bulletin of the School of Oriental Studies*, ix (1937-9), 903-27, and x (1939-42), 107-32. Additions to Sachau's edition of al-Bīrūnī's *Chronology* can be found in K. Garbers, *Eine Ergänzung zur Sachauschen Ausgabe von al-Bīrūnīs Chronologie orientalischer Völker*, in *Der Islam*, xxx (1952), 39-80.

**[col. 271a]** Conversions of dates have traditionally been performed with the *Wüstenfeld-Mahler'sche Vergleichungstabellen*, revised by B. Spuler and J. Mayr, Wiesbaden 1961, or with R. Schram, *Kalendariographische und chronologische Tafeln*, Leipzig 1908. Now programs for various types of computers are available, partially from the Internet. The DOS programme *CALH* by B. van Dalen (version 1.2, 1997) includes most of the calendars and epochs described in this article.

# INDEXES

## Index of Subjects

*abjad* notation: VIII 408, note 8
*aḥwāl* (of the planets): VII 31
algebra: IV 197
almanac: VII 31
arc of visibility: VII 25
arithmetic: IV 197
ascendance (*suʿūd*): VII 20
ascendant: I 90; VII 18, note 4, 21, 27–30
ascensional difference: I 101, 106
*aṣl* (days since a calendar epoch): VII 16; IX 9–11, 13, 14
astrological houses: IV 209; VII 18
astrology: I 90; IV 209–10; VII 18, 28–30
astronomical tables: *passim*

bias (of a statistical estimator): I 115–16, 120–25, 127–9; III 187–8

calendars:
Akbar (Moghul emperor): V note 4
Arabic: IV 201, 208; V 42, 47–8; VII 16; VIII 419; IX 2, 4–5, 9–15
Byzantine (*rūmī*) or Syrian: IV 201; VII 16–17; VIII 419; IX 2, 4, 5–6, 10–15
Chinese-Uighur: IX 2, 7–8
Coptic / Egyptian: VII 16; IX 2–3, 6
(ancient) Egyptian: IX 2–3, 6
Indian: IX 2, 9
Jewish: IX 2, 8–9
Julian: V 42 (see also under Byzantine or Syrian)
Maliki / Jalālī: IX 2, 6–7
Muʿtaḍid: VII 16; IX 2, 8
Persian: IV 201; V note 4; VI 7, note 2; VII 16–17, 22; IX 2, 4, 5, 10–15
calendar conversion: VII 17; IX 11–13, 14–16
central limit theorem: I 133–6; III 187
centre of the world (*qubba*): VII 30
chords / table of chords: II 115, 127; IV 205; VII 18, 20; VIII 414
chronological tables: IX 14–15
chronology: IV 201; VII 15–17; IX *passim*
climate: VI note 6; VII 20
colour of eclipse: VII 26
comet: VI 6
completed period (chronology, mean motion tables): V 44; VI 6, note 1; IX 9
computational error: I 117; IV 224; V 49
computational techniques: see *method of computation*
computer programs (including the author's TA and MM): I 86; IV 230, 233; V 48; VII note 3; IX 16
conditions for the use of statistical estimators (especially independence and uniform distribution of tabular errors): I 97–8, 100–101, 105, 110–11, 119; II 109, 127, 145, note 34, note 38; III 189; IV 231–2
confidence interval: I 88, 94, 97, 100, 104–5, 110–11, 113, 116, 124–6, 130, note 85, note 87; II 108, 126, 128, 133, 136, 139; III 176, 180–82, 185, 189; IV 232–3, 240, 245;
conjunction (of sun and moon): IV 208; VII 21, 27
conversion factor (for the equation of time): II 104, 107, 114, 116, 124, 126, 135, 139; IV 216, 218–19, 226, 241, 245
cosine of the obliquity: I 89
cosmology: VII 30–31
cotangent: IV 207; VII 18; VIII 415–17
current months / days (chronology, mean motion tables): see *incomplete period*

daily mean motion: V 47–53; VI 6–12
declination: I 90, 98, 101; II 113, note 16; IV 203, 245; VII 20, 24; VIII 411–13, note 5
decrements of the planetary equation of anomaly: VI 25–9, note 11
derivative (of a mathematical function): IV 230
descendance (*hubūṭ*): VII 20
difference between two longitudes: V 44–7, note 7; VI 6, 15
difference in the lunar equation of anomaly: VI 21–3
difference in the planetary equation of anomaly: VI 25, note 9
displaced planetary equation / displacement: IV 236–8, 246; VI 16–18, note 7, 21–3, 25–6, 31–2

# Index of Personal Names

The Arabic names are primarily arranged by *nisba*.

# Index of Titles

## Index of Localities


Akhlāṭ: VII 19
Alexandria: II 109–10, 112
Amber: V 42
Āmid: VII 19
Arīn ( Ujjayn): IV 201
Arzan: VII 19

Badlīs: VII 19
Baghdad: I 107; III 172, 181–2; IV 196–7, 209; VI 3–4, 24; VII 9–10, 19–20, 27–8, 30–31, note 11
Benares: V 42

Cairo: VI 4
China: VI 5, 9; addition to I 142, note 48
Constantinople: VI 2, 5
Cordoba: IV 196, 198, 208–9

Damascus: VII 10
Delhi: V 42, 44–7, note 7; VI 5–7, 13–15, note 6, 31
Diyār Bakr: VII 19

Egypt: I note 52

Ghazna: VI 4; VII 16
Gīlān (region in Iran): III 172

Ḥarrān: VII 28

India: VI 1
Iran (Persia): I 107, note 57; VI 4
Isfahan: I note 48
Is'ird: VII 10

Jaipur: V 42, 46, note 7
Jerusalem: I note 48

Khilāṭ: VII 19
Khorezm: IV 197; VII 16

Mardin: I 106–7, note 56, note 57; VI 4
Marw: VI 4
Mathura (India): V 42
Mayyāfāriqīn: VII 10, 19
Mosul: VII 19, 26

Nanjing: addition to I 142, note 48
Newminster: IV note 9, note 10

Paris: V 42–7, note 7
Raqqa: VII 10, 19, 28; VIII 405

Quṭrubbul (suburb of Baghdad): IV 197

Raqqa: VI 4; VII 10, 19, 28; VIII 419

Samarqand: I 98
Samarra: VI 3; VII 10
Shiraz: I note 51; III 181
Shirwan: VI 4, 15, note 6, 19, 31
Si'ird: VII 10

Tabriz: VI 2, note 6
Tibet: I 98; VI 5

Ujjain: IV 201; V 42

Yazd: VI note 7
Yemen: VI 2, 5, 15, 31


## Index of Manuscripts


Berlin, Staatsbibliothek
We. 90 (=Ahlwardt 5750, *Zīj* of Ḥabash al-Ḥāsib): I 112, note 70; VI 3, 11, 19; VII 11; VIII note 5; IX 3
Or. qu. 101 (=Ahlwardt 5751, *Jāmi' Zīj* by Kūshyār ibn Labbān): I note 24; III 172–3 *et passim*

Cairo, Dār al-Kutub
mīqāt 188/2 (*Jāmi' Zīj*): III 172
Ṭal'at mīqāt 138 (*Shāmil Zīj*): I note 49
Taymūr riyāda
99 (Ibn Masrūr's commentary on al-Khwārizmī's *zīj*): IV 199
296/1 (*Shāmil Zīj*): I note 49
Cambridge
Gonville and Caius College
456 (Ibn al-Muthannā''s commentary in Latin): IV 199
University Library
Browne O.1 (*Mufrad Zīj*): VIII 422
Gg. 3.27/2 (*Muẓaffarī Zīj*): VI 5
Kk.I.1 (*Toledan Tables*): VIII 407 *etc.*
Chartres, Bibliothèque publique
214 (al-Khwārizmī's *zīj* in Latin): IV 199, 218

Dublin, Chester Beatty Library
Arabic 4076 (*Zīj* of al-Abharī): I 106–7, note 53

T - #0471 - 101024 - C0 - 224/149/21 - PB - 9781138382657 - Gloss Lamination